Steel Design for Engineers and Architects

Steel Design for Engineers and Architects

Second Edition

David A. Fanella
Rene Amon
Bruce Knobloch
Atanu Mazumder

VNR VAN NOSTRAND REINHOLD
―――――――――― New York

Copyright © 1992 by Van Nostrand Reinhold

Library of Congress Catalog Card Number 91-41835
ISBN 0-442-00927-5

All rights reserved. Certain portions of this work © 1982 by Van Nostrand Reinhold. No part of this work covered by the copyright hereon may be reproduced or used in any form or by any means—graphic, electronic, or mechanical, including photocopying, recording, taping, or informational storage and retrieval systems—without written permission of the publisher.

Manufactured in the United States of America

Published by Van Nostrand Reinhold
115 Fifth Avenue
New York, NY 10003

Chapman and Hall
2-6 Boundary Row
London, SE 1 8HN

Thomas Nelson Australia
102 Dodds Street
South Melbourne 3205
Victoria, Australia

Nelson Canada
1120 Birchmount Road
Scarborough, Ontario M1K 5G4, Canada

16 15 14 13 12 11 10 9 8 7 6 5 4 3 2 1

Library of Congress Cataloging-in-Publication Data

Steel design for engineers and architects.—2nd ed./David A. Fanella . . . [et al.]
    p. cm.
  Rev. ed. of: Steel design for engineers and architects/Rene Amon, Bruce Knobloch. Atanu Mazumder. © 1982.
  Includes index.
  ISBN 0-442-00927-5
  1. Building, Iron and steel. 2. Steel, Structural. I. Fanella, David Anthony. II. Amon, Rene. Steel design for engineers and architects.
TA684.S7543 1992
624.1′821—dc20
                                      91-41835
                                           CIP

TO
Our Families

Contents

Preface to the Second Edition / xi

Preface to the First Edition / xiii

Symbols and Abbreviations / xv

Introduction / xxv

    0.1 Steel as a Building Material / xxv
    0.2 Loads and Safety Factors / xxviii

1. Tension Members / 1

    1.1 Tension Members / 1
    1.2 Gross, Net, and Effective Net Sections / 1
    1.3 Allowable Tensile Stresses / 12
    1.4 AISC Design Aids / 19
    1.5 Slenderness and Elongation / 23
    1.6 Pin-Connected Members and Eyebars / 25
    1.7 Built-Up Members / 29
    1.8 Fatigue / 34

2. Members Under Flexure: 1 / 39

    2.1 Members Under Flexure / 39
    2.2 Determining the Allowable Bending Stress / 39
    2.3 Continuous Beams / 59
    2.4 Biaxial Bending / 64
    2.5 Shear / 67
    2.6 Holes in Beams / 72
    2.7 Beams with Concentrated Loads / 75
    2.8 Design of Bearing Plates / 80

2.9 Deflections / 84
2.10 Allowable Loads on Beams Tables / 88

3. Members Under Flexure: 2 / 93

3.1 Cover-Plated Beams / 93
3.2 Built-Up Members / 106
3.3 Design of Built-Up Beams / 107
3.4 Design of Plate Girders / 123
3.5 Open-Web Steel Joists / 132

4. Columns / 136

4.1 Columns / 136
4.2 Effective Lengths / 143
4.3 Allowable Axial Stress / 149
4.4 Design of Axially Loaded Columns / 153
4.5 Combined Stresses / 162
4.6 Column Base Plates / 178

5. Bolts and Rivets / 186

5.1 Riveted and Bolted Connections / 186
5.2 Rivets / 186
5.3 Bolts / 187
5.4 Design of Riveted or Bolted Connections / 189
5.5 Fasteners For Horizontal Shear in Beams / 215
5.6 Block Shear Failure / 219
5.7 AISCM Design Tables / 221

6. Welded Connections / 231

6.1 Welded Connections / 231
6.2 Weld Types / 231
6.3 Inspection of Welds / 233
6.4 Allowable Stresses on Welds / 234
6.5 Effective and Required Dimensions of Welds / 234
6.6 Design of Simple Welds / 238
6.7 Repeated Stresses (Fatigue) / 255
6.8 Eccentric Loading / 257
6.9 AISCM Design Tables / 266

7. Special Connections / 274

7.1 Beam-Column Connections / 274
7.2 Single-Plate Shear Connections / 277

7.3 Tee Framing Shear Connections / 279
7.4 Design of Moment Connections / 282
7.5 Moment-Resisting Column Base Plates / 302
7.6 Field Splices of Beams and Plate Girders / 311
7.7 Hanger Type Connections / 317

8. Torsion / 324

8.1 Torsion / 324
8.2 Torsion of Axisymmetrical Members / 324
8.3 Torsion of Solid Rectangular Sections / 330
8.4 Torsion of Open Sections / 335
8.5 Shear Center / 345
8.6 Torsion of Closed Sections / 350
8.7 Membrane Analogy / 358

9. Composite Design / 363

9.1 Composite Design / 363
9.2 Effective Width of Concrete Slab / 365
9.3 Stress Calculations / 367
9.4 Shear Connectors / 377
9.5 Formed Steel Deck / 385
9.6 Cover Plates / 392
9.7 Deflection Computations / 396
9.8 AISCM Composite Design Tables / 399

10. Plastic Design of Steel Beams / 420

10.1 Plastic Design of Steel Beams / 420
10.2 Plastic Hinges / 427
10.3 Beam Analysis by Virtual Work / 431
10.4 Plastic Design of Beams / 438
10.5 Additional Considerations / 446
10.6 Cover Plate Design / 446

Appendix A Repeated Loadings (Fatigue) / 459

Appendix B Highway Steel Bridge Design / 471

B.1 Overview of Highway Steel Bridge Design and the AASHTO Code / 471
B.2 Design of a Simply Supported Composite Highway Bridge Stringer / 480

Index / 499

Preface to the Second Edition

In 1989, the American Institute of Steel Construction published the ninth edition of the *Manual of Steel Construction* which contains the "Specification for Structural Steel Buildings—Allowable Stress Design (ASD) and Plastic Design." This current specification is completely revised in format and partly in content compared to the last one, which was published in 1978. In addition to the new specification, the ninth edition of the *Manual* contains completely new and revised design aids.

The second edition of this book is geared to the efficient use of the aforementioned manual. To that effect, all of the formulas, tables, and explanatory material are specifically referenced to the appropriate parts of the AISCM. Tables and figures from the *Manual*, as well as some material from the *Standard Specifications for Highway Bridges*, published by the American Association of State Highway and Transportation Officials (AASHTO), and from the *Design of Welded Structures*, published by the James F. Lincoln Arc Welding Foundation, have been reproduced here with the permission of these organizations for the convenience of the reader.

The revisions which led to the second edition of this book were performed by the first two authors, who are both experienced educators and practitioners.

Two major new topics can be found in Appendices A and B: design for repeated stresses (fatigue) and highway steel bridge design, respectively. Within the body of the text, the following additions have been included: composite design with formed metal deck; single-plate and tee framing shear connections; and beam bearing plates. The remainder of the topics have been modified, adjusted, and in some cases expanded to satisfy the requirements given in the ninth edition of the *Manual*. A solutions manual for all of the problems to be solved at the end of each chapter is available to the instructor upon adoption of the text for classroom use.

A one quarter course could include all of the material in Chapters 1 through 6 with the exception of Sects. 3.2 through 3.4. Chapters 7 through 10, as well as the sections omitted during the first quarter can be covered during a second quarter course. Any desired order of presentation can be used for the material in the second quarter course since these chapters are totally independent of each other. Similarly, Appendices A or B can be substituted in whole or in part for one or more of Chapters 7 through 10.

A one semester course could include all of the material in Chapters 1 through 6 as well as Chapter 9 or 10. As was noted above, any parts of Chapters 7 through 10, Appendix A, or Appendix B can be substituted as desired.

The authors would like to thank Mr. Andre Witschi, S.E., P.E. who is Chief Structural Engineer at Triton Consulting Engineers for his help in reviewing the material in the Appendices. Thanks are also due to Mr. Tom Parrott and all of the secretarial staff of the School of Architecture at the University of Illinois at Chicago for all of their generous assistance. Finally, we would like to thank our families for their patience and understanding.

DAVID A. FANELLA
RENE AMON

Chicago, Illinois

Preface to the First Edition

This is an introductory book on the design of steel structures. Its main objective is to set forth steel design procedures in a simple and straightforward manner. We chose a format such that very little text is necessary to explain the various points and criteria used in steel design, and we have limited theory to that necessary for understanding and applying code provisions. This book has a twofold aim: it is directed to the practicing steel designer, whether architect or engineer, and to the college student studying steel design.

The practicing structural engineer or architect who designs steel structures will find this book valuable for its format of centralized design requirements. It is also useful for the veteran engineer who desires to easily note all the changes from the seventh edition of the *AISC Steel Design Manual* and the logic behind the revisions. Yet, the usefulness as a textbook is proven by field-testing. This was done by using the appropriate chapters as texts in the following courses offered by the College of Architecture of the University of Illinois at Chicago Circle: Steel Design; Additional Topics in Structures; and Intermediate Structural Design (first-year graduate). The text was also field-tested with professionals by using the entire book, less Chapter 8, as text for the steel part of the Review for the State of Illinois Structural Engineers' Licence Examination, offered by the University of Illinois at Chicago Circle. The text, containing pertinent discussion of the numerous examples, was effectively combined with supplemental theory in classroom lectures to convey steel design requirements. Unsolved problems follow each chapter to strengthen the skill of the student.

The American Institute of Steel Construction (AISC) is the authority that codifies steel construction as applied to buildings. Its specifications are used by most governing bodies, and, therefore, the material presented here has been written with the AISC specifications in mind. We assumed—indeed, it is necessary—that the user of this book have an eighth edition of the *AISC Manual of Steel Construction*. Our book complements, but does not replace, the ideas presented in the *Manual*. Various tables from the *Manual* have been reprinted here, with permission, for the convenience of the user.

We assume that the user is familiar with methods of structural analysis. It is the responsibility of the designer to assure that proper loadings are used and proper details employed to implement the assumed behavior of the structure.

In this book, the chapters follow a sequence best suited for a person familiar with steel design. For classroom use, the book has been arranged for a two-quarter curriculum as follows:

First quarter—Chapters 1 through 6. Section 3.2 and Examples 3.7 through 3.11 should be omitted and left for discussion in the second quarter. Furthermore, examples in Section 3.1 could be merely introduced initially and discussed fully after Chapters 5 and 6 have been covered. This amount of material, if properly covered, will enable the student to understand the fundamentals of steel design and to comprehensively design simple steel structures.

Second quarter—Chapters 7 through 10 and Section 3.2. This material is complementary to steel design. An instructor can follow any desired order because these chapters are totally independent of each other.

This book can also be used for a one-semester course in steel design by covering Chapters 1 through 6, as discussed above for First quarter, and Chapters 9 and 10. An instructor may also want to include some material from Chapter 7.

In conclusion, we would like to thank all our friends, families, and colleagues who provided help and understanding during the writing of this manuscript. Without their help and cooperation, we might never have made this book a reality.

Rene Amon
Bruce Knobloch
Atanu Mazumder

Chicago, Illinois

Symbols and Abbreviations

A	Cross-sectional area, in.^{2}
	Gross area of an axially loaded compression member, in.^{2}
A_b	Nominal body area of a fastener, in.^{2}
A_c	Actual area of effective concrete flange in composite design, in.^{2}
A_{ctr}	Concrete transformed area in compression, in.^{2}
	$= \left(\dfrac{b}{n}\right) t$
A_e	Effective net area of an axially loaded tension member, in.^{2}
A_f	Area of compression flange, in.^{2}
A_{fe}	Effective tension flange area, in.^{2}
A_{fg}	Gross beam flange area, in.^{2}
A_{fn}	Net beam flange area, in.^{2}
A_g	Gross area of member, in.^{2}
A_n	Net area of an axially loaded tension member, in.^{2}
A_s	Area of steel beam in composite design, in.^{2}
A_{st}	Cross-sectional area of a stiffener or pair of stiffeners, in.^{2}
A_t	Net tension area, in.^{2}
A_v	Net shear area, in.^{2}
A_w	Area of girder web, in.^{2}
A_1	Area of steel bearing concentrically on a concrete support, in.^{2}
A_2	Maximum area of the portion of the supporting surface that is geometrically similar to and concentric with the loaded area, in.^{2}
AASHTO	American Association of State Highway and Transportation Officials
AISCM	American Institute of Steel Construction *Manual* of *Steel Construction, Allowable Stress Design*
AISCS	American Institute of Steel Construction Specifications
B	Allowable load per bolt, kips
	Width of column base plate

xvi SYMBOLS AND ABBREVIATIONS

B_c	Load per bolt, including prying action, kips
B_x, B_y	Area of column divided by its appropriate section modulus
C	Coefficient for determining allowable loads in kips for eccentrically loaded connections
C_a	Coefficient used in Table 3 of Numerical Values
	Constant used in calculating moment for end-plate design: 1.13 for 36-ksi and 1.11 for 50-ksi steel
C_b	Bending coefficient dependent upon moment gradient

$$= 1.75 + 1.05 \left(\frac{M_1}{M_2}\right) + 0.3 \left(\frac{M_1}{M_2}\right)^2$$

Coefficient used in calculating moment for end-plate design
$= \sqrt{b_f/b_p}$

C_c	Column slenderness ratio separating elastic and inelastic buckling
C_m	Coefficient applied to bending term in interaction equation for prismatic members and dependent upon column curvature caused by applied moments
C_v	Ratio of "critical" web stress, according to the linear buckling theory, to the shear yield stress of web material
C_1	Coefficient for web tear-out (block shear)
	Increment used in computing minimum spacing of oversized and slotted holes
C_2	Coefficient for web tear-out (block shear)
	Increment used in computing minimum edge distance for oversized and slotted holes
D	Factor depending upon type of transverse stiffeners
	Outside diameter of tubular member, in.
	Number of $\frac{1}{16}$-inches in weld size
	Clear distance between flanges of a built-up bridge section, in.
DL	Dead load
	Subscript indicating dead load
E	Modulus of elasticity of steel (29,000 ksi)
E_c	Modulus of elasticity of concrete, ksi
F	Maximum induced stress in the bottom flange of a bridge stringer due to wind loading when the top flange is continuously supported, psi
F_a	Axial compressive stress permitted in a prismatic member in the absence of bending moment, ksi
F_b	Bending stress permitted in a prismatic member in the absence of axial force, ksi
F_b'	Allowable bending stress in compression flange of plate girders as reduced for hybrid girders or because of large web depth-to-thickness ratio, ksi

SYMBOLS AND ABBREVIATIONS

F_{cb}	Stress in the bottom flange of a bridge stringer due to wind loading, psi
F'_e	Euler stress for a prismatic member divided by factor of safety, ksi
F_p	Allowable bearing stress, ksi
F_t	Allowable axial tensile stress, ksi
F_u	Specified minimum tensile strength of the type of steel or fastener being used, ksi
F_v	Allowable shear stress, ksi
F_w	Weld capacity, kips/in.
F_y	Specified minimum yield stress of the type of steel being used, ksi. As used in the Manual, "yield stress" denotes either the specified minimum yield point (for those steels that have a yield point) or specified minimum yield strength (for those steels that do not have a yield point)
F'_y	The theoretical maximum yield stress (ksi) based on the width-thickness ratio of one-half the unstiffened compression flange, beyond which a particular shape is not "compact." See AISC Specification Sect. B5.1. $$= \left[\frac{65}{b_f/2t_f}\right]^2$$
F'''_y	The theoretical maximum yield stress (ksi) based on the depth-thickness ratio of the web below which a particular shape may be considered "compact" for any condition of combined bending and axial stresses. See AISC Specification Sect. B5.1. $$= \left[\frac{257}{d/t_w}\right]^2$$
F_{yc}	Specified minimum column yield stress, ksi
F_{yf}	Specified minimum yield stress of flange, ksi
F_{yst}	Specified minimum stiffener yield stress, ksi
F_{yw}	Specified minimum yield stress of beam web, ksi
G	Shear modulus of elasticity of steel (11,200 ksi)
H	Force in termination region of cover plate, kips
H_s	Length of a stud shear connector after welding, in.
I	Moment of inertia of a section, in.^{4}
	Impact factor
	Subscript indicating impact
I_{eff}	Effective moment of inertia of composite sections for deflection computations, in.^{4}
I_s	Moment of inertia of steel beam in composite construction, in.^{4}
I_{tr}	Moment of inertia of transformed composite section, in.^{4}
I_x	Moment of inertia of a section about the $X-X$ axis, in.^{4}
I_y	Moment of inertia of a section about the $Y-Y$ axis, in.^{4}

J	Torsional constant of a cross-section, in.^{4}
	Polar moment of inertia of a bolt or weld group, in.^{4}
K	Effective length factor for a prismatic member
L	Span length, ft
	Length of connection angles, in.
	Unbraced length of tensile members, in.
	Unbraced length of member measured between centers of gravity of the bracing members, in.
	Plate length, in.
	Column length, ft.
LL	Live load
	Subscript indicating live load
L_c	Maximum unbraced length of the compression flange at which the allowable bending stress may be taken at $0.66F_y$ or as determined by AISC Specification Eq. (F1-3) or Eq. (F2-3), when applicable, ft
L_u	Maximum unbraced length of the compression flange at which the allowable bending stress may be taken at $0.6F_y$, ft
L_v	Span for maximum allowable web shear of uniformly loaded beam, ft
M	Moment, kip-ft
	Maximum factored bending moment, kip-ft
M_1	Smaller moment at end of unbraced length of beam-column, kip-ft
M_2	Larger moment at end of unbraced length of beam-column, kip-ft
M_D	Moment produced by dead load, kip-ft
M_L	Moment produced by live load, kip-ft
	Moment produced by loads imposed after the concrete has achieved 75% of its required strength, kip-ft
M_R	Beam resisting moment, kip-ft
M_{cb}	Moment induced in the bottom flange of a bridge stringer due to wind loading, lb-ft
M_e	Extreme fiber bending moment in end-plate design, kip-in.
M_p	Plastic moment, kip-ft
M_{range}	Range of the applied moment, kip-ft
N	Length of base plate, in.
	Length of bearing of applied load, in.
N_e	Length at end bearing to develop maximum web shear, in.
N_r	Number of stud shear connectors on a beam in one transverse rib of a metal deck, not to exceed 3 in calculations
N_1	Number of shear connectors required between point of maximum moment and point of zero moment

N_2	Number of shear connectors required between concentrated load and point of zero moment
P	Applied load, kips
	Force transmitted by a fastener, kips
	Factored axial load, kips
	Normal force, kips
	Force in the concrete slab in composite bridge sections, pounds
P_1	Force in the concrete slab in composite bridge sections, pounds $= A_s F_y$
P_2	Force in the concrete slab in composite bridge sections, pounds $= 0.85 f'_c b t_s$
P_R	Beam reaction divided by the number of bolts in high-strength bolted connection, kips
P_b	Plate bearing capacity in single-plate shear connections, kips
	Tee stem bearing capacity in tee framing shear connections, kips
P_{bf}	Factored beam flange or connection plate force in a restrained connection, kips
P_{cr}	Maximum strength of an axially loaded compression member or beam, kips
P_e	Euler buckling load, kips
P_{ec}	Effective horizontal bolt distance used in end-plate connection design, in.
P_f	Distance between top or bottom of top flange to nearest bolt, in.
P_{fb}	Force, from a beam flange or moment connection plate, that a column will resist without stiffeners, as determined using Eq. (K1-1), kips
P_{wb}	Force, from a beam flange or moment connection plate, that a column will resist without stiffeners, as determined using Eq. (K1-8), kips
P_{wi}	Force, in addition to P_{wo}, that a column will resist without stiffeners, from a beam flange or moment connection plate of one inch thickness, as derived from Eq. (K1-9), kips
P_{wo}	Force, from a beam flange or moment connection plate of zero thickness, that a column will resist without stiffeners, as derived from Eq. (K1-9), kips
P_y	Plastic axial load, equal to profile area times specified minimum yield stress, kips
P_{range}	Applied load range, kips
Q	Prying force per fastener, kips
	First statical moment of the area used for horizontal shear computations, in.^{3}

xx SYMBOLS AND ABBREVIATIONS

R Maximum end reaction for $3\frac{1}{2}$ in. of bearing, kips
Reaction or concentration load applied to beam or girder, kips
Radius, in.
Constant used in determining the maximum induced stress due to wind in bottom flange of a bridge stringer

R_1 A constant used in web yielding calculations, from Eq. (K1-3), kips
$= 0.66 \, F_y \, t_w \, (2.5k)$

R_2 A constant used in web yielding calculations, from Eq. (K1-3), kips/in.
$= 0.66 \, F_y \, t_w$

R_3 A constant used in web crippling calculations, from Eq. (K1-5), kips
$= 34 \, t_w^2 \sqrt{F_{yw} t_f / t_w}$

R_4 A constant used in web crippling calculations, from Eq. (K1-5), kips/in.
$= 34 \, t_w^2 \left[3 \left(\dfrac{1}{d} \right) \left(\dfrac{t_w}{t_f} \right)^{1.5} \right] \sqrt{F_{yw} t_f / t_w}$

R_{BS} Resistance to web tear-out (block shear), kips
R_{PG} Plate girder bending strength reduction factor
R_e Hybrid girder factor
R_i Increase in reaction R in kips for each additional inch of bearing
R_{ns} Net shear fracture capacity of the plate in single-plate shear connections, kips
Net shear fracture capacity of the tee stem in tee framing shear connections, kips

R_o Plate capacity in yielding in single-plate shear connections, kips
Tee stem capacity in yielding in tee framing shear connection, kips

R_v Shear capacity of the net section of connection angles
Allowable shear or bearing value for one fastener, kips

S Elastic section modulus, in.^{3}
Stringer spacing, ft.

SD Superimposed dead load
Subscript indicating superimposed dead load

S_{eff} Effective section modulus corresponding to partial composite action, in.^{3}

S_r Range of horizontal shear at the slab-girder interface in composite bridge sections, kips/in.

S_s Section modulus of steel beam used in composite design, referred to the bottom flange, in.^{3}

S_t Section modulus of transformed composite cross-section, referred to the top of concrete, in.^{3}

S_{tr}	Section modulus of transformed composite cross section, referred to the bottom flange; based upon maximum permitted effective width of concrete flange, in.^{3}
S_x	Elastic section modulus about the $X - X$ axis, in.^{3}
S_u	Ultimate strength of a shear connector in composite bridge sections, pounds
T	Horizontal force in flanges of a beam to form a couple equal to beam end moment, kips
	Bolt force, kips
	Distance between web toes of fillet at top and at bottom of web, in.
	$= d - 2k$
T_b	Specified pretension of a high-strength bolt, kips
U	Factor for converting bending moment with respect to $Y - Y$ axis to an equivalent bending moment with respect to $X - X$ axis
	$= \dfrac{F_{bx} S_x}{F_{by} S_y}$
	Reduction coefficient used in calculating effective net area
V	Maximum web shear, kips
	Statical shear on beam, kips
	Shear produced by factored loading, kips
V_h	Total horizontal shear to be resisted by connectors under full composite action, kips
V_h'	Total horizontal shear provided by the connectors providing partial composite action, kips
V_r	Range of shear due to live loads and impact, kips
W	Total uniform load, including weight of beam, kips
Y	Ratio of yield stress of web steel to yield stress of stiffener steel
Z	Plastic section modulus, in.^{3}
Z_x	Plastic section modulus with respect to the major $(X - X)$ axis, in.^{3}
Z_y	Plastic section modulus with respect to the minor $(Y - Y)$ axis, in.^{3}
Z_r	Allowable range of horizontal shear on an individual shear connector in composite bridge sections, pounds
a	Distance from bolt line to application of prying force Q, in.
	Clear distance between transverse stiffeners, in.
	Dimension parallel to the direction of stress, in.
a'	Distance beyond theoretical cut-off point required at ends of welded partial length cover plate to develop stress, in.
b	Actual width of stiffened and unstiffened compression elements, in.
	Dimension normal to the direction of stress, in.

xxii SYMBOLS AND ABBREVIATIONS

	Fastener spacing vertically, in.
	Distance from the bolt centerline to the face of tee stem or angle leg in determining prying action, in.
	Effective concrete slab width based on AISC Specification Sect. I1, in.
b_e	Effective width of stiffened compression element, in.
b_f	Flange width of rolled beam or plate girder, in.
b_{fb}	Beam flange width in end-plate design, in.
b_p	End-plate width, in.
b_{tr}	Transformed concrete slab width, in.
d	Depth of column, beam or girder, in.
	Nominal diameter of a fastener, in.
	Stud diameter, in.
d_b	Bolt diameter, in.
d_c	Web depth clear of fillets, in.
d_h	Diameter of hole, in.
e	Eccentricity or distance from point of load application to bolt line
e_o	Distance from outside face of web to the shear center of a channel section, in.
f	Axial compression stress on member based on effective area, ksi
f_a	Computed axial stress, ksi
f_b	Computed bending stress, ksi
f'_c	Specified compression strength of concrete, ksi
f_p	Actual bearing pressure on support, ksi
f_t	Computed tensile stress, ksi
f_v	Computed shear stress, ksi
f_{vs}	Shear between girder web and transverse stiffeners kips per linear inch of single stiffener or pair of stiffeners
g	Transverse spacing locating fastener gage lines, in.
h	Clear distance between flanges of a beam or girder at the section under investigation, in.
h_r	Nominal rib height for steel deck, in.
k	Distance from outer face of flange to web toe of fillet of rolled shape or equivalent distance on welded section, in.
k_v	Shear buckling coefficient for girder webs
l	For beams, distance between cross sections braced against twist or lateral displacement of the compression flange, in.
	For columns, actual unbraced length of member, in.
	Unsupported length of a lacing bar, in.
	Length of weld, in.
	Largest laterally unbraced length along either flange at the point of load, in.

SYMBOLS AND ABBREVIATIONS

l_b	Actual unbraced length in plane of bending, in.
l_v	Distance from centerline of fastener hole to free edge of part in the direction of the force, in.
l_h	Distance from centerline of fastener hole to end of beam web, in.
l_w, l_1, l_2	Weld lengths, in.
m	Factor for converting bending to an approximate equivalent axial load in columns subjected to combined loading conditions
	Cantilever dimension of base plate, in.
n	Number of fasteners in one vertical row
	Cantilever dimension of base plate, in.
	Modular ratio (E/E_c)
n'	An equivalent cantilever dimension of a base plate, in.
q	Allowable horizontal shear to be resisted by a shear connector, kips
r	Governing radius of gyration, in.
r_T	Radius of gyration of a section comprising the compression flange plus $\frac{1}{3}$ of the compression web area, taken about an axis in the plane of the web, in.
r_x	Radius of gyration with respect to the $X - X$ axis, in.
r_y	Radius of gyration with respect to the $Y - Y$ axis, in.
r'_y	Radius of gyration with respect to $Y - Y$ axis of double angle member, in.
s	Longitudinal center-to-center spacing (pitch) of any two consecutive holes, in.
t	Thickness of a connected part, in.
	Wall thickness of a tubular member, in.
	Angle thickness, in.
	Compression element thickness, in.
	Thickness of concrete in compression, in.
t_b	Thickness of beam flange or moment connection plate at rigid beam-to-column connection, in.
t_c	Thickness of flange used in prying, in.
t_f	Flange thickness, in.
t_{fb}	Thickness of beam flange in end-plate connection design, in.
t_o	Thickness of concrete slab above metal deck, in.
t_p	End-plate thickness, in.
t_s	Stiffener plate thickness, in.
	Thickness of concrete slab in composite bridge sections, in.
t_w	Web thickness, in.
t_{wc}	Column web thickness, in.
w	Length of channel shear connectors, in.
	Plate width (distance between welds), in.
	Weld size, in.

w_r	Average width of rib or haunch of concrete slab on formed steel deck, in.
x	Subscript relating symbol to strong axis bending
y	Subscript relating symbol to weak axis bending
y_s	Distance from neutral axis of steel section to bottom of steel, in.
y_{tr}	Distance from neutral axis of composite beam to bottom of steel, in.
α	Constant used in equation for hybrid girder factor R_e, Ch. G $= 0.6\, F_{yw}/F_b \leq 1.0$
	Moment ratio used in prying action formula for end-plate design
β	Ratio S_{tr}/S_s or S_{eff}/S_s
Δ	Beam deflection, in.
	Displacement of the neutral axis of a loaded member from its position when the member is not loaded, in.
δ	Ratio of net area (at bolt line) to the gross area (at the face of the stem on angle leg)
υ	Poisson's ratio, may be taken as 0.3 for steel
ϕ	Reduction factor $= 0.85$
	Angle of rotation, rad.
	Symbol representing the diameter of a circular element
kip	1,000 lb
ksi	Expression of stress in kips per sq. in.

Introduction

0.1 STEEL AS A BUILDING MATERIAL

From the dawn of history, man has been searching for the perfect building material to construct his dwellings. Not until the discovery of iron and its manufacture into steel did he find the needed material to largely fulfill his dreams. All other building materials discovered and used in construction until then proved to be either too weak (wood), too bulky (stone), too temporary (mud and twigs), or too deficient in resisting tension and fracture under bending (stone and concrete). Other than its somewhat unusual ability to resist compression and tension without being overly bulky, steel has many other properties that have made it one of the most common building materials today. Brief descriptions of some of these properties follow. Some of the other concepts not discussed here have been introduced in treatments of the strength of materials and similar courses and will not be repeated.

High strength. Today steel comes in various strengths, designated by its yield stress F_y or by its ultimate tensile stress F_u. Even steel of the lowest strengths can claim higher ratios of strength to unit weight or volume than any of the other common building materials in current use. This allows steel structures to be designed for smaller dead loads and larger spans, leaving more room (and volume) for use.

Ease of erection. Steel construction allows virtually all members of a structure to be prefabricated in the shops, leaving only erection to be completed in the field. As most structural components are standard rolled shapes, readily available from suppliers, the time required to produce all the members for a structure can often be shortened. Because steel members are generally standard shapes having known properties and are available through various producers, analyzing, remodeling, and adding to an existing structure are very easily accomplished.

xxvi INTRODUCTION

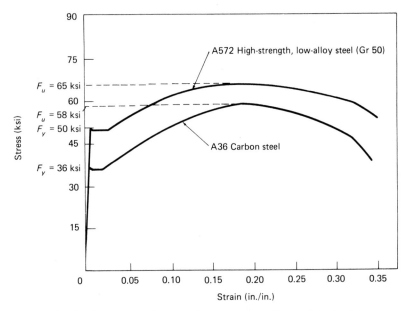

Fig. 0.1. Typical stress-strain curves for two common classes of structural steel.

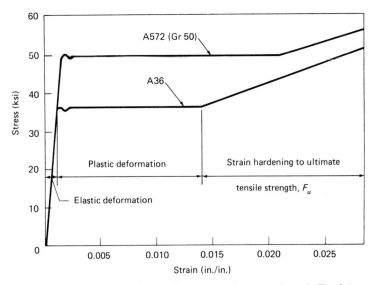

Fig. 0.2. Initial portions of the typical stress-strain curves shown in Fig. 0.1.

INTRODUCTION xxvii

Uniformity. The properties of steel as a material and as structural shapes are so rigidly controlled that engineers can expect the members to behave reasonably as expected, thus reducing overdesign due to uncertainties. Figure 0.1 shows typical stress-strain curves for two types of structural steel. The initial portions of these curves are shown in Fig. 0.2.

Ductility. The property of steel that enables it to withstand extensive deformations under high tensile stresses without failure, called *ductility*, gives steel structures the ability to resist sudden collapse. This property is extremely

Fig. 0.3. Fabrication of special trusses at a job site to be used for the wall and roof structural system for the atrium lobby of an office building. Note that the trusses, made from tubular sections, are partially prefabricated in the shop to transportable lengths and assembled at the site to form each member.

valuable when one considers the safety of the occupants of a building subject to, for instance, a sudden shock, such as an earthquake.

Some of the other advantages of structual steel are (1) speed of erection, (2) weldability, (3) possible reuse of structural components, (4) scrap value of unreusable components, and (5) permanence of the structure with proper maintenance. Steel also has several disadvantages, among which are (1) the need to fireproof structural components to meet local fire codes, (2) the maintenance costs to protect the steel from excessive corrosion, and (3) its susceptibility to buckling of slender members capable of carrying its axial loads but unable to prevent lateral displacements. Engineers should note that under high temperatures, such as those reached during building fires, the strength of structural steel is severely reduced, and only fireproofing or similar protection can prevent the structural members from sudden collapse. Heavy timber structural members usually resist collapse much longer than unprotected structural steel. The most common methods of protecting steel members against fire are a sprayed-on coating (about 2 in.) of a cementitious mixture, full concrete embedment, or encasement by fire-resistant materials, such as gypsum board.

0.2 LOADS AND SAFETY FACTORS

Components of a structure must be designed to resist applied loads without excessive deformations or stresses. These loads are due to the dead weight of the structure and its components, such as walls and floors; snow; wind; earthquakes; and people and objects supported by the structure. These loads can be applied to a member along its longitudinal axis (axially), causing it to elongate or shorten depending on the load; perpendicular to its axis (transversely), causing it to flex in a bending mode; by a moment about its axis (torsionally), causing the member to twist about that axis; or by a combination of any two or all three. It is very important for the engineer to recognize all the loads acting on each and every element of a structure and on the entire structure as a whole and to determine which mode they are applied in and the combinations of loads that critically affect the individual components and the entire structure. The study of these loads and their effects is primarily the domain of structural analysis.

Loads are generally categorized into two types, dead and live. Loads that are permanent, steady, and due to gravity forces on the structural elements (dead weight) are called *dead loads*. Estimating the magnitudes of dead loads is usually quite accurate, and Table 0.1 can be used for that purpose. *Live loads*, however, are not necessarily permanent or steady and are due to forces acting on a structure's superimposed elements, such as people and furniture, or due to wind, snow, earthquakes, etc. Unlike dead loads, live loads cannot be accu-

Fig. 0.4. A typical steel-frame high-rise building under construction. Note how, in steel construction, the entire structure can be framed and erected before other trades commence work, thus reducing conflict and interference among various trades.

rately predicted, but can only be estimated. To relieve the engineer of the burden of estimating live loads, building codes often dictate the magnitude of the loads, based on structure type and occupancy. National research and standardization organizations, such as the American Society of Civil Engineers (ASCE),[1] and other national and city building codes dictate the magnitude of wind, snow, and earthquake loads based on extensive research data. It is important to determine which codes govern in a particular situation and then to find the applicable loads which must act on the structure. Some average values for dead and live loads for buildings that can be used for preliminary design are shown in Table 0.1.

A structure cannot be designed just to resist the estimated dead loads and the estimated or code-specified live loads. If that were allowed, the slightest variation of loads toward the high side would cause the structure or member to deform unacceptably (considered failure). To avoid this, the stresses in the members are knowingly kept to a safe level below the ultimate limit. This safe level is usually specified to be between one-half and two-thirds of the yield

---

[1] "Minimum Design Loads for Buildings and Other Structures," American Society of Civil Engineers (ASCE 7-88; formerly ANSI A58.1-1982), New York, 1990.

Table 0.1 Approximate Values of Some Common Loads in Building Design

Material	Weight	Units
Dead Loads (Weights) of Some Common Building Materials[a]		
Plain concrete (normal weight)	145	pcf
Reinforced concrete (normal weight)	150	pcf
Lightweight concrete	85–130	pcf
Masonry (brick, concrete block)	120–145	pcf
Earth (gravel, sand, clay)	70–120	pcf
Steel	490	pcf
Stone (limestone, marble)	165	pcf
Brick walls		
4 in.	40	psf
8 in.	80	psf
12 in.	120	psf
Hollow concrete block walls		
Heavy aggregate		
4 in.	30	psf
6 in.	43	psf
8 in.	55	psf
12 in.	80	psf
Light aggregate		
4 in.	21	psf
6 in.	30	psf
8 in.	38	psf
12 in.	55	psf
Wood (seasoned)	25–50	pcf
Live Loads[b]		
Rooms (residences, hotels, etc.)	40	psf
Offices	50	psf
Corridors	80–100	psf
Assembly rooms, lobbies, theaters	100	psf
Wind (depends on location, terrain, and height above the ground)	15–60	psf
Snow (depends on location and roof type)	10–80	psf

[a]See pages 6-7 through 6-9 of the AISCM for weights and specific gravities of other materials.
[b]See AISCS A4 for specific rules about loading and AISCS A5.2 for provisions concerning wind and earthquake loads.

stress level, which means that from one-half to one-third of a member's capacity is kept on reserve for uncertainties in loading, material properties, and workmanship. This reserve capacity is the *safety factor*.

In the United States, the American Institute of Steel Construction (AISC) recommends what safety factors should be used for every type of structural steel

Fig. 0.5. Wide-flange beam and girder floor system with sprayed-on cementitious fireproofing. Note that openings in the webs are reinforced because they are large and occur at locations of large shear.

component for buildings. These safety factors are usually determined from experiments conducted or approved by the AISC. Most municipalities in the United States have local building codes that require that the AISC specifications be met. Structures other than buildings are designed according to other specifications, such as the American Association of State Highway and Transportation Officials (AASHTO) for highway bridges and the American Railway Engineering Association (AREA) for railway bridges. Throughout this book, the AISC specifications will be referred to as AISCS. These specifications can be found in the AISC *Manual of Steel Construction, Allowable Stress Design*, from here on called AISCM.

The design philosophy of the AISCS can be stated as follows: All structural members and connections must be proportioned so that the maximum stresses due to the applied loads do not exceed the allowable stresses given in Chapters D through K of the specification (AISCS A5.1). These allowable stresses are typically a function of the yield stress F_y or the ultimate stress F_u of the steel divided by an appropriate factor of safety (as noted above). The allowable stresses given in the AISCS may be increased by one-third if the stresses are produced by wind or seismic loads acting alone or in combination with the dead and live loads, provided that the required section is satisfactory for only dead and live loads without the increase (AISCS A5.2). The increase mentioned above does not apply, however, to the allowable stress ranges for fatigue loading given in Appendix K4 of the AISCS. It is important to note that when computing the maximum bending stresses for simply supported beams or girders, the effective length of the span should be taken as the distance between the centers of gravity of the supporting members (AISCS B8).

The steels which are approved for use are given in AISCS A3.1.a. Each of the steels listed has an American Society for Testing and Materials (ASTM) designation. The various types of steels and their corresponding values of F_y and F_u are listed in Table 1, p. 1-7 of the AISCM. One of the steels which is often used in building design is ASTM A36 ($F_y = 36$ ksi, $F_u = 58$ ksi). For this reason, unless specifically stated otherwise, the reader should assume A36 steel for all examples and problems to be solved throughout this book. Additionally, because of their popularity, the same assumption should be made for E70 electrodes when dealing with welds.

1
Tension Members

1.1 TENSION MEMBERS[1]

A tension member is defined as an element capable of resisting tensile loads along its longitudinal axis. Classic examples are bottom chords of trusses and sag rods (Fig. 1.1). For the most part, the shape of the cross section has little effect on the tensile capacity of a member. The net cross-sectional area will be uniformly stressed except at points of load applications and their vicinity (St. Venant's principle). If fasteners are used, it may become necessary to design for stress concentrations near the fasteners, referred to as *shear lag*. Other stress buildups, in the form of bending stresses, will develop if the centers of gravity of the connected members do not line up. This effect is usually neglected, however, in statically loaded members (AISCS J1.9).

Allowable stresses are computed for both gross member area and effective net area. The gross area stress is designed to remain below the yield stress, at which point excessive deformations will occur, and the effective net area is designed to prevent local fracture.

To account for the effective net area, it is necessary to use reduction coefficients for tension members that are not connected through all elements of the cross section. This provision is intended to account for the phenomenon of shear lag. For example, the angle in Fig. 1.2 is connected through one leg. The shear stress being transferred through its bolts will concentrate at the connection. The effect of shear lag will diminish, however, as the number of fasteners increases.

1.2 GROSS, NET, AND EFFECTIVE NET SECTIONS

The gross section of a member, A_g, is defined as the sum of the products of the thickness and the gross width of each element as measured normal to the axis

---

[1]This chapter deals with members subjected to pure tensile stresses only. For the case of members subjected to tension and flexure, refer to Chap. 4, Sect. 4.5.

2   STEEL DESIGN FOR ENGINEERS AND ARCHITECTS

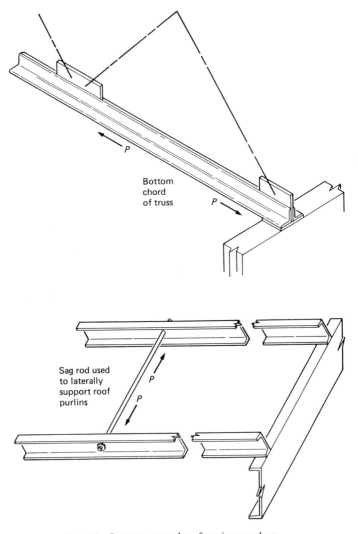

Fig. 1.1. Common examples of tension members.

of the member (AISCS B1). This is the cross-sectional area of the member with no parts removed.

The net width multiplied by the member thickness is the net area. Net width is determined by deducting from the gross width the sum of all holes in the section cut. In AISCS B2, the code states: "The width of a bolt or rivet shall be taken as $\frac{1}{16}$ in. larger than the nominal dimension of the hole." AISC Table J3.1 lists the diameter of holes as a function of fastener size. For standard holes,

TENSION MEMBERS 3

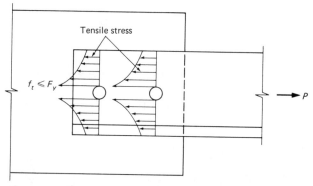

Fig. 1.2. Shear lag concept. Because the stress is transferred through the bolts, a concentration of tensile stress occurs at the bolt holes. As the number of bolts increases, the magnitude of shear lag decreases (see AISCS Commentary B3).

Fig. 1.3. John Hancock Building, Chicago. (*Courtesy of U.S. Steel*)

4 STEEL DESIGN FOR ENGINEERS AND ARCHITECTS

the hole diameter is $\frac{1}{16}$ in. larger than the nominal fastener size. Thus, a value of fastener diameter plus $\frac{1}{8}$ in. ($d + \frac{1}{16} + \frac{1}{16}$) must be used in computing net sections.

Example 1.1. Determine the net area of a $4 \times 4 \times \frac{1}{2}$ angle with one line of 3-$\frac{3}{4}$-in. bolts as shown.

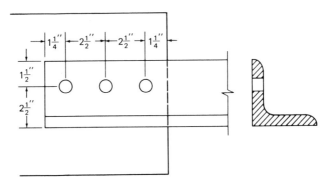

Solution. The net area is equal to the gross area less the sum of the nominal hole dimension plus $\frac{1}{16}$ in. (AISCS B2).

$A_g = 3.75$ in.^{2}

$A_n = 3.75$ in.$^2 - [\frac{3}{4}$ in. $+ (\frac{1}{16}$ in. $+ \frac{1}{16}$ in.$)] \times \frac{1}{2}$ in. $= 3.31$ in.^{2}

If there is a chain of holes on a diagonal or forming a zigzag pattern, as in Fig. 1.4, the net width is taken as the gross width minus the diameter of all the holes in the chain and then adding for each gage space in the chain the quantity

$$\frac{s^2}{4g} \qquad (1.1)$$

where s is the longitudinal spacing (pitch), in inches, of two consecutive holes and g is the transverse spacing (gage), in inches, of the same two holes (AISCS B2).

The critical net section is taken at the chain that yields the least net width. In no case, however, shall the net section, when taken through riveted or bolted splices, gusset plates, or other connection fittings subject to a tensile force exceed 85% of the gross section (AISCS B3).

For determining the areas of angles, the gross width is the sum of the widths

TENSION MEMBERS 5

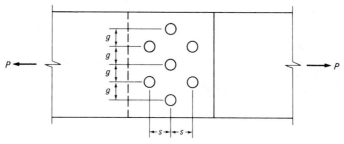

Fig. 1.4. Gage and pitch spacing.

of the legs less the angle thickness (AISCS B1). The gage for holes in opposite legs, as shown in Fig. 1.5, is the sum of the gages from the back of the angle less the angle thickness (AISCS B2).

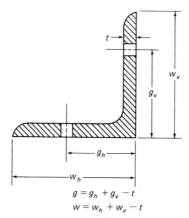

$$g = g_h + g_v - t$$
$$w = w_h + w_v - t$$

Fig. 1.5. Measurement of gage dimension and gross width for angles.

Example 1.2. Determine the net area of the plate below if the holes are for $\frac{7}{8}$-in. bolts.

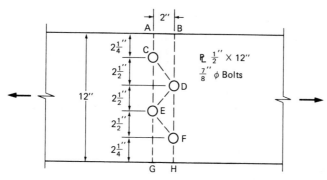

6 STEEL DESIGN FOR ENGINEERS AND ARCHITECTS

Solution. To find the net section, consult the AISC Code Section B2. The net width must be calculated by considering all possible lines of failure and deducting the diameters of the holes in the chain. Then for each diagonal path, the quantity $(s^2/4g)$ is added, where s = longitudinal spacing (pitch) of any two consecutive holes and g = transverse spacing (gage) of the same two holes. The critical net section is the chain that gives the least net width. The net critical width is then multiplied by the thickness to obtain the net area (AISCS B2).

Chain	Width $-$ Holes $+ \dfrac{s^2}{4g}$ for each diagonal path
ACEG	$12 - 2 \times (\frac{7}{8} + \frac{1}{8}) = 10.0''$
BDFH	$12 - 2 \times (\frac{7}{8} + \frac{1}{8}) = 10.0''$
ACDEG	$12 - 3 \times (\frac{7}{8} + \frac{1}{8}) + (2^2/4(2.5)) + (2^2/4(2.5)) = 9.8''$
ACDFH	$12 - 3 \times (\frac{7}{8} + \frac{1}{8}) + (2^2/4(2.5)) = 9.4''$
ACDEFH	$12 - 4 \times (\frac{7}{8} + \frac{1}{8}) + 3 \times (2^2/4(2.5)) = 9.2''$

$$\text{Critical section} = 9.2 \text{ in.} \times \tfrac{1}{2} \text{ in.} = 4.6 \text{ in.}^2$$

$$\therefore A_n = 4.6 \text{ in.}^2$$

Example 1.3. For the two lines of bolt holes shown, determine the pitch (s) that will give a net section along ABCDEF equal to a net section through two holes. Holes are for $\frac{3}{4}$-in. bolts.

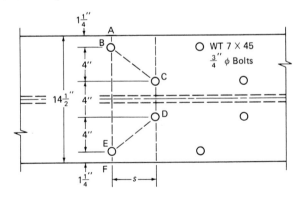

Solution. The section through four holes plus the quantity $2 \times (s^2/4g)$ must equal the gross section minus two holes.

$$14.5 - 4 \times \left(\frac{3}{4} + \frac{1}{8}\right) + 2 \times \frac{s^2}{4 \times 4} = 14.5 - 2 \times \left(\frac{3}{4} + \frac{1}{8}\right)$$

$$11.0 + \frac{s^2}{8} = 12.75$$

$$s^2 = 14 \quad s = 3.74 \text{ in.}$$

Use pitch of $3\frac{3}{4}$ in.

Example 1.4. A single angle tension member $6 \times 4 \times \frac{3}{4}$ has gage lines in the legs as shown. Determine the pitch (s) for $\frac{3}{4}$-in. rivets, so that the reduction in area is equivalent to two holes in line.

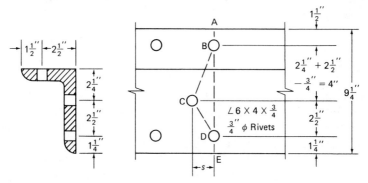

Solution. Because of the angle thickness, the gross width of the angle must be $(6 + 4 - \frac{3}{4}) = 9.25$ in.

Path ABDE = path ABCDE

$$\text{Path ABDE} = 9.25 - 2 \times \left(\frac{3}{4} + \frac{1}{8}\right) = 7.5 \text{ in.}$$

$$\text{Path ABCDE} = 9.25 - 3 \times \left(\frac{3}{4} + \frac{1}{8}\right) + \frac{s^2}{4 \times 4} + \frac{s^2}{4 \times 2.5}$$

$$= 6.625 + \frac{s^2}{16} + \frac{s^2}{10}$$

$$7.5 = 6.625 + \frac{s^2}{16} + \frac{s^2}{10}$$

$$0.875 = \frac{(5 + 8)s^2}{80}$$

$$s^2 = 5.385 \quad s = 2.32 \text{ in.}$$

Use pitch of $2\frac{3}{8}$ in.

8 STEEL DESIGN FOR ENGINEERS AND ARCHITECTS

The calculation of the effective net area A_e is given in AISCS B3 for bolted, riveted, and welded connections.

If the tensile force is transmitted by bolts or rivets through some, but not all, of the cross-sectional elements of the member, then A_e is given by (B3-1):

$$A_e = UA_n \tag{1.2}$$

where A_n is the net area of the member as defined previously and U is a reduction coefficient. Values of U based on the requirements set forth in AISCS B3 can be found in Table 1.1.

When a tensile force is transmitted by welds through some, but not all, of the cross-sectional elements of the member, then (B3-2) should be used to determine the value of A_e:

$$A_e = UA_g \tag{1.3}$$

where A_g is the gross cross-sectional area of the member. When transverse welds are used to transmit the tensile load to some, but not all, of the cross-sectional elements of W, M, or S shapes and WT sections cut from these shapes, then the effective area A_e is defined in AISCS B3 as the area of the directly connected elements only. An example of this provision is shown in Fig. 1.6.

For the case when the tensile load is transmitted to a plate by longitudinal welds along both edges at the end of the plate, the effective net area A_e is given by (B3-2). The values of the reduction coefficients U to be used are given in Fig. 1.7. Note that the length, l, of the longitudinal welds shall not be less than the width of the plate for any case.

Although not stated in AISCS B3, the AISCS Commentary B3 gives a conservative method to determine U for situations not covered in the specification.

Table 1.1. Values for the Reduction Coefficient U Based on AISCS B3.

Shape[a]	No. of Fasteners Per Line	U
1) W, M, or S shapes where $b_f \geq \frac{2}{3} d$. Tee sections from above shapes Connection to the flanges.	3 or more	0.90
2) W, M, or S shapes not meeting above conditions and all other shapes including built-up sections	3 or more	0.85
3) All members	2	0.75

[a] b_f = flange width; d = member depth.

TENSION MEMBERS 9

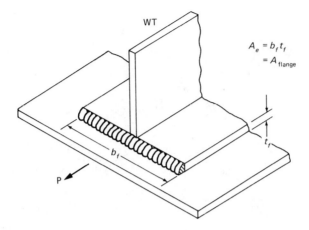

Fig. 1.6. Calculation of A_e for a member with a transverse weld.

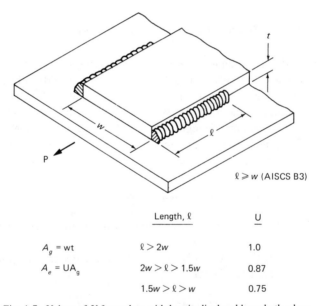

	Length, ℓ	U
$A_g = wt$	$\ell > 2w$	1.0
$A_e = UA_g$	$2w > \ell > 1.5w$	0.87
	$1.5w > \ell > w$	0.75

Fig. 1.7. Values of U for a plate with longitudinal welds on both edges.

Figure 1.8 illustrates this method for the case of a rolled member with longitudinal welds along each edge of the flange only. Note that this method will yield more conservative values of U than those which are specifically given in AISCS B3 for both bolted and welded connections.

For the case of shapes which are connected by both longitudinal welds and

10 STEEL DESIGN FOR ENGINEERS AND ARCHITECTS

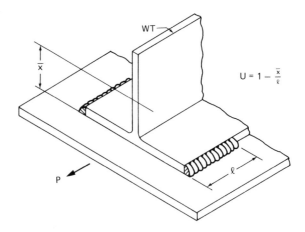

$\bar{x}$ = distance from the centroid of the shape to the plane of the connection, in.

ℓ = weld length, in.

Fig. 1.8. Determination of U as defined in AISCS Commentary B3.

transverse welds at the ends, it is the understanding of the authors that the effective net area A_e can be taken as $0.85 A_g$ (i.e., $U = 0.85$). For plates, however, U should equal 1 since the tensile load is transmitted directly to the total cross-sectional area of the member. When a tensile load is transmitted by longitudinal or a combination of longitudinal and transverse welds through one leg of a single-angle tensile member, Sect. 2 of the Specification for Allowable Stress Design of Single-Angle Members requires that $A_e = 0.85 A_g$ (see eqn. (2-1), p. 5-310).

Example 1.5. What is the effective net area of the angle in Example 1.1?

Solution. The effective net area is the product of the net area and the required reduction coefficient (AISCS B3).

$$A_e = U \times A_n$$

$$U = 0.85 \text{ (from Table 1.1)}$$

$$A_n = 3.31 \text{ in.}^2 \text{ (from Example 1.1)}$$

$$A_e = 0.85 \times 3.31 \text{ in.}^2 = 2.81 \text{ in.}^2$$

Example 1.6. Determine A_e for the WT 7 × 13 shown below.

TENSION MEMBERS 11

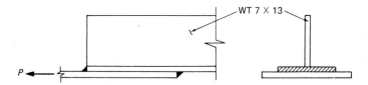

Solution. A_e = area of directly connected elements (AISCS B3)

$$= b_f \times t_f$$
$$= 5.025 \text{ in.} \times 0.42 \text{ in.} = 2.11 \text{ in.}^2$$

Example 1.7. Determine A_e for the 1 × 5 in. plate shown below if (a) $l = 7$ in., (b) $l = 8.5$ in., (c) $l = 11$ in.

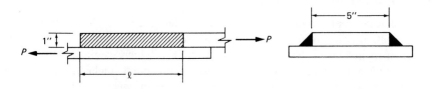

Solution. $A_e = UA_g$ (B3-2)

$$A_g = 1 \text{ in.} \times 5 \text{ in.} = 5 \text{ in.}^2$$

$w = 5$ in., $1.5w = 7.5$ in., $2w = 10$ in.

a) Since $1.5w > l > w$ in this case, then $U = 0.75$ (see Fig. 1.7) and $A_e = 0.75 \times 5 = 3.75$ in.^{2}
b) Since $2w > l > 1.5w$ in this case, then $U = 0.87$ and $A_e = 0.87 \times 5 = 4.35$ in.^{2}
c) Since $l > 2w$, then $U = 1.0$ and $A_e = 5$ in.^{2}

Example 1.8. Determine A_e for the WT 5 × 15 shown below.

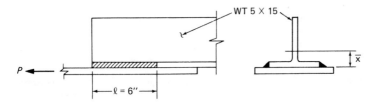

12 STEEL DESIGN FOR ENGINEERS AND ARCHITECTS

Solution. Since this problem is not specifically referred to in AISCS B3, use the procedure given in the commentary (see Fig. 1.8):

$$U = 1 - \frac{\bar{x}}{l}$$

$\bar{x} = 1.1$ in. (see AISCM Properties Section)

$l = 6$ in.

$$U = 1 - \frac{1.1}{6} = 0.82$$

$$A_e = UA_g = 0.82 \times 4.42 = 3.61 \text{ in.}^2$$

1.3 ALLOWABLE TENSILE STRESSES

A tension member can fail in either of two modes: excessive elongation of the gross section or localized fracture of the net section.

As an applied tensile force increases, the strain will increase linearly until the stress reaches its yield stress F_y (Fig. 1.9). At this point, inelastic strain will develop and continue in the ultimate stress (F_u) region, where additional stress capacity is realized. Once the yield stress has been reached and inelastic elongation occurs, the member's usefulness is diminished. Furthermore, the failure of other members in the structural system may result. To safeguard against yield failure, the AISCS D1 states that stress on the gross member section (except

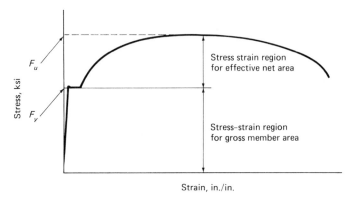

Fig. 1.9. Stress-strain diagram of mild steel. The design of tension members is based on strain (rate of elongation) on the gross member area and stress on the effective net area.

pin-connected members) shall not exceed

$$F_t = 0.60F_y \quad (1.4)$$

Localized fracture will occur at the net section of least resistance. The value of the load may be less than that required to yield the gross area. Therefore, the stress on the effective net section, as defined in AISCS D1, shall not be greater than

$$F_t = 0.50F_u \quad (1.5)$$

Hence, a factor of safety of 1.67 against yielding of the entire member and 2.0 against fracture of the weakest effective net area has been established.

For pin-connected members, the allowable stress on its net section is given in AISCS D3.1:

$$F_t = 0.45F_y \quad (1.6)$$

Table 1.2. Allowable Stresses for Tension Members.

F_y	On Net Area $0.45\,F_y$	On Gross Area $0.60\,F_y$
36	16.2	22.0
42	18.9	25.2
45	20.3	27.0
50	22.5	30.0
55	24.8	33.0
60	27.0	36.0
65	29.3	39.0
90	40.5	54.0
100	45.0	60.0

F_u	On Nominal Rod Area $0.33\,F_u$	On Eff. Net Area $0.50\,F_u$
58.0	19.1	29.0
60.0	19.8	30.0
65.0	21.5	32.5
70.0	23.1	35.0
75.0	24.8	37.5
80.0	26.4	40.0
100.0	33.0	50.0
110.0	36.3	55.0

14 STEEL DESIGN FOR ENGINEERS AND ARCHITECTS

The design of threaded rods incorporates the use of the nominal area of the rod, that is, the area corresponding to its gross diameter. To allow for the reduced area through the threaded part, the allowable stress for threaded bars is now limited to (see Table J3.2):

$$F_t = 0.33F_u \tag{1.7}$$

Computed values of allowable yield stress and allowable ultimate stress are provided in Table 1.2. These values can also be obtained from Tables 1 and 2 in the AISC Numerical Values Section, pp. 5-117 and 5-118.

Example 1.9. Determine the member capacity for the section shown in Example 1.1.

Solution. From Ex. 1.1, $A_g = 3.75$ in.^{2}

$$A_n = 3.31 \text{ in.}^2$$

From Ex. 1.5, $A_e = 2.81$ in.^{2}

$$P_{max} = \text{lesser of} \begin{cases} 0.6F_y A_g = 22 \times 3.75 = 82.5 \text{ k} \\ 0.5F_u A_e = 29 \times 2.81 = 81.5 \text{ k (governs)} \end{cases}$$

Example 1.10. What tensile load can a $4 \times 4 \times \frac{3}{8}$ angle carry with the connections shown?

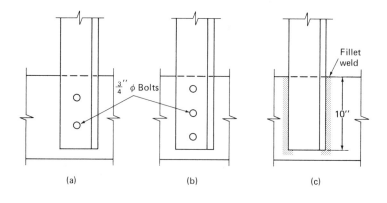

(a) (b) (c)

Solution.

For a 4 × 4 × ⅜ angle

$$A_g = 2.86 \text{ in.}^2$$

$$A_n = 2.86 \text{ in.}^2 - [\tfrac{3}{8} \text{ in.} \times (\tfrac{3}{4} \text{ in.} + \tfrac{1}{16} \text{ in.} + \tfrac{1}{16} \text{ in.})] = 2.53 \text{ in.}^2$$

$$A_e = U \times A_n$$

a)
$$U = 0.75 \text{ (from Table 1.1)}$$

$$A_e = 0.75 \times 2.53 \text{ in.}^2 = 1.90 \text{ in.}^2$$

$$P_{\max} = \text{lesser of} \begin{cases} 0.6F_y A_g = 22 \times 2.86 = 62.9 \text{ k} \\ 0.5F_u A_e = 29 \times 1.90 = 55.1 \text{ k (governs)} \end{cases}$$

b)
$$U = 0.85$$

$$A_e = 0.85 \times 2.53 \text{ in.}^2 = 2.15 \text{ in.}^2$$

$$P_{\max} = \text{lesser of} \begin{cases} 22 \times 2.86 = 62.9 \text{ k} \\ 29 \times 2.15 = 62.4 \text{ k (governs)} \end{cases}$$

c)

From eqn. (2-1), p. 5-310,

$$A_e = 0.85 \times 2.86 \text{ in.}^2 = 2.43 \text{ in.}^2$$

$$P_{\max} = \text{lesser of} \begin{cases} 22 \times 2.86 = 62.9 \text{ k (governs)} \\ 29 \times 2.43 = 70.5 \text{ k} \end{cases}$$

From AISCS Commentary B3 (C-B1-1),

$$U = 1 - \frac{\bar{x}}{l} = 1 - \frac{1.14}{10} = 0.89; \quad A_e = 0.89 \times 2.86 = 2.55 \text{ in.}^2$$

$$P_{\max} = \text{lesser of} \begin{cases} 22 \times 2.86 = 62.9 \text{ k (governs)} \\ 29 \times 2.55 = 73.8 \text{ k} \end{cases}$$

16 STEEL DESIGN FOR ENGINEERS AND ARCHITECTS

Example 1.11. Determine the maximum tensile load a $\frac{3}{4} \times 7$-in. plate connection fitting can carry if it has welded connections and punched holes as shown. Use A572 Gr 50 steel.

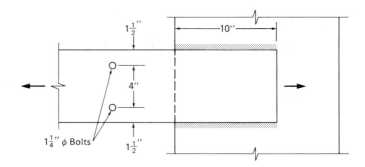

Solution. The lesser of the critical loads which can be transmitted through the bolts (P_1) or through the longitudinal welds (P_2) will govern.

- For the bolted connection:

$$A_n = 5.25 - 2 \times [0.75 \times (1.25 + \tfrac{1}{8})] = 3.19 \text{ in.}^2$$

$$A_{n,\max} = 0.85 A_g = 4.46 \text{ in.}^2 \text{ for short connection fittings}$$

Therefore, $A_n = 3.19$ in.^{2} (governs)
Since $U = 1$, $A_e = A_n = 3.19$ in.^{2}

$$P_1 = \text{lesser of} \begin{cases} 0.6 F_y A_g = 30 \times 5.25 = 157.5 \text{ k} \\ 0.5 F_u A_e = 32.5 \times 3.19 = 103.7 \text{ k (governs)} \end{cases}$$

- For the welded connection:

$$A_e = U A_g$$

$l = 10$ in., $w = 7$ in., $1.5w = 10.5$ in.; therefore, $U = 0.75$
$A_e = 0.75 \times 5.25 = 3.94$ in.^{2}

$$P_2 = \text{lesser of} \begin{cases} 30 \times 5.25 = 157.5 \text{ k} \\ 32.5 \times 3.94 = 128.1 \text{ k (governs)} \end{cases}$$

Since $P_1 = 103.7$ k is less than $P_2 = 128.1$ k, the maximum load the plate connection can resist is $P_{\max} = 103.7$ k.

Example 1.12. Calculate the allowable tensile load in the $\frac{1}{2}$-in. × 14-in. plate. The holes are for $\frac{3}{4}$-in. bolts; use A36 steel.

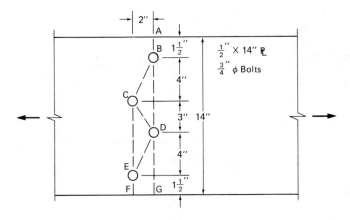

Solution.

Chain   Width − Holes + $\dfrac{s^2}{4g}$ for each diagonal path (AISCS B2)

ABDG $14 - 2 \times (\frac{3}{4} + \frac{1}{8}) = 12.25$ in.

ABDEF $14 - 3 \times (\frac{3}{4} + \frac{1}{8}) + 2^2/4(4) = 11.625$ in.

ABCEF $14 - 3 \times (\frac{3}{4} + \frac{1}{8}) + 2^2/4(4) = 11.625$ in.

ABCDG $14 - 3 \times (\frac{3}{4} + \frac{1}{8}) + 2^2/4(4) + 2^2/4(3) = 11.96$ in.

ABCDEF $14 - 4 \times (\frac{3}{4} + \frac{1}{8}) + 2 \times (2^2/4(4)) + 2^2/4(3) = 11.33$ in.

Critical section = 11.33 in. × $\frac{1}{2}$ in. = 5.67 in.^{2} = A_n.

$$A_e = 1.0 \, A_n = 5.67 \text{ in.}^2$$

$$P_{max} = \text{lesser of} \begin{cases} 22 \times 7.0 = 154 \text{ k (governs)} \\ 29 \times 5.67 = 164.4 \text{ k} \end{cases}$$

Example 1.13. A single angle tension member L 6 × 6 × $\frac{1}{2}$ has two gage lines in each leg as shown. Determine the allowable tension load that can be carried. Holes are for $\frac{3}{4}$-in. bolts.

18 STEEL DESIGN FOR ENGINEERS AND ARCHITECTS

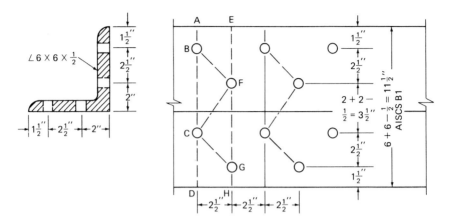

Solution.

Path ABCD or EFGH $\left(6 + 6 - \dfrac{1}{2}\right) - 2 \times \left(\dfrac{3}{4} + \dfrac{1}{8}\right) = 9.75$ in.

Path ABFCGH $\left(6 + 6 - \dfrac{1}{2}\right) - 4 \times \left(\dfrac{3}{4} + \dfrac{1}{8}\right) + 2 \times \dfrac{(2.5)^2}{4 \times 2.5}$

$+ \dfrac{(2.5)^2}{4 \times \left(2 + 2 - \dfrac{1}{2}\right)} = 9.70$ in.

Path ABFCD $\left(6 + 6 - \dfrac{1}{2}\right) - 3 \times \left(\dfrac{3}{4} + \dfrac{1}{8}\right) + \dfrac{(2.5)^2}{4 \times 2.5}$

$+ \dfrac{(2.5)^2}{4 \times \left(2 + 2 - \dfrac{1}{2}\right)} = 9.95$ in.

Path ABFGH $\left(6 + 6 - \dfrac{1}{2}\right) - 3 \times \left(\dfrac{3}{4} + \dfrac{1}{8}\right) + \dfrac{(2.5)^2}{4 \times 2.5}$

$= 9.5$ in. (governs)

$A_g = 5.75$ in.^{2}

$A_n = 9.5$ in. $\times \tfrac{1}{2}$ in. $= 4.75$ in.^{2}

$A_e = 1.0 \times 4.75 = 4.75$ in.^{2}

$P_{max} = $ lesser of $\begin{cases} 22 \times 5.75 = 126.5 \text{ k (governs)} \\ 29 \times 4.75 = 137.8 \text{ k} \end{cases}$

Example 1.14. A threaded rod cut from A36 stock is to be used as a tie bar carrying a 10-kip tensile force. Determine the diameter required.

Solution. The allowable tensile stress can be found in AISC Table J3.2, and is repeated in Table 1.2 for convenience.

$$F_t = 0.33 F_u = 19.1 \text{ ksi}$$

The stress is to be calculated on the nominal body area

$$A = \frac{P}{F_t} = \frac{10 \text{ k}}{19.1 \text{ ksi}} = 0.52 \text{ in.}^2$$

$$\frac{\pi d^2}{4} > 0.52 \text{ in.}^2$$

$$d^2 > 0.66 \text{ in.}^2$$

$$d > 0.81 \text{ in.}$$

Use minimum $\frac{7}{8}$-in. ϕ rod.

1.4 AISC DESIGN AIDS

The AISCM provides a chart to determine net areas for double angles in tension. Common double angles used as tension members are given on p. 4-96 for two, four, and six holes out. To use, simply find the angle designation in the left column and find the number of holes out in the fastener size in the top row. To find A_n, as determined in accordance with AISCS B1 and B2, carry the angle designation across and the fastener diameter down until the two lines intersect. The appropriate U value is then applied to determine the effective net section. Values for single angles can also be determined by assuming one hole out instead of two for double angles, two holes out instead of four, and three holes out instead of six, and noting that the area will be *one-half* of the value listed.

AISCM (p. 4-98) also provides a table for area reductions due to holes. The thickness of steel along the left margin is matched with the hole diameter, and the value determined is the area of the hole.

Values for $s^2/4g$ can be determined from the chart on p. 4-99 labeled "Net Section of Tension Members." Entering the chart on the side with the gage value g, carry the line across until it intersects the curve for the appropriate pitch s. The value found vertically above or below that point is $s^2/4g$.

20 STEEL DESIGN FOR ENGINEERS AND ARCHITECTS

Fig. 1.10. Wind bracing for an office building, each made from four angles. Note the stiffeners on the beam web to resist the large concentrated reactions from the braces.

Example 1.15. Using the AISC Design Aids, find the net area for two 4 × 4 × $\frac{3}{8}$ angles with two rows of $\frac{3}{4}$-in. bolts in each angle.

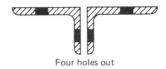

Four holes out

Solution. The 4 × 4 × $\frac{3}{8}$-in. angle can be found in the "Connections" section of AISCM. With two $\frac{3}{4}$-in. bolts in each of the angles, there are a total of four holes out. Finding the correct column for $\frac{3}{4}$-in. fasteners with four holes out, follow down until it intersects with the row of L 4 × 4 × $\frac{3}{8}$ in. The value for the net area is 4.41 in.^{2}

$$A_n = 4.41 \text{ in.}^2$$

Example 1.16. Using the AISC chart, find the net area for an angle 6 × 4 × $\frac{3}{8}$ with two rows of $\frac{3}{4}$-in. bolts.

Solution. The angle $6 \times 4 \times \frac{3}{8}$ can be considered as one-half of a double angle. Therefore, use the column for four holes out with a double angle. Reading down the column for $\frac{3}{4}$-in. fasteners and across for the L $6 \times 4 \times \frac{3}{8}$ in. yields a value of 5.91 in.^{2}. Because this gives the net area for two angles, the value must be divided by 2, which gives a net area for a single angle as 2.95 in.^{2}

$$A_n = 2.95 \text{ in.}^2$$

Example 1.17. Two $6 \times 4 \times \frac{1}{2}$ angles are connected with long legs back to back. Assume two rows of $\frac{3}{4}$-in. bolts are used in the long legs and one row of $\frac{3}{4}$-in. bolts is used in the short legs. Determine the maximum tensile load that can be carried.

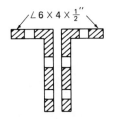

Solution. Two methods will be shown, one using the tables in the AISCM.

a) Using the AISCM, determine the number of holes out, and read down the column of the row corresponding to the angle designation.
From the AISCM

$$A_g = 2 \times 4.75 \text{ in.}^2 = 9.50 \text{ in.}^2$$

$$A_n = 6.88 \text{ in.}^2 \quad (\text{p. 4-96})$$

$$U = 1.0$$

$$A_e = 1.0 \times A_n = 6.88 \text{ in.}^2$$

$$P_{max} = \text{lesser of } \begin{cases} 22 \times 9.5 = 209 \text{ k} \\ 29 \times 6.88 = 199.5 \text{ k (governs)} \end{cases}$$

b) $A_n = 2 \times [4.75 \text{ in.}^2 - 3 \times \frac{1}{2} \text{ in.} \times (\frac{3}{4} \text{ in.} + \frac{1}{8} \text{ in.})] = 6.88 \text{ in.}^2$
The rest of the computation is the same as part a.

Example 1.18. Using the AISC charts for tension member net areas, determine the net areas and maximum tensile load for
a) $6 \times 6 \times \frac{3}{4}$ angle connected by two lines of 3-$\frac{3}{4}$-in. bolts in one leg.
b) $6 \times 6 \times \frac{3}{4}$ angle connected by one line of 3-$\frac{3}{4}$-in. bolts in each leg, and connected with lug angles.

22 STEEL DESIGN FOR ENGINEERS AND ARCHITECTS

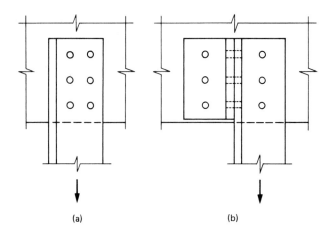

(a)   (b)

Solution.

$$A_g = 8.44 \text{ in.}^2$$

a) Using the column for $\frac{3}{4}$-in. fasteners with four holes out (two holes out in a single angle), and the row for L $6 \times 6 \times \frac{3}{4}$

$$A_n = 14.3 \text{ in.}^2/2 = 7.15 \text{ in.}^2$$

$$A_e = U \times A_n$$

$$U = 0.85$$

$$A_e = 0.85 \times 7.15 \text{ in.}^2 = 6.08 \text{ in.}^2$$

$$A_g = 8.44 \text{ in.}^2$$

$$P_{max} = \text{lesser of} \begin{cases} 22 \times 8.44 = 185.7 \text{ k} \\ 29 \times 6.08 = 176.3 \text{ k (governs)} \end{cases}$$

b) The load is transferred through both legs of the angle, and therefore no reduction need be taken:

$$A_n = 14.3 \text{ in.}^2/2 = 7.15 \text{ in.}^2$$

$$U = 1.0$$

$$A_e = 7.15 \text{ in.}^2$$

$$P_{max} = \text{lesser of} \begin{cases} 22 \times 8.44 = 185.7 \text{ k (governs)} \\ 29 \times 7.15 = 207.4 \text{ k} \end{cases}$$

1.5 SLENDERNESS AND ELONGATION

To prevent lateral movement or vibrations, the AISC recommends limits to the slenderness ratio l/r for tension members other than rods. Although not essential to structural integrity, AISCS B7 recommends that the slenderness ratio l/r be limited to 300, where l is the length of the member in inches and r is the least radius of gyration, equal to $\sqrt{I/A}$.

Provided tension members are designed within stated allowable stresses, elongations of tension members should not be critical. Should the elongation of a member be desired, however, it can be calculated in the elastic range ($f_t \leq F_y$) by

$$\Delta = \frac{Pl}{AE} \qquad (1.8)$$

where l is the member length in inches and E is the modulus of elasticity. For this calculation, the area should be taken as the gross area, though at net sections the strain value will locally be greater.

Example 1.19. A WT 8 × 13 structural tee is used as a main tension member with a length of 20 ft. Determine if the member is within recommended AISC limits to the slenderness ratio.

Solution. (AISCS B7)

$$\frac{l}{r} \leq 300$$

The member length in inches is 20 ft × 12 in./ft = 240 in.

Checking properties for designing, $r_{xx} = 2.47$ in., $r_{yy} = 1.12$ in. Use $r_{yy} = 1.12$ in. (least value):

$$\frac{l}{r} = \frac{240 \text{ in.}}{1.12 \text{ in.}} = 214$$

$$214 < 300 \quad \text{ok}$$

The member satisfies the AISC recommended slenderness limit.

24 STEEL DESIGN FOR ENGINEERS AND ARCHITECTS

Example 1.20. A 5 × 3 × $\frac{5}{16}$-in. angle is used as a bracing member carrying tension. The member is to be within recommended AISC slenderness ratio limits. Determine the maximum length of the member to be within AISC limits.

Solution. (AISCS B7)

$$\frac{l}{r} \leq 300$$

$$l \leq 300 \times r$$

$$r_{xx} = 1.61 \text{ in.}; \quad r_{yy} = 0.853 \text{ in.}; \quad r_{zz} = 0.658 \text{ in.}$$

Use $r_{zz} = 0.658$ in.

$$l < 300 \times 0.658 \text{ in.} = 197.4 \text{ in.} = 16.45 \text{ ft}$$

To satisfy recommended AISC slenderness limits, the length of the member cannot exceed 16 ft, 5 in.

Example 1.21. Design the 12 ft WT 4 structural tee shown to carry 60 kips and satisfy the recommended slenderness ratio. Use $\frac{7}{8}$-in. bolts.

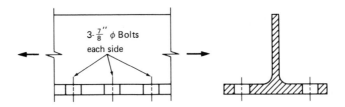

Solution.

$$\text{Gross area required} = \frac{60 \text{ k}}{22.0 \text{ ksi}} = 2.73 \text{ in.}^2$$

$$\text{Effective net area required} = \frac{60 \text{ k}}{29.0 \text{ ksi}} = 2.07 \text{ in.}^2$$

Using the tables for structural tees cut from W shapes, find the most economical section by choosing one with the required area and the least weight.

Try WT 4 × 10.5

$A_g = 3.08$ in.$^2 > 2.73$ in.^{2} ok

$A_n = 3.08$ in.$^2 - 2 \times [0.400$ in. $\times (\frac{7}{8}$ in. $+ \frac{1}{8}$ in.$)] = 2.28$ in.^{2}

$U = 0.90$ ($b_f = 5.27$ in. $> \frac{2}{3}(4.14$ in.$) = 2.76$ in.)

$A_e = U \times A_n = 0.90 \times 2.28$ in.$^2 = 2.05$ in.$^2 < 2.07$ in.^{2} N.G.

Try WT 4 × 12

$A_g = 3.54$ in.$^2 > 2.73$ in.^{2} ok

$A_n = 3.54$ in.$^2 - 2 \times \left[0.400 \text{ in.} \times \left(\frac{7}{8} \text{ in.} + \frac{1}{8} \text{ in.} \right) \right] = 2.74$ in.^{2}

$A_e = 0.90 \times 2.74$ in.$^2 = 2.47$ in.$^2 > 2.07$ in.^{2} ok

$r_{xx} = 0.999$ in.; $r_{yy} = 1.61$ in.

$\dfrac{l}{r} = \dfrac{12 \text{ ft} \times 12 \text{ in/ft}}{0.999 \text{ in.}} = 144 < 300$ ok

Use WT 4 × 12.

1.6 PIN-CONNECTED MEMBERS AND EYEBARS

Eyebar members and pin-connected plates are designed to carry the tensile load through the bar and transfer the load through the pinhole to the pin. The allowable stress in the eyebar is $F_t = 0.45F_y$ and is taken across the member net area. Figures 1.11 and 1.12 show the requirements for eyebars and pin-connected plates, as stated in AISCS D3. When exposed to weather, the pins and the eyebar may have a tendency to rust, which could freeze the joint and cause some distress. To avoid this, many designers prefer using plates of A588 steel (Gr 50) and stainless steel pins.

26 STEEL DESIGN FOR ENGINEERS AND ARCHITECTS

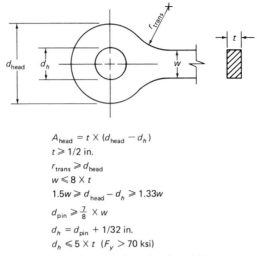

$A_{head} = t \times (d_{head} - d_h)$
$t \geq 1/2$ in.
$r_{trans} \geq d_{head}$
$w \leq 8 \times t$
$1.5w \geq d_{head} - d_h \geq 1.33w$
$d_{pin} \geq \frac{7}{8} \times w$
$d_h = d_{pin} + 1/32$ in.
$d_h \leq 5 \times t$ $(F_y > 70$ ksi$)$
Note: Eyebar shall be of uniform thickness, without reinforcement at the pinholes.

Fig. 1.11. Design requirements of eyebars according to AISCS D3.3.

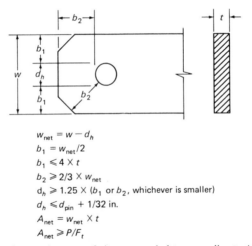

$w_{net} = w - d_h$
$b_1 = w_{net}/2$
$b_1 \leq 4 \times t$
$b_2 \geq 2/3 \times w_{net}$
$d_h \geq 1.25 \times (b_1$ or b_2, whichever is smaller$)$
$d_h \leq d_{pin} + 1/32$ in.
$A_{net} = w_{net} \times t$
$A_{net} \geq P/F_t$

Fig. 1.12. Design requirements of pin-connected plates according to AISCS D3.2.

Example 1.22. Design an eyebar to carry a tensile load of 150 kips.

Solution. (AISCS D3)

$$F_t = 0.45 F_y \quad \text{(AISCS D3.1)}$$

$$F_t = 16.2 \text{ ksi}$$

$$A_n \geq \frac{150 \text{ k}}{16.2 \text{ ksi}} = 9.26 \text{ in.}^2$$

Try plate $1\frac{1}{4}$ in. $\times$ $7\frac{1}{2}$ in.; $A_g = 9.38$ in.^{2} > 150 k$/0.6 \times 36 = 6.9$ in.^{2} ok
(AISCS D3.1)

$$w \leq 8t$$

$$7\frac{1}{2} \text{ in.} < 8 \times 1\frac{1}{4} \text{ in.} = 10 \text{ in.} \quad \text{ok}$$

$$d_{\text{pin}} \geq \tfrac{7}{8} w$$

$$d_{\text{pin}} \geq \tfrac{7}{8} \times 7\tfrac{1}{2} \text{ in.} = 6\tfrac{18}{32} \text{ in.} \quad (6.56 \text{ in.})$$

Use $6\frac{5}{8}$ in. diameter pin

$$d_n = d_{\text{pin}} + \tfrac{1}{32} \text{ in.} = 6\tfrac{21}{32} \text{ in.} \quad (6.66 \text{ in.})$$

$$1.5w \geq d_{\text{head}} - d_n \geq 1.33w$$

$$1.5w = 1.5 \times 7.5 \text{ in.} = 11.25 \text{ in.}$$

$$1.33w = 1.33 \times 7.5 \text{ in.} = 9.98 \text{ in.}$$

$$6.66 + 11.25 = 17.91 \text{ in.} \geq d_{\text{head}} \geq 6.66 + 9.98 = 16.64 \text{ in.}$$

Try $d_{\text{head}} = 17.25$ in.

$$A_n = \left(17\tfrac{1}{4} \text{ in.} - 6\tfrac{21}{32} \text{ in.}\right) \times 1\tfrac{1}{4} \text{ in.} = 13.24 \text{ in.}^2 > 9.26 \text{ in.}^2 \quad \text{ok}$$

$$r_{\text{trans}} \geq d_{\text{head}}$$

Use $17\frac{1}{4}$-in. radius
Use eyebar as shown.

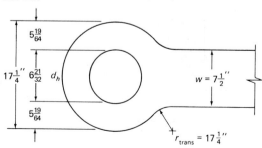

28 STEEL DESIGN FOR ENGINEERS AND ARCHITECTS

Example 1.23. Design a pin-connected plate to carry a tensile load of 150 kips. Assume the pin diameter to be $5\frac{1}{4}$ in.

Solution. (AISCS D3)

$$F_t = 0.45 F_y = 16.2 \text{ ksi} \quad (\text{AISCS D3.1})$$

$$A_n \geq \frac{150 \text{ k}}{16.2 \text{ ksi}} = 9.26 \text{ in.}^2$$

$$b_1 \leq 4t$$

Try $t = 1\frac{1}{8}$ in.

$$b_1 \leq 4t = 4\frac{1}{2} \text{ in.}$$

Try $b_1 = 4\frac{1}{8}$ in.

$$A_n = (2 \times 4\frac{1}{8} \text{ in.}) \times 1\frac{1}{8} \text{ in.} = 9.28 \text{ in.}^2 > 9.26 \text{ in.}^2 \quad \text{ok}$$

$$d_n \geq 1.25 \times 4\frac{1}{8} \text{ in.} = 5.16 \text{ in.}$$

$$d_n \geq 5\frac{1}{4} \text{ in.} + \frac{1}{32} \text{ in.} = 5\frac{9}{32} \text{ in.} > 5.16 \text{ in.} \quad \text{ok}$$

$$b_2 \geq \frac{2}{3} w_{net} = \frac{2}{3} \times (2 \times 4\frac{1}{8} \text{ in.}) = 5\frac{1}{2} \text{ in.}$$

Use pin-connected plate as shown.

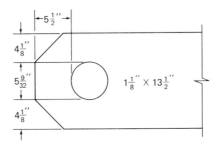

1.7 BUILT-UP MEMBERS

Requirements for built-up tension members are discussed in AISCS D2. For two plates or a plate and a rolled shape, the longitudinal spacing of rivets, bolts, or intermittent fillet welds shall not exceed 24 times the thickness of the thinner plate nor 12 in. for painted members or unpainted members not subject to corrosion; or 14 times the thickness of the thinner plate, nor 7 in. for unpainted members of weathering steel subject to atmospheric corrosion. The longitudinal spacing of rivets, bolts, or intermittent welds connecting two or more rolled

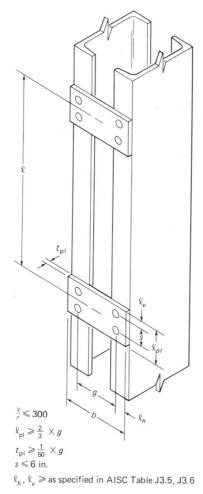

$\frac{\ell}{r} \leqslant 300$

$\ell_{pl} \geqslant \frac{2}{3} \times g$

$t_{pl} \geqslant \frac{1}{50} \times g$

$s \leqslant 6$ in.

$\ell_h, \ell_v \geqslant$ as specified in AISC Table J3.5, J3.6

Fig. 1.13. Spacing requirements of tie plates for built-up tension members.

30 STEEL DESIGN FOR ENGINEERS AND ARCHITECTS

shapes shall not exceed 24 in. For members separated by intermittent fillers, connections must be made at intervals such that the slenderness ratio of either component between the fasteners does not exceed 300.

Perforated cover plates or tie plates without lacing may be used on the open sides of the built-up tension members (Fig. 1.13). Such tie plates must be designed to satisfy the criteria below. The spacing shall be such that the slenderness ratio of any component in the length between tie plates will not exceed 300.

Example 1.24. A 30-ft pinned member is to consist of four equal leg angles arranged as shown. The tensile load is to be 150 kips. One $\frac{5}{8}$ in. bolt will be used in each angle leg at the location of every tie plate.

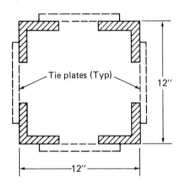

Solution.

$$\text{Gross area required} = \frac{150 \text{ k}}{22.0 \text{ ksi}} = 6.82 \text{ in.}^2$$

$$\text{Gross area for one angle} = \frac{6.82 \text{ in.}^2}{4} = 1.7 \text{ in.}^2$$

$$\text{Effective net area required} = \frac{150 \text{ k}}{29.0 \text{ ksi}} = 5.17 \text{ in.}^2$$

$$\text{Effective net area for one angle} = \frac{5.17 \text{ in.}^2}{4} = 1.29 \text{ in.}^2$$

TENSION MEMBERS

Try $L\ 2\frac{1}{2} \times 2\frac{1}{2} \times \frac{1}{2}$:

$A_g = 2.25$ in.$^2 > 1.7$ in.^{2} ok

$r_{zz} = 0.487$ in.

$A_n = 2.25 - 2 \times \left[\left(\dfrac{5}{8} + \dfrac{1}{8}\right) \times \dfrac{1}{2}\right] = 1.5$ in.^{2}

$A_e = 1 \times A_n = 1.5$ in.$^2 > 1.29$ in.^{2} ok

$I = I_o + Ad^2 = 4 \times [1.23 + 2.25(6 - 0.806)^2] = 247.7$ in.^{4}

$r = \sqrt{\dfrac{I}{A}} = \sqrt{\dfrac{247.7}{4 \times 2.25}} = 5.25$ in.

$\dfrac{l}{r} = \dfrac{30 \times 12}{5.25} = 68.6 < 300$ ok

Maximum spacing of tie plates $= l_{max} = \dfrac{300 \times 0.487}{12} = 12.1$ ft

Use 10 ft 0 in. spacing (third points).

Plate length $= l_{pl} \geq \frac{2}{3} g$

Use $g = 12$ in. $- (2 \times 1\frac{3}{8}$ in.$) = 9.25$ in. (see Fig. 1.13 and p. 1-52 of the AISCM for the usual gages for angles).

$l_{pl} \geq \frac{2}{3}(9.25$ in.$) = 6.167$ in. Use $6\frac{1}{4}$ in.

Plate thickness $= t \geq g/50 = 9.25$ in./50 $= 0.185$ in. Use $\frac{3}{16}$ in.

Minimum width of tie plates $= g + 2l_{h,min}$

$l_{h,min} = 1\frac{1}{8}$ in. (see Table J3.5)

Minimum width $= 9.25$ in. $+ (2 \times 1\frac{1}{8}$ in.$) = 11.5$ in.

32 STEEL DESIGN FOR ENGINEERS AND ARCHITECTS

Use tie plates $\frac{3}{16} \times 6\frac{1}{4} \times 11\frac{1}{2}$ in. at a maximum spacing of 10 ft 0 in.

Example 1.25. Design the most economical W 6 or W 8 shape to carry a 100-kip tensile load. The length is to be 20 ft, and two rows of 3-$\frac{3}{4}$-in. bolts will be used in each flange.

Solution.

$$\text{Gross area required} = \frac{100 \text{ k}}{22.0 \text{ ksi}} = 4.55 \text{ in.}^2$$

$$\text{Effective net area required} = \frac{100 \text{ k}}{29.0 \text{ ksi}} = 3.45 \text{ in.}^2$$

Try W 6 × 16

$A = 4.74$ in.^{2}, $t_f = 0.405$ in., $b_f = 4.03$ in., $d = 6.28$ in.

$A_n = 4.74$ in.$^2 - 4 \times [(\frac{3}{4}$ in. $+ \frac{1}{8}$ in.$) \times 0.405$ in.$] = 3.32$ in.^{2}

$b_f = 4.03$ in. $< \frac{2}{3}(6.28$ in.$) = 4.19$ in. $\therefore U = 0.85$

$A_e = U \times A_n = 0.85 \times 3.32$ in.$^2 = 2.82$ in.$^2 < 3.45$ in.^{2} N.G.

Try W 8 × 18

$A = 5.26$ in.^{2}, $t_f = 0.33$ in., $b_f = 5.25$ in., $d = 8.14$ in.

$A_n = 5.26$ in.$^2 - 4 \times \left[\left(\frac{3}{4}\text{ in.} + \frac{1}{8}\text{ in.}\right) \times 0.33 \text{ in.}\right] = 4.11$ in.^{2}

$b_f = 5.25$ in. $< \frac{2}{3}(8.14$ in.$) = 5.42$ in. $\therefore U = 0.85$

$A_e = U \times A_n = 0.85 \times 4.11$ in.$^2 = 3.49$ in.$^2 > 3.45$ in.^{2} ok

$l = 240$ in.; $r_{yy} = 1.23$ in.

$$\frac{l}{r} = \frac{240 \text{ in.}}{1.23 \text{ in.}} = 195 < 300 \quad \text{ok}$$

Use W 8 × 18.

TENSION MEMBERS 33

Example 1.26. Design the most economical common single angle to carry a 50-kip tensile load. The angle is to be 15 ft long and is to be connected by four $\frac{3}{4}$-in. bolts in one line in one leg.

Solution.

$$\text{Gross area required} = \frac{50 \text{ k}}{22.0 \text{ ksi}} = 2.27 \text{ in.}^2$$

$$\text{Effective net area required} = \frac{50 \text{ k}}{29.0 \text{ ksi}} = 1.72 \text{ in.}^2 \quad (U = 0.85)$$

In determining the most economical angle, choose the angle with the least area that can carry the load. Therefore, the easiest method of investigating different angles is to make a table, as shown below. Note that the only angle which will not satisfy the strength requirements is L $3\frac{1}{2} \times 3 \times \frac{3}{8}$.

Lightest Angles Available	Wt k/ft	Gross Area (in.^{2})	Area of One Hole (in.^{2})*	Effective Net Area (in.^{2})
$3\frac{1}{2} \times 3 \times \frac{3}{8}$	7.9	2.30	0.33	1.67
$3\frac{1}{2} \times 3\frac{1}{2} \times \frac{3}{8}$	8.5	2.48	0.33	1.83
$4 \times 4 \times \frac{5}{16}$	8.2	2.40	0.27	1.81
$5 \times 3 \times \frac{5}{16}$	8.2	2.40	0.27	1.81

*$(\frac{3}{4} + \frac{1}{8}$ in.$) \times$ thickness.

Try L $3\frac{1}{2} \times 3\frac{1}{2} \times \frac{3}{8}$:

$$r_{zz} = 0.687 \text{ in.}$$

$$\frac{l}{r} = \frac{15 \text{ ft} \times 12 \text{ in.}/\text{ft}}{0.687 \text{ in.}} = 262 < 300 \quad \text{ok, but not the lightest section.}$$

Try L $5 \times 3 \times \frac{5}{16}$:

$$r_{zz} = 0.658 \text{ in.}$$

$$\frac{l}{r} = \frac{15 \text{ ft} \times 12 \text{ in.}/\text{ft}}{0.658 \text{ in.}} = 273.6 \quad \text{ok}$$

34 STEEL DESIGN FOR ENGINEERS AND ARCHITECTS

Try L 4 × 4 × $\tfrac{5}{16}$:

$$r_{zz} = 0.791 \text{ in.}$$

$$\frac{l}{r} = \frac{15 \text{ ft} \times 12 \text{ in.}/\text{ft}}{0.791 \text{ in.}} = 227.6 \quad \text{ok}$$

Use L 4 × 4 × $\tfrac{5}{16}$ or L 5 × 3 × $\tfrac{5}{16}$.

1.8 FATIGUE

Occasionally, it becomes necessary to design for fatigue if frequent variations or reversals in stress occur. An example of a system that encounters fluctuations is found in bridge structures. AISCS K4 and Appendix K4 give fatigue provisions. See Appendix A in this text for a discussion on fatigue.

PROBLEMS TO BE SOLVED

1.1. Using AISCS, determine the tensile capacity for the following tension members connected by welds:

a) One angle 4 × 6 × $\tfrac{3}{8}$, A36 steel, 6-in. outstanding leg
b) One angle 4 × 6 × $\tfrac{3}{8}$, A572 Gr 50, 4-in. outstanding leg
c) One W 8 × 24, A36 steel
d) One C 8 × 11.5, A572 Gr 50 steel

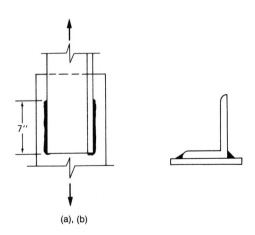

(a), (b)

TENSION MEMBERS 35

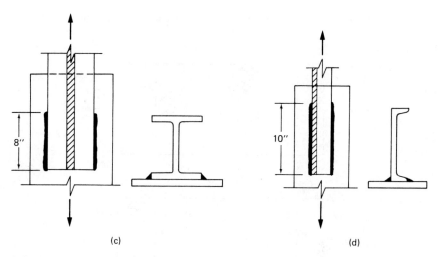

(c)   (d)

1.2. Determine the capacity of the members connected as shown. Use $\frac{7}{8}$-in. diameter bolts.

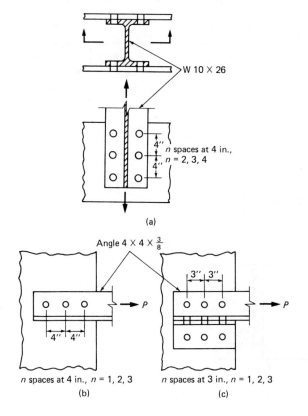

36 STEEL DESIGN FOR ENGINEERS AND ARCHITECTS

1.3. Determine the tensile capacity of $\frac{1}{2}$-in.-thick plates connected as shown. Use $\frac{7}{8}$-in. diameter bolts.

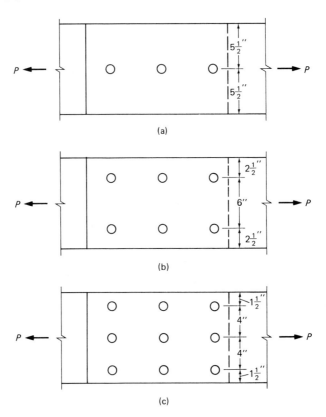

1.4. Calculate $P_{\text{allowable}}$ for the plates shown if $t = \frac{1}{2}$ in., and bolt diameter = $\frac{3}{4}$ in. Use A572 Gr 50 steel.

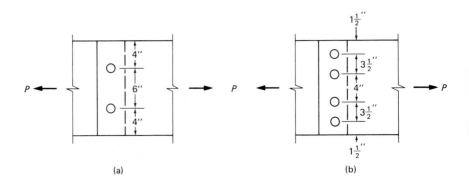

1.5. Determine the tensile capacity of the plate shown if $F_y = 36$ ksi, $t = \frac{1}{2}$ in., and bolt diameter is $\frac{3}{4}$ in.

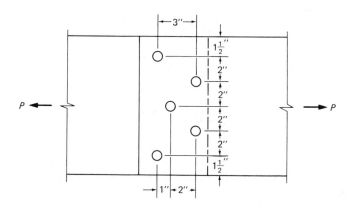

1.6. Determine the tensile capacity of a $7 \times 4 \times \frac{3}{4}$ angle as shown if bolt diameter $= 1$ in. Holes in the 4-in. leg have a $1\frac{1}{2}$-in. edge distance. Use A572 Gr 50 steel.

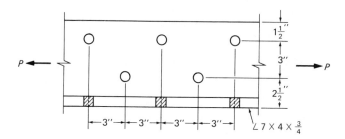

1.7. Calculate the pitch s, such that the net area is equal to the gross area less the area of two holes. What is the allowable tensile force if bolt diameters is $\frac{7}{8}$ in. and A572 Gr 50 steel is used?

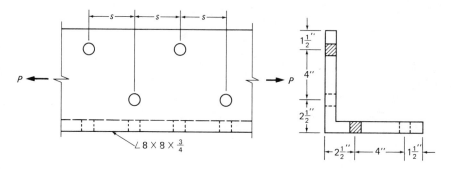

1.8. Determine the pitch s, such that the loss in area is equal to the loss of two holes. Use A36 steel and $\frac{3}{4}$-in. diameter bolts for the $7 \times 4 \times \frac{3}{4}$ angle.

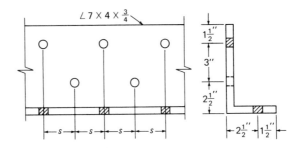

1.9. What is the maximum recommended length for a $4 \times 4 \times \frac{1}{4}$ angle tension member?

1.10. A 30-ft-long angle of equal legs is to carry 150 kips. Design the tension member so that it is within the recommended slenderness ratio. Assume the angle will have longitudinal welds on one leg only.

1.11. Design a round tension bar to carry 100 kips. Use threaded ends and A572 Gr 50 steel.

1.12. Design an eyebar to carry 200 kips. Use A36 steel.

1.13. Design a pin-connected plate to carry 200 kips. Assume 6-in. ϕ pin and A36 steel.

1.14. Design the 14×14-in. tension member shown to carry 200 kips if $L = 50$ ft, $F_y = 36$ ksi, and bolt diameter $= \frac{3}{4}$ in. Use $3\frac{1}{2}$-in. leg maximum.

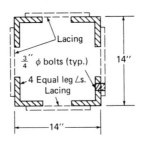

2
Members under Flexure: 1

2.1 MEMBERS UNDER FLEXURE

Flexural members are generally defined as structural members that support transverse loads. This chapter will cover the design of simple flexural members only.

According to Chapter F of the AISC, two types of flexural members are considered: beams and plate girders. Beams are distinguished from plate girders when the web slenderness ratio h/t_w is less than or equal to $760/\sqrt{F_b}$, where h is the clear distance between the flanges, t_w is the thickness of the web, and F_b is the allowable bending stress in ksi. The discussion of plate girders is given in Chapter 3.

Beams can usually be categorized into the following:
- beams per se
- joists, which are closely spaced beams supporting floors and roofs of buildings
- lintels, which span openings in walls, such as doors and windows
- spandrel beams, which support the exterior walls of buildings and, in some cases, part of the floor loads
- girders, which are generally large beams carrying smaller ones

Beams can be designed as simply supported, fixed-ended, partially fixed, or continuous. It is very important that the designer verify that the support conditions of the beam be detailed to satisfy design assumptions for the member to behave as predicted in its analysis.

2.2 DETERMINING THE ALLOWABLE BENDING STRESS

Bending stress in a beam is determined by the flexure formula

$$f_b = \frac{Mc}{I} \qquad (2.1)$$

40 STEEL DESIGN FOR ENGINEERS AND ARCHITECTS

Fig. 2.1. A wide-flange beam and girder floor system.

where M is the bending moment, c is the distance of the extreme fibers of the beam from the neutral axis, and I is the moment of inertia of the cross section. Because the section modulus of a beam is defined as the value I/c, the flexure formula becomes

$$f_b = \frac{M}{S} \tag{2.2}$$

where S is the section modulus.

Example 2.1. Calculate the maximum bending stress f_b due to a 170-ft-k moment about the strong axis on a:[1]

a) W 12 × 65 wide-flange beam
b) W 18 × 65 wide-flange beam

---

[1] d, I, and S can be obtained in the Dimensions and Properties Section of the AISCM, pp. 1-10ff.

Solution.

a) For a W 12 × 65, $S_x = 87.9$ in.^{3}

$$f_{bx} = \frac{170 \text{ ft-k} \times 12 \text{ in./ft}}{87.9 \text{ in.}^3} = 23.2 \text{ ksi}$$

b) For a W 18 × 65, $S_x = 117$ in.^{3}

$$f_{bx} = \frac{170 \text{ ft-k} \times 12 \text{ in./ft}}{117 \text{ in.}^3} = 17.4 \text{ ksi}$$

Example 2.2. Determine the bending stress on a W 12 × 79 subjected to a moment of 80 ft-k about a) the strong axis b) the weak axis.

Solution. For a W 12 × 79, $S_x = 107$ in.^{3}, $S_y = 35.8$ in.^{3}

a) $$f_{bx} = \frac{80 \text{ ft-k} \times 12 \text{ in./ft}}{107 \text{ in.}^3} = 8.97 \text{ ksi}$$

b) $$f_{by} = \frac{80 \text{ ft-k} \times 12 \text{ in./ft}}{35.8 \text{ in.}^3} = 26.8 \text{ ksi}$$

The allowable bending stresses are given in AISCS F1, F2, and F3 for strong axis bending of I-shaped members and channels; weak axis bending of I-shaped members, solid bars, and rectangular plates; and bending of box members, rectangular tubes, and circular tubes, respectively. The following discussion will present these allowable stresses in detail.

Allowable Stress: Strong Axis Bending of I-Shaped Members and Channels (AISCS F1)

For a compact member whose compression flange is adequately braced, the allowable bending stress F_b is given by (F1-1):

$$F_b = 0.66F_y \tag{2.3}$$

Compact members are defined as those capable of developing their full plastic moment before localized buckling occurs, provided that the conditions in AISCS F1.1 are satisfied. The yield stresses beyond which a shape is not compact are

42 STEEL DESIGN FOR ENGINEERS AND ARCHITECTS

labeled F'_y and F'''_y and are given in the properties tables in the AISCM.[2] The limits for compactness are given in Table B5.1. Almost all of the W and S shapes are compact for A36 steel, and only some are not compact for 50 ksi steel.

Members bent about their major axis and having an axis of symmetry may fail by buckling of their compression flange and twisting about the longitudinal axis. To avoid this, the member must be "laterally braced" within certain intervals to resist such buckling. When a transverse load is applied to a beam, the compression flange behaves in the same manner as a column. As the length of the member increases, the flange tends to buckle. The resulting displacements in the weaker axis will induce torsion and may ultimately cause failure.

What constitutes lateral support is at times a matter of judgment. A beam flange encased in a concrete slab is fully laterally supported. Cross beams framing into the sides of beams provide lateral support *if* an adequate connection is made to the compression flange. However, care must be taken to provide rigidity to the cross beams. It may be necessary to provide diagonal bracing in one section to resist movement in both directions. Bracing as shown in Fig. 2.2 will provide rigidity for several bays.

Metal decking, in some cases, does not constitute lateral bracing. With adequate connections, up to full lateral support may be assumed. Cases of partial support are usually transformed to full support by a multiple of the actual spacing. For instance, decking that is tack-welded every 4 ft may be considered to provide a third of the full lateral support, yielding an equivalent full lateral support every 12 ft.

In most situations, the compression flange is laterally supported, and therefore $F_b = 0.66F_y$, as given in eqn. (2.3). At times, however, it is not possible to brace the compression flange. In such cases, $F_b = 0.66F_y$, provided that the interval of lateral support L_b is less than the smaller of the values of L_c as given

---

[2]The stresses F'_y and F'''_y are defined in the Symbols Section of the ninth edition of the AISCS (preceding the Index). For the benefit of the reader, these definitions are as follows:

F'_y The theoretical maximum yield stress (ksi) based on the width-thickness ratio of one-half the unstiffened compression flange, beyond which a particular shape is not "compact."

$$= \left[\frac{65}{b_f/2t_f}\right]^2$$

F'''_y The theoretical maximum yield stress (ksi) based on the depth-thickness ratio of the web below which a particular shape may be considered "compact" for any condition of combined bending and axial stresses.

$$= \left[\frac{257}{d/t_w}\right]^2$$

MEMBERS UNDER FLEXURE: 1   43

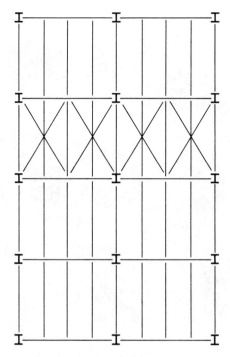

Fig. 2.2. Lateral bracing of a floor or roof framing system. Bracing in one bay can offer rigidity for several bays.

in (F1-2):

$$L_c = \frac{76b_f}{\sqrt{F_y}} \quad \text{or} \quad \frac{20{,}000}{(d/A_f)F_y}, \quad \text{whichever is smaller} \quad (2.4)$$

The AISC, in Section F1.3, specifies the stresses that can be used for certain unbraced lengths. Considering lateral instability, $F_b = 0.66F_y$, provided that the unsupported length of a beam is less than or equal to L_c. The value $F_b = 0.60F_y$ may be used when the unsupported length falls between L_c and another established length L_u.

The value of L_u is the length which provides the equality for the largest F_b in eqns. (F1-6) or (F1-7) and (F1-8) as applicable:

When

$$\sqrt{\frac{102 \times 10^3 C_b}{F_y}} \leq \frac{l}{r_T} \leq \sqrt{\frac{510 \times 10^3 C_b}{F_y}}:$$

44 STEEL DESIGN FOR ENGINEERS AND ARCHITECTS

Fig. 2.3. Lateral bracing for a floor system made with channel sections.

$$F_b = \left[\frac{2}{3} - \frac{F_y(l/r_T)^2}{1530 \times 10^3 C_b}\right] F_y \leq 0.60 F_y \quad (2.5)$$

When

$$\frac{l}{r_T} \geq \sqrt{\frac{510 \times 10^2 C_b}{F_y}}:$$

$$F_b = \frac{170 \times 10^3 C_b}{(l/r_T)^2} \leq 0.60 F_y \quad (2.6)$$

For any value of l/r_T:

$$F_b = \frac{12 \times 10^3 C_b}{ld/A_f} \leq 0.60 F_y \quad (2.7)$$

where all quantities are defined in the AISCM. In general, for any unsupported length L_b greater than L_u, the largest F_b, as determined by eqns. (2.5) or (2.6)

MEMBERS UNDER FLEXURE: 1   45

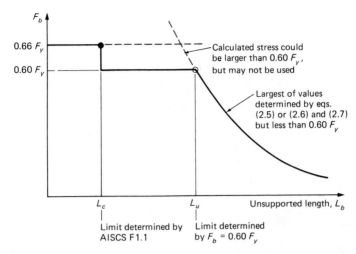

Fig. 2.4. Bending stress versus unsupported length. In the range of relatively short, unsupported lengths, the allowable stress is given as a step function. For longer, unsupported lengths, the stress is determined by a hyperbolic-type function.

and (2.7), is the one that governs. Figure 2.4 indicates the values of the allowable stress for different unsupported lengths.

For noncompact members as, defined in AISCS F1.2, with their compression flanges braced at a distance less than L_c, the allowable bending stress is given in (F1-3):

$$F_b = F_y \left[0.79 - 0.002 \frac{b_f}{2t_f} \sqrt{F_y} \right] \qquad (2.8)$$

Note that for noncompact members with compression flanges braced at a distance greater than L_c, the discussion given above for the allowable bending stresses of compact members is also valid. The study of more slender beams (i.e., those sections that exceed the noncompact limits of Table B5.1) is beyond the scope of this text. The use of design aids that are available in the AISCM will be discussed in the following examples.

Examples 2.3 and 2.7. Select the most economical[3] W sections for the beams shown. Assume full lateral support and compactness. Neglect the weight of the beam.

---

[3]Usually, "most economical" means the lightest section. However, in some cases (e.g., conditions of clearance), it may be different.

46 STEEL DESIGN FOR ENGINEERS AND ARCHITECTS

Example 2.3

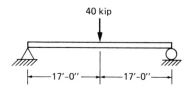

Solution 2.3.

$$M = \frac{PL}{4} = \frac{40 \text{ k} \times 34 \text{ ft}}{4} = 340 \text{ ft-k}$$

$F_b = 0.66 \, F_y$ for full lateral support (AISCS F1.1)

$F_b = 0.66 \times 36$ ksi $= 24.0$ ksi   (for A36 steel; Numerical Values Table 1, p. 5-117)

$$S_{req} = \frac{M}{F_b} = \frac{340 \text{ ft-k} \times 12 \text{ in.}/\text{ft}}{24 \text{ ksi}} = 170 \text{ in.}^3$$

W shapes that are satisfactory (from properties tables)

W 12 × 136	$S = 186$ in.^{3}
W 14 × 109	$S = 173$
W 16 × 100	$S = 175$
W 18 × 97	$S = 188$
W 21 × 83	$S = 171$
W 24 × 76	$S = 176$
W 27 × 84	$S = 213$
W 30 × 99	$S = 269$
W 33 × 118	$S = 359$
W 36 × 135	$S = 439$

The last number represents the member weight per foot length. Therefore, W 24 × 76 is the lightest satisfactory section.

The AISCM has provided design aids which facilitate the selection of an economical W or M shape in accordance with AISCS F1 (see AISCM, p. 2-4). If the beam satisfies the requirements of compactness and lateral support as given in AISCS F1.1 (i.e., $F_b = 0.66F_y$), then the most economical beam can be obtained for steels having $F_y = 36$ ksi or $F_y = 50$ ksi (shaded in gray) by utilizing the Allowable Stress Design Selection Table, which begins on p. 2-7. Once the maximum moment for the beam has been determined, enter the table and choose the first beam that appears in boldface type with a moment capacity M_R greater than or equal to the maximum applied moment. This is the most economical beam for this moment. Note that all of the beams above (and, possibly, some of the beam below) have sufficient capacity but are not the most economical. Similarly, the required section modulus S_x can be determined (i.e., $S_x = M_{max}/(0.66F_y)$), and the table can be used to obtain an economical beam in the same way as for M_R. It is important to note that if the stress F_b is different from $0.66F_y$, then the method described previously that uses the required section modulus S_x can only be used to obtain an economical beam size.

For this case, a W 24 × 76 is the lightest section which satisfies the moment requirements (i.e., M_R = 348 ft-k is larger than the applied moment M = 340 ft-k). If there was a restriction on the depth of this beam equal to 20 in., then a W 18 × 97, W 16 × 100, or W 14 × 109 could be used. Obviously, the W 18 × 97 would be the most economical section in this case.

Example 2.4

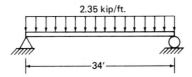

Solution 2.4.

$$M_{max} = \frac{wL^2}{8} = \frac{2.35 \text{ k} \times (34 \text{ ft})^2}{8} = 339.6 \text{ ft-k}$$

$$F_b = 0.66 \, F_y = 24.0 \text{ ksi}$$

$$S_{req} = \frac{M}{F_b} = \frac{339.6 \text{ ft-k} \times 12 \text{ in.}/\text{ft}}{24 \text{ ksi}} = 169.8 \text{ in.}^3$$

W shapes that are satisfactory are the same W shapes that are satisfactory for Example 2.3.

The lightest section is W 24 × 76, S_x = 176 in.^{3}

Alternative approach

$$S_{req} = 169.8 \text{ in.}^3$$

From S_x tables (Part 2), W 24 × 76 (in bold print) has an S_x of 176 in.^{3} and for F_y = 36 ksi can carry a moment of 348 ft-k.

Use W 24 × 76

Example 2.5

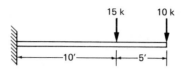

Solution 2.5.

$$M_{max} = (15 \text{ k} \times 10 \text{ ft}) + (10 \text{ k} \times 15 \text{ ft}) = 300 \text{ ft-k}$$

$$F_b = 24.0 \text{ ksi}$$

$$S_{req} = \frac{M}{F_b} = \frac{300 \text{ ft-k} \times 12 \text{ in./ft}}{24.0 \text{ ksi}} = 150 \text{ in.}^3$$

Through investigation, W 24 × 68 is the lightest section, $S = 154$ in.³

Alternative approach

$$M_{req} = 300 \text{ ft-k}$$

From the AISCM, W 24 × 68 (in bold print) has a moment capacity of 305 ft-k.

W 24 × 68 is satisfactory and is the lightest section.

Example 2.6

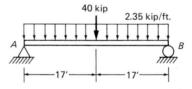

Solution 2.6. To determine the maximum moment, either (a) draw shear and moment diagrams, (b) use the method of sections, or (c) use superposition principles.

a) Total load = 40 k + (2.35 k/ft × 34 ft) = 119.9 k

$$R_A = R_B = \frac{119.9 \text{ k}}{2} = 59.95 \text{ k—say 60 k}$$

Shear just to the left of the 40-k load

$$60 \text{ k} - (2.35 \text{ k/ft} \times 17 \text{ ft}) = 20 \text{ k}$$

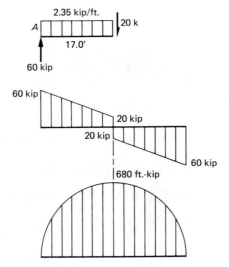

M_{max} at center = area under the shear diagram

$$M_{\text{max}} = \frac{60.0 \text{ k} + 20.0 \text{ k}}{2} \times 17 \text{ ft} = 680 \text{ ft-k}$$

b) $R_A = R_B = 60.0$ k

M_{max} at center = $(60.0 \text{ k} \times 17 \text{ ft}) - ((2.35 \text{ k/ft} \times 17 \text{ ft})$
$\times (17 \text{ ft}/2)) = 680$ ft-k

c) By inspection, the maximum moment is at the center.

Moment due to 40-kip load = $PL/4$

$$\frac{40 \text{ k} \times 34 \text{ ft}}{4} = 340 \text{ ft-k}$$

Moment due to distributed load = $wL^2/8$

$$\frac{2.35 \text{ k/ft} \times (34 \text{ ft})^2}{8} = 339.6 \text{ ft-k—say } 340 \text{ ft-k}$$

Total moment = 340 ft-k + 340 ft-k = 680 ft-k.

Using the AISCM beam chart, select a W 33 × 118 (in bold print) as the lightest satisfactory section with the moment capacity of 711 ft-k for A36 steel.

Example 2.7

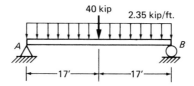

Solution 2.7. Redesign the beam in Example 2.6 for 50-ksi steel.

$$M_{max} = 680 \text{ ft-k} \quad \text{(from Example 2.6)}$$

$$F_b = 0.66 \, F_y = 0.66 \times 50 \text{ ksi} = 33.0 \text{ ksi}$$

$$S_{req} = \frac{M}{F_b} = \frac{680 \text{ ft-k} \times 12 \text{ in./ft}}{33.0 \text{ ksi}} = 247 \text{ in.}^3$$

Using the AISCM beam chart for $F_y = 50$ ksi, select a W 30 × 99 as the lightest satisfactory section, $S = 269$ in.^{3}

Example 2.8. Determine the most economical wide-flange section for the loading case shown. Assume full lateral support. Include the weight of the beam.

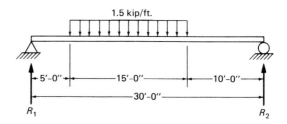

Solution. Reactions, shears, and moments can be determined from the beam diagrams and formulas (AISCM p. 2-296). Use diagram 1 for beam weight and diagram 4 for imposed loading.

- Imposed Loading

$$M_{max} = R_1 \left(a + \frac{R_1}{2w} \right) \text{ at } x = a + \frac{R_1}{w}$$

$$R_1 = \frac{wb}{2L}(2c + b)$$

$$a = 5 \text{ ft}, \quad b = 15 \text{ ft}, \quad c = 10 \text{ ft}, \quad L = 30 \text{ ft}$$

$$w = 1.5 \text{ k/ft}$$

$$R_1 = \frac{1.5 \text{ k/ft} \times 15 \text{ ft}}{2 \times 30 \text{ ft}} (2(10 \text{ ft}) + 15 \text{ ft}) = 13.13 \text{ k}$$

$$M_{max} = 13.13 \text{ k}\left(5 \text{ ft} + \frac{13.13 \text{ k}}{2(1.5 \text{ k/ft})}\right) = 123.12 \text{ ft-k}$$

$$x = 5 \text{ ft} + \frac{13.13 \text{ k}}{1.5 \text{ k/ft}} = 13.75 \text{ ft}$$

- Beam Loading

  Assume beam weight is 40 lb/ft

  Determine moment at point of maximum applied moment.

$$x = 13.75 \text{ ft}$$

$$M_x = \frac{wx}{2}(L - x) = \frac{0.04 \text{ k/ft} \times 13.75 \text{ ft}}{2}(30.0 \text{ ft} - 13.75 \text{ ft}) = 4.47 \text{ ft-k}$$

- Total Maximum Moment

$$123.12 \text{ ft-k} + 4.47 \text{ ft-k} = 127.6 \text{ ft-k}$$

Consulting the allowable stress design selection table (AISCM, Part 2), find the lightest beam with a resisting moment (M_R) greater than or equal to 127.6 ft-k ($F_y = 36$ ksi).

Use W 16 × 40 ($M_R = 128$ ft-k).

Example 2.9. Determine the distributed load that can be carried by a W 14 × 22 beam spanning 11 ft with an unsupported length of 5 ft, 6 in.

Solution. For a W 14 × 22,

$$L_c = 5.3 \text{ ft}; \quad L_u = 5.6 \text{ ft}; \quad S = 29.0 \text{ in.}^3 \quad \text{(from Part 2 of AISCM)}$$

Because the unsupported $L = 5.5$ ft is between L_c and L_u,

$$F_b = 0.6 F_y = 22.0 \text{ ksi} \quad \text{(AISCS F1.3)}$$

$$M = S \times F_b = 29.0 \text{ in.}^3 \times 22 \text{ ksi} \times \frac{1}{12 \text{ in./ft}} = 53 \text{ ft-k}$$

$$M = \frac{wL^2}{8}, \quad w = \frac{8M}{L^2} = \frac{8 \times 53 \text{ ft-k}}{(11.0)^2} = 3.5 \text{ k/ft}$$

52 STEEL DESIGN FOR ENGINEERS AND ARCHITECTS

Subtracting the weight of the beam,

$$w = 3.5 \text{ k/ft} - 0.022 \text{ k/ft} = 3.48 \text{ k/ft}$$

Example 2.10. Determine the distributed load that can be carried by a W 10 × 22 beam spanning 16 ft, 6 in. with an unsupported length of 5 ft, 6 in. F_y = 50 ksi.

Solution. For a W 10 × 22,

$$L_c = 5.2 \text{ ft}; \quad L_u = 6.8 \text{ ft}; \quad S = 23.2 \text{ in.}^3$$

Because the unsupported $L = 5.5$ ft is between L_c and L_u,

$$F_b = 0.6 F_y = 0.6 \times 50 \text{ ksi} = 30 \text{ ksi}$$

$$M = S \times F_b = 23.2 \text{ in.}^3 \times 30 \text{ ksi} \times \frac{1}{12 \text{ in./ft}} = 58.0 \text{ ft-k}$$

$$w = \frac{8M}{L^2} = \frac{8 \times 58.0 \text{ ft-k}}{(16.5)^2} = 1.70 \text{ k/ft}$$

Subtracting the weight of the beam,

$$w = 1.70 \text{ k/ft} - 0.02 \text{ k/ft} = 1.68 \text{ k/ft}$$

Example 2.11. Determine the distributed load that can be carried by a W 16 × 31 beam spanning 30 ft with an unsupported length of 15 ft, 0 in.

Solution. For a W 16 × 31,

$$L_c = 5.8 \text{ ft}; \quad L_u = 7.1 \text{ ft}; \quad S = 47.2 \text{ in.}^3$$

Because the unsupported $L = 15$ ft is greater than both L_c and L_u, F_b must be determined by AISCS Section F1.3.

Referring to the inequalities of AISC Section F1.3, we need to determine the relative magnitude of l/r_T:

$$r_T = 1.39 \text{ in.}$$

$$\frac{l}{r_T} = \frac{15 \text{ ft} \times 12 \text{ in./ft}}{1.39 \text{ in.}} = 129.5$$

The value of C_b is

$$1.75 + 1.05 \, (M_1/M_2) + 0.3 \, (M_1/M_2)^2$$

which depends on the magnitude and sign of end moments and can be taken conservatively equal to 1 (AISCS F1.3 footnote).

$$\sqrt{\frac{102 \times 10^3 \, C_b}{F_y}} = \sqrt{\frac{102 \times 10^3}{36}} = 53.2$$

$$\sqrt{\frac{510 \times 10^3 \, C_b}{F_y}} = \sqrt{\frac{510 \times 10^3}{36}} = 119.0$$

$$\frac{l}{r_T} > 119.0$$

$$F_b = \frac{170 \times 10^3 \, C_b}{(l/r_T)^2} = 10.14 \text{ ksi} \qquad \text{AISCS (F1-7)}$$

or

$$F_b = \frac{12 \times 10^3 \, C_b}{(ld/A_f)} = \frac{12 \times 10^3 \times 1}{180 \times (15.88/2.43)} = 10.21 \text{ ksi} \quad \text{AISCS (F1-8)}$$

The greater value is to be used.

$$F_b = 10.21 \text{ ksi}$$

$$M = S \times F_b = \frac{47.2 \text{ in.}^3 \times 10.21 \text{ ksi}}{12 \text{ in./ft}} = 40.2 \text{ ft-k}$$

$$w = \frac{8M}{l^2} = \frac{8 \times 40.2 \text{ ft-k}}{(30 \text{ ft})^2} = 0.36 \text{ k/ft}$$

Subtracting the beam weight to determine the beam capacity,

$$w = 0.36 \text{ k/ft} - 0.031 \text{ k/ft} = 0.33 \text{ k/ft}$$

Example 2.12. Select the lightest wide-flange section for the beam shown if lateral support is provided at the ends only. Include the weight of the beam. Use the ''Allowable Moments in Beams'' charts.

54 STEEL DESIGN FOR ENGINEERS AND ARCHITECTS

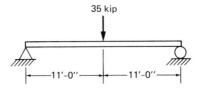

Solution. The AISCM has provided the Allowable Moments in Beams charts starting on p. 2-149 for the economical selection of a beam which includes lateral instability. The charts for $F_y = 36$ ksi and $F_y = 50$ ksi (gray border) assume a conservative value of $C_b = 1$ (see the definition of C_b and the corresponding footnote on p. 5-47). To select a member for a given moment and unbraced length, enter an appropriate chart at the required resisting moment (ordinate) and proceed to the right to meet the vertical line corresponding to the unbraced length (abscissa). Any beam located above and to the right of the intersection is satisfactory, with the beam shown in the solid line immediately to the right and above being the lightest satisfactory section. Beams indicated by broken lines satisfy the requirements of unbraced length and moment but are not the lightest sections available for those conditions. The values corresponding to L_c are indicated in the charts by solid circles (dots), and values corresponding to L_u are indicated by open circles.

In this case,

$$\text{Unsupported length } L = 22.0 \text{ ft}$$

$$\text{Beam weight (assumed)} = 80 \text{ lb/ft}$$

$$M = \frac{PL}{4} + \frac{wL^2}{8} = \frac{35.0 \text{ k} \times 22.0 \text{ ft}}{4} + \frac{0.08 \text{ k/ft} \times (22.0 \text{ ft})^2}{8}$$

$$= 197.3 \text{ ft-k}$$

A study of the chart on p. 2-168 reveals that a W 21 × 83 could be a solution; however, the line for this beam is broken in this area. Further inspection of the chart on p. 2-166 indicates that the lightest beam is a W 14 × 74, which for an unsupported length of 22 ft can carry a moment of 205.2 ft-k.

The bending stress is

$$f_b = \frac{197.3 \text{ ft-k} \times 12 \text{ in./ft}}{112 \text{ in.}^3} = 21.1 \text{ ksi}$$

Example 2.13. Select the lightest wide-flange section for the beam shown if lateral support is provided only at the point loads and beam ends. Use the "Allowable Moments in Beams" charts for $F_y = 50$ ksi.

MEMBERS UNDER FLEXURE: 1   55

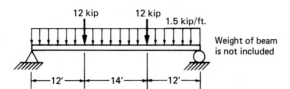

Solution.

$$\text{Maximum unsupported length } L = 14.0 \text{ ft}$$

$$\text{Beam weight (assumed)} = 90 \text{ lb/ft}$$

$$M_{max} = 1.59 \text{ k/ft} \times \frac{(38 \text{ ft})^2}{8} + (12 \text{ k} \times 12 \text{ ft}) = 431 \text{ ft-k}$$

From the "Allowable Moments in Beams" chart, W 27 × 84 is satisfactory (p. 2-198).

Use W 27 × 84.

Bending stress

$$f_b = \frac{431 \text{ ft-k} \times 12 \text{ in./ft}}{213 \text{ in.}^3} = 24.28 \text{ ksi}$$

Example 2.14. A W 27 × 84 carries a uniformly distributed load w over its entire span length of 24 ft. Calculate F_b and w_{all} (including the weight of the beam) if the maximum unbraced length L_b is a) 24 ft, b) 16 ft, c) 12 ft, d) 8 ft.

Solution. For W 27 × 84, $S = 213$ in.^{3}, $L_c = 10.5$ ft, $L_u = 11.0$ ft. Inspection of the Allowable Moments in Beams charts yields the following results:

Case	L_b (ft)	M_{all} (ft-k)	$F_b = \frac{M_{all}}{S}$ (ksi)	$w_{all} = \frac{8 \times M_{all}}{L^2}$ (kip/ft)
a	24	225.0	$\frac{225.0 \times 12}{213} = 12.68$	$\frac{8 \times 225.0}{(24)^2} = 3.13$
b	16	337.0	$\frac{337.0 \times 12}{213} = 18.99$	$\frac{8 \times 337.0}{(24)^2} = 4.68$
c	12	376.0	$\frac{376.0 \times 12}{213} = 21.18$	$\frac{8 \times 376.0}{(24)^2} = 5.22$
d	8	426.0	$\frac{426.0 \times 12}{213} = 24.00$	$\frac{8 \times 426.0}{(24)^2} = 5.92$

Example 2.15. Calculate F_b and w_{all} for the beam in Example 2.14 if A572 Gr 50 steel is used.

56 STEEL DESIGN FOR ENGINEERS AND ARCHITECTS

Solution. $F_y = 50$ ksi, $S = 213$ in.^{3}, $L_c = 8.0$ ft, $L_u = 9.4$ ft. Again, use the Allowable Moments in Beams charts (gray border in this case).

Case	L_b (ft)	M_{all} (ft-k)	$F_b = \dfrac{M_{all}}{S}$ (ksi)	$w_{all} = \dfrac{8 \times M_{all}}{L^2}$ (ksi)
a	24	225.0^a	$\dfrac{225.0 \times 12}{213} = 12.68$	$\dfrac{8 \times 225.0}{(24)^2} = 3.13$
b	16	420.0	$\dfrac{420.0 \times 12}{213} = 23.66$	$\dfrac{8 \times 420.0}{(24)^2} = 5.83$
c	12	493.0	$\dfrac{493.0 \times 12}{213} = 27.77$	$\dfrac{8 \times 493.0}{(24)^2} = 6.85$
d	8	586.0	$\dfrac{586.0 \times 12}{213} = 33.01$	$\dfrac{8 \times 586.0}{(24)^2} = 8.14$

^{a}From extrapolation. This moment and allowable stress can be verified with the aid of the governing eqn. (F1-7).

Comparing the results from Example 2.14 with those obtained from Example 2.15, it can readily be seen that for long unsupported lengths, the values of F_b are identical regardless of the value of F_y. This is due to the fact that F_b depends mostly on the modulus of elasticity E in these situations (i.e., failure of the beam is caused by buckling of the compression flange). It can also be seen that for unsupported lengths of L_c or less, the allowable stress F_b depends on the yield stress F_y of the beam.

Case a $F_b = 12.68$ ksi (A36) $= 12.68$ ksi   (A572, Gr 50)

Case d $F_b = 24.00$ ksi (A36) vs. 33.01 ksi   (A572, Gr 50)

Allowable Stress: Weak Axis Bending of I-Shaped Members, Solid Bars, and Rectangular Plates (AISCS F2)

For compact doubly symmetric I- and H-shape members, solid round and square bars, and solid rectangular sections, the allowable bending stress is given in eqn. (F2-1):

$$F_b = 0.75 F_y \tag{2.9}$$

For noncompact members where the width-thickness ratio as defined in AISCS B5 is equal to $95/\sqrt{F_y}$, and for members which are not covered in AISCS F3, the allowable stress is given in eqn. (F2-2):

$$F_b = 0.60 F_y \tag{2.10}$$

MEMBERS UNDER FLEXURE: 1   57

When the width-thickness ratio of the member falls between $65/\sqrt{F_y}$ and $95/\sqrt{F_y}$, F_b is given by the transition formula (F2-3):

$$F_b = F_y\left[1.075 - 0.005\left(\frac{b_f}{2t_f}\right)\sqrt{F_y}\right] \qquad (2.11)$$

Note that lateral support for the member types noted above is not required. As before, a discussion of slender members is beyond the scope of this text.

Example 2.16. Determine the lightest channel which can support a uniformly distributed load of 0.3 kip/ft (includes beam weight) on a span of 12 ft. The channel is bent about the weak axis. Use A572 Gr 50 steel.

Solution.

$$\text{Maximum moment} = \frac{0.3 \text{ kip/ft} \times (12 \text{ ft})^2}{8} = 5.4 \text{ ft-k}$$

$$F_b = 0.6 \times 50 \text{ ksi} = 30 \text{ ksi} \quad (\text{eqn. 2.10})$$

$$S_y = \frac{5.4 \text{ ft-k} \times 12 \text{ in./ft}}{30 \text{ ksi}} = 2.16 \text{ in.}^3$$

The following channels are acceptable:

$$\text{C } 15 \times 33.9, \ S_y = 3.11 \text{ in.}^3$$
$$\text{MC } 10 \times 22, \ S_y = 2.8 \text{ in.}^3$$
$$\text{MC } 8 \times 21.4, \ S_y = 2.74 \text{ in.}^3$$
$$\text{MC } 6 \times 18, \ S_y = 2.48 \text{ in.}^3 \quad \text{(lightest)}$$

Example 2.17. Determine the moment capacity of the following members bent about their weak axes (A588 Gr 50 steel): a) W 10 × 88, b) W 14 × 90.

Solution.

a)

$$\text{Width-thickness ratio} = \frac{b_f}{2 \times t_f} = \frac{10.265 \text{ in.}}{2 \times 0.99 \text{ in.}} = 5.18 \quad (\text{AISCS B5.1})$$

$$\frac{65}{\sqrt{50}} = 9.19 > 5.18.$$

58 STEEL DESIGN FOR ENGINEERS AND ARCHITECTS

Therefore, $F_b = 0.75 \times 50$ ksi $= 37.5$ ksi.

$$S_y = 34.8 \text{ in.}^3$$

$$\text{Moment capacity} = \frac{34.8 \text{ in.}^3 \times 37.5 \text{ ksi}}{12 \text{ in.}/\text{ft}} = 109 \text{ ft-k}$$

b)

$$\text{Width-thickness ratio} = \frac{b_f}{2 \times t_f} = \frac{14.520 \text{ in.}}{2 \times 0.71 \text{ in.}} = 10.2$$

$$\frac{65}{\sqrt{50}} = 9.19 \qquad \frac{95}{\sqrt{50}} = 13.44$$

Using transition formula (F2-3):

$$F_b = 50 \text{ ksi} \times (1.075 - 0.005 \times 10.2\sqrt{50}) = 35.7 \text{ ksi}$$

$$S_y = 49.9 \text{ in.}^3$$

$$\text{Moment capacity} = \frac{49.9 \text{ in.}^3 \times 35.7 \text{ ksi}}{12 \text{ in.}/\text{ft}} = 148.5 \text{ ft-k}$$

Allowable Stress: Bending of Box Members, Rectangular Tubes, and Circular (AISCS F3)

For compact members bent about their strong or weak axes, the allowable bending stress is given in (F3-1):

$$F_b = 0.66 F_y \qquad (2.12)$$

By definition, a compact box-shaped member is one which satisfies the requirements in Table B5.1; also, it must have a depth not greater than six times its width and a flange thickness not greater than two times its web thickness. The lateral support requirements are given in (F3-2):

$$L_c = \left(1{,}950 + 1{,}200 \frac{M_1}{M_2}\right) \frac{b}{F_y} \qquad (2.13)$$

If the section is noncompact, $F_b = 0.6 F_y$, as given in (F3-3).

2.3 CONTINUOUS BEAMS

The AISC allows beams that span continuously over an interior support or are rigidly framed to columns to assume a slight moment redistribution. AISCS F1.1 states that, except for hybrid girders and members with $F_y > 65$ ksi, continuous or rigidly framed compact beams and girders may be designed for $\frac{9}{10}$ of the negative moments due to gravity loading if the positive moments are increased by $\frac{1}{10}$ of the average negative moments (see Fig. 2.5). These provi-

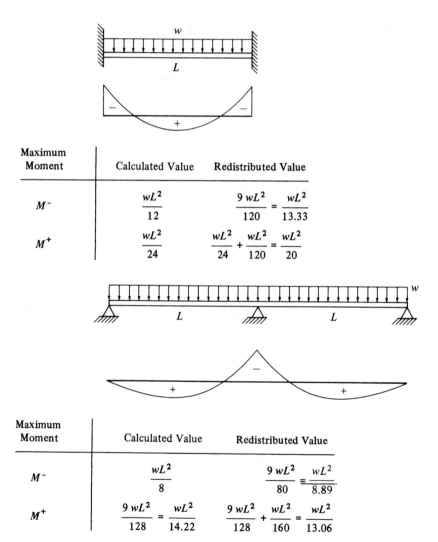

Fig. 2.5. Moment redistribution of rigid and continuous beams.

60 STEEL DESIGN FOR ENGINEERS AND ARCHITECTS

sions do not apply for cantilevers. If the negative moment is resisted by a rigid column-beam connection, the $\frac{1}{10}$ reduction may be used in proportioning the column for the combined axial and bending loads, provided that the stress f_a does not exceed 0.15 F_a. Note that moment redistribution usually provides a savings in material quantities.

Example 2.18. A beam spanning two 30-ft bays supports a load of 1.7 k/ft. Assuming full lateral support, design the beam

a) for calculated positive and negative moments
b) using the moment redistribution allowed by AISCS.

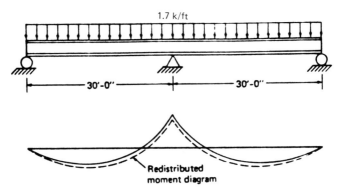

Solution.

$$M^+ = \tfrac{9}{128} wL^2 \text{ at 11 ft-3 in. from outside supports}$$

$$M^- = \tfrac{1}{8} wL^2 \text{ at interior support.}$$

$$M^+ = \tfrac{9}{128} (1.7 \text{ k/ft}) \times (30.0 \text{ ft})^2 = 107.6 \text{ ft-k}$$

$$M^- = \tfrac{1}{8} (1.7 \text{ k/ft}) \times (30.0 \text{ ft})^2 = 191.25 \text{ ft-k}$$

a) $M_{max} = 191.25$ ft-k

$$S = M_{max}/F_b = \frac{191.25 \text{ ft-k} \times 12 \text{ in.}/\text{ft}}{24.0 \text{ ksi}} = 95.6 \text{ in.}^3$$

From the allowable stress design selection table

Use W 18 × 55 $S_x = 98.3$ in.^{3}

b) For AISC redistribution, the negative moment becomes $\frac{9}{10}$ that produced by gravity, and the positive moment increases $\frac{1}{10}$ of the average negative moments

$$M^- = \frac{9}{10} \times 191.25 \text{ ft-k} = 172.1 \text{ ft-k}$$

$$M^+ = 107.6 \text{ ft-k} + \left(\frac{1}{10} \times \frac{191.25 \text{ ft-k}}{2}\right) = 117.2 \text{ ft-k}$$

$$S = M_{max}/F_b = \frac{172.1 \text{ ft-k} \times 12 \text{ in.}/\text{ft}}{24.0 \text{ ksi}} = 86.1 \text{ in.}^3$$

From the allowable stress design selection table

Use W 18 × 50 $S_x = 88.9$ in.^{3}

Example 2.19. Design a continuous beam for the condition shown.

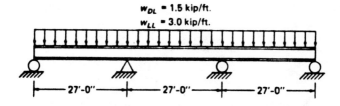

Solution. The analysis can be done easily with the aid of AISC beam diagrams and formulas. Design the beam to carry the dead load uniformly plus the maximum moment that can occur according to variable live load conditions (AISCM Part 2). Referring to AISCM p. 2-308, the following moments can be obtained:

62 STEEL DESIGN FOR ENGINEERS AND ARCHITECTS

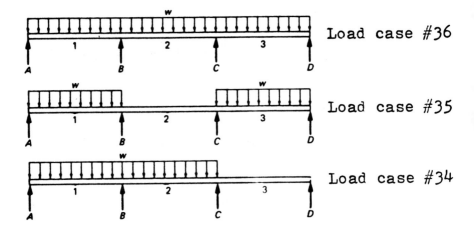

For dead load, use load case #36:

$$M_1 = M_3 = 0.08(1.5 \text{ k/ft} \times (27.0 \text{ ft})^2) = 87.5 \text{ ft-k}$$

$$M_B = M_C = -0.1(1.5 \text{ k/ft} \times (27.0 \text{ ft})^2) = -109.4 \text{ ft-k}$$

It can be seen that the maximum midspan moment occurs in the outer bays and that the maximum interior support moment occurs with one bay unloaded. For midspan live load moment, use load case #35 (outer bays):

$$M_1 = 0.1013(3.0 \text{ k/ft} \times (27.0 \text{ ft})^2) = 221.5 \text{ ft-k}$$

For support moment, use load case #34 (at support B):

$$M_B = -0.1167(3.0 \text{ k/ft} \times (27.0 \text{ ft})^2) = -255.2 \text{ ft-k}$$

The beam is symmetrical, and moments would be the same for the opposite side.

Adding dead load moment and live load moments

$$M_1 = 87.5 \text{ ft-k} + 221.5 \text{ ft-k} = 309.0 \text{ ft-k}$$

$$M_B = (-109.4 \text{ ft-k}) + (-255.2 \text{ ft-k}) = -364.6 \text{ ft-k}$$

Note that the moments M_1 for the dead load and M_1 for the live load do not occur at the same location on the beam; however, we are justified in directly superimposing these two moments, since this result is very close to (actually, slightly greater than) the exact value of the total moment at this point. The following calculation for the exact moment at point 1 (i.e., 0.45×27 ft = 12.15 ft to the right of support A) illustrates the above remarks.

Dead load (load case #36): $M_B = -109.4$ ft-k (see above)
Live load (load case #35): $M_B = -0.050$ wl^2 = $-0.05 \times 3 \times (27)^2$
$$= -109.35 \text{ ft-k}$$
Total $M_B = -109.4 - 109.35 = -218.75$ ft-k

$$R_A = (3.0 + 1.5) \times \frac{27}{2} - \frac{218.75}{27} = 60.75 - 8.10 = 52.65 \text{ k}$$

Total moment at point 1:

$$M_1 = \frac{(52.65)^2}{2 \times (3.0 + 1.5)} = 308 \text{ ft-k, which is less than the total moment of 309 ft-k obtained from direct superposition (less than a 0.5\% difference)}$$

Using moment redistribution

$M_B = 0.9 \times -364.6$ ft-k $= -328.1$ ft-k

$M_1 = 309.0$ ft-k $+ \dfrac{39.46 \text{ ft-k}}{2} = 327.2$ ft-k (nearly equal moments)

$$S = M/F_b = \frac{328.1 \text{ ft-k} \times 12 \text{ in./ft}}{24.0 \text{ ksi}} = 164.05 \text{ in.}^3$$

Use W 24 × 76 $S_x = 176.0$ in.^{3}

If moment redistribution was not considered, maximum moment = M_B = -364.6 ft-k and

$$S = \frac{364.6 \text{ ft-k} \times 12 \text{ in./ft}}{24.0 \text{ ksi}} = 182.3 \text{ in.}^3$$

Use W 24 × 84 ($S_x = 196.0$ in.^{3}).

2.4 BIAXIAL BENDING

All cross sections have two principal axes passing through the centroid for which, about one axis, the moment of inertia is maximum and, about the other axis, the moment of inertia is minimum. If loads passing through the shear center are perpendicular to one axis, simple bending occurs about that axis. Members loaded such that bending occurs simultaneously about both principal axes are subject to biaxial bending. The total bending stress at any point in the cross section of such a member is

$$f_b = \frac{M_1}{I_1} m + \frac{M_2}{I_2} n \tag{2.14}$$

where the subscripts refer to the principal axes, m is the distance to the point measured perpendicular to axis 1, and n is the distance to the point measured perpendicular to axis 2. The moment of inertia I_1 and I_2 about the principal axes can be determined by

$$I_1 = \frac{I_x + I_y}{2} + \sqrt{\left(\frac{I_x + I_y}{2}\right)^2 + I_{xy}^2}$$

$$I_2 = \frac{I_x + I_y}{2} - \sqrt{\left(\frac{I_x - I_y}{2}\right)^2 + I_{xy}^2}$$

where I_x and I_y are the moments of inertia through the centroid about the x and y axes, respectively, and I_{xy} is the product of inertia of the cross section. The product of inertia of a section is the geometrical characteristic of the section defined by the integral $I_{xy} = \int_A xy \, dA$. If orthogonal axes x and y, or one of them, are axes of symmetry, the product of inertia with respect to such axes is equal to zero. In such cases, the principal axes coincide with the x and y centroidal axes of the member.

In the case of loads applied perpendicular to symmetrical sections, the stresses and deflections may be calculated separately for bending about each axis and superimposed. Loads applied that are not perpendicular to either principal axis can be broken down into components that are perpendicular to the principal axes. The extreme bending stress due to biaxial bending is then

$$f_b = \frac{M_x y}{I_x} + \frac{M_y x}{I_y} \tag{2.15}$$

and the resultant deflection is

$$\Delta_T = [(\Delta_x)^2 + (\Delta_y)^2]^{1/2} \tag{2.16}$$

MEMBERS UNDER FLEXURE: 1

Shapes that do not have an axis of symmetry have the principal axes inclined to the x and y centroidal axes. For these nonsymmetrical shapes, the total stress at any point in the cross section can be determined by

$$f_b = \frac{M_x - M_y(I_{xy}/I_y)}{I_x - (I_{xy}^2/I_y)} y + \frac{M_y - M_x(I_{xy}/I_y)}{I_y - (I_{xy}^2/I_x)} x$$

where M_x and M_y are bending moments caused by loads perpendicular to the x and y axes, and I_{xy} is the product of inertia of the cross section, referred to the x and y axes.

In most cases, steel flexural members have different allowable bending stresses with respect to their major and minor axes. The use of the interaction equation below is then required to limit the stress levels in each plane (see AISCS H1 or H2 with $f_a = 0$):

$$\frac{f_{bx}}{F_{bx}} + \frac{f_{by}}{F_{by}} \le 1.0 \qquad (2.17)$$

The allowable stresses given in Sect. 2.2 above are applicable (AISCS F1, F2, F3).

Example 2.20. The wide-flange beam shown is loaded through the center subjecting the beam to a 150-ft-k moment. Design the lightest W 14 section by breaking down the moment to components of the principal axes and solving for biaxial bending. Assume full lateral support.

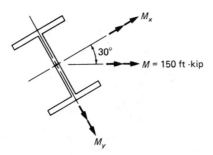

Solution. The moment is broken down by the components as shown

$$M_x = 150 \text{ ft-k } (\cos 30°) = 130.0 \text{ ft-k}$$

$$M_y = 150 \text{ ft-k } (\sin 30°) = 75.0 \text{ ft-k}$$

66 STEEL DESIGN FOR ENGINEERS AND ARCHITECTS

For member under combined stress

$$\frac{f_{bx}}{F_{bx}} + \frac{f_{by}}{F_{by}} \leq 1.0 \quad \text{(AISCS H1)}$$

Try W 14 × 109 $S_x = 173 \text{ in.}^3$ $S_y = 61.2 \text{ in.}^3$

$$F_{bx} = 24.0 \text{ ksi}$$
$$F_{by} = 27.0 \text{ ksi}$$

$$f_{bx} = M_x/S_x = \frac{130.0 \text{ ft-k} \times 12 \text{ in./ft}}{173 \text{ in.}^3} = 9.02 \text{ ksi}$$

$$f_{by} = M_y/S_y = \frac{75.0 \text{ ft-k} \times 12 \text{ in./ft}}{61.2 \text{ in.}^3} = 14.71 \text{ ksi}$$

$$\frac{9.02 \text{ ksi}}{24.0 \text{ ksi}} + \frac{14.71 \text{ ksi}}{27.0 \text{ ksi}} = 0.376 + 0.545 = 0.921 \leq 1.0 \quad \text{ok}$$

If the next lightest section, W 14 × 99, were checked, the interaction equation would yield 1.018 > 1.0 (N.G.).

Therefore, W 14 × 109 is the lightest W 14 section.

Example 2.21. Design a C15 channel spanning 20 ft with full lateral support as a purlin subject to biaxial bending. Assume the live load of 100 lb/ft, dead load of 40 lb/ft, and wind load of 80 lb/ft to act through the shear center, thus eliminating torsion in the beam.

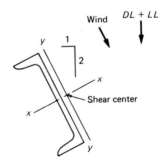

Solution. The total wind load is carried in the x axis of the channel, because the wind acts perpendicular to the roof surface. The live load and dead load will be broken down into x and y components.

MEMBERS UNDER FLEXURE: 1   67

Gravity loads w = 40 lb/ft (DL) + 100 lb/ft (LL)
= 140 lb/ft

$$w_x = 80 \text{ lb/ft (wind)} + \frac{2}{\sqrt{5}} \, 140 \text{ lb/ft} = 205 \text{ lb/ft}$$

$$w_y = \frac{1}{\sqrt{5}} \, 140 \text{ lb/ft} = 63 \text{ lb/ft}$$

$$M_x = w_x L^2/8 = \frac{0.205 \text{ k/ft} \times (20.0 \text{ ft})^2}{8} = 10.25 \text{ ft-k}$$

$$M_y = w_y L^2/8 = \frac{0.063 \text{ k/ft} \times (20.0 \text{ ft})^2}{8} = 3.15 \text{ ft-k}$$

For members subject to combined stresses, the bending stresses must be proportioned such that

$$\frac{f_{bx}}{F_{bx}} + \frac{f_{by}}{F_{by}} \le 1.0$$

$$F_{bx} = 22.0 \text{ ksi} \quad (\text{AISCS F1.3})$$

$$F_{by} = 22.0 \text{ ksi} \quad (\text{AISCS F2.2})$$

Try C 15 × 33.9 $S_x = 42.0 \text{ in.}^3$ $S_y = 3.11 \text{ in.}^3$

$$f_{bx} = M_x/S_x = \frac{10.25 \text{ ft-k} \times 12 \text{ in./ft}}{42.0 \text{ in.}^3} = 2.9 \text{ ksi}$$

$$f_{by} = M_y/S_y = \frac{3.15 \text{ ft-k} \times 12 \text{ in./ft}}{3.11 \text{ in.}^3} = 12.2 \text{ ksi}$$

$$\frac{2.9 \text{ ksi}}{22.0 \text{ ksi}} + \frac{12.2 \text{ ksi}}{22.0 \text{ ksi}} = 0.69 < 1.0 \quad \text{ok}$$

Use C 15 × 33.9 channel.

(Note: Increase in allowable stress due to wind not considered.)

2.5 SHEAR

For a beam subjected to a positive bending moment, the lower fibers of the member are elongated and the upper fibers are shortened, while, at the neutral

68 STEEL DESIGN FOR ENGINEERS AND ARCHITECTS

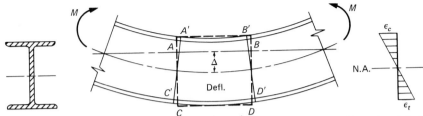

Fig. 2.6. Deformation of a beam in bending. Due to a moment, a section of beam in original position $ABCD$ will deflect a distance Δ. In doing so, compressive strains will develop on the loaded side of the neutral axis, and tensile strains on the opposite side. Under elastic deformation, the strains are linear, with no strain occurring at the neutral axis. The original section $ABCD$ assumes a new position $A'B'C'D'$.

axis, the length of the fibers remain unchanged (see Fig. 2.6). Due to these varying deformations, individual fibers have a tendency to slip on the adjacent ones. If a beam is built by merely stacking several boards on top of each other and then transversely loaded, it will take the configuration shown in Fig. 2.7(a). If the boards are connected, as in Fig. 2.7(b), the tendency to slip will be resisted by the shearing strength of the connectors. For single-element beams, the tendency to slip is resisted by the shearing strength of the material.

From mechanics of materials, the longitudinal shearing stress in a beam is given by the formula

$$f_v = \frac{VQ}{It} \qquad (2.18)$$

where V is the vertical shear force, Q is the moment of the areas on one side of the sliding interface taken about the centroid of the cross section, I is the moment of inertia of the section, and t is the width of the section where the shearing stress is investigated. For rectangular beams, the maximum shearing stress is given by $f_v = 3V/2A$, where A is the area of the cross section. For rolled and fabricated shapes, the AISC allows the longitudinal shearing stress

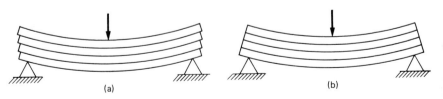

Fig. 2.7. Built-up beam in flexure: (a) transversely unconnected, (b) transversely connected.

to be determined by the formula

$$f_v = \frac{V}{A_{web}} \qquad (2.19)$$

where A_{web} is the product of the overall depth of the section d and the thickness of the web t_w (AISCS F4). However, this value is 10 to 15% less than the maximum shearing stress, as determined by the more accurate eqn. (2.18). The allowable shearing stress, as determined by the AISC on the gross section of a member, is given by $F_v = 0.40\ F_y$ (AISCS (F4-1)), provided that $h/t_w \le 380/\sqrt{F_y}$. Except for short spans with heavy loading, or heavy concentrated loads close to supports, shear seldom governs in the design of beams.

Example 2.22. Determine the maximum shearing stress for the following sections when the external shear force $V = 75$ kips.

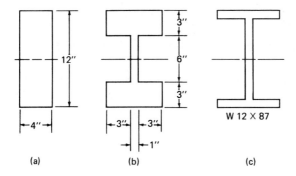

(a)   (b)   (c)

Solution.

a)
$$f_{v\,max} = \frac{3V}{2A}$$

$$A = 4 \text{ in.} \times 12 \text{ in.} = 48 \text{ in.}^2$$

$$f_{v\,max} = \frac{3V}{2A} = \frac{3 \times 75 \text{ k}}{2 \times 48 \text{ in.}^2} = 2.34 \text{ ksi}$$

b)
$$f_{v\,max} = \frac{VQ}{It}$$

70 STEEL DESIGN FOR ENGINEERS AND ARCHITECTS

By inspection, it is seen that the neutral axis is located at the center, 6 in. from the bottom.

$$I = \Sigma \frac{bh^3}{12} + (Ad^2)$$

$$= 2 \times \left[\frac{7 \times (3)^3}{12} + (3 \times 7)(6 - 1.5)^2\right] + \frac{1 \times (6)^3}{12} = 900 \text{ in.}^4$$

Because $f_{v\max}$ is at the neutral axis,

$$Q = (3 \times 7)(6 - 1.5) \times (3 \times 1)(1.5) = 99.0 \text{ in.}^3$$

$$t = 1 \text{ in.}$$

$$f_{v\max} = \frac{VQ}{It} = \frac{75 \text{ k} \times 99 \text{ in.}^3}{900 \text{ in.}^4 \times 1 \text{ in.}} = 8.25 \text{ ksi}$$

Verify with $f_v = \dfrac{V}{A_{\text{web}}}$ (AISCS F4)

$$f_v = \frac{75 \text{ k}}{12 \text{ in.} \times 1 \text{ in.}} = 6.25 \text{ ksi}$$

As the web and the flanges are very thick, there is approximately one-third difference in shear values.

c) $\quad f_v = \dfrac{V}{A_{\text{web}}}$

$$A_{\text{web}} = d \times t = 12.53 \text{ in.} \times 0.515 \text{ in.} = 6.45 \text{ in.}^2$$

$$f_v = \frac{V}{A_{\text{web}}} = \frac{75 \text{ k}}{6.45 \text{ in.}^2} = 11.62 \text{ ksi}$$

Verify with $f_v = \dfrac{VQ}{It}$

$$Q = 0.81 \text{ in.} \times 12.125 \text{ in.} \times \left(\frac{12.53 \text{ in.}}{2} - \frac{0.810 \text{ in.}}{2}\right)$$

$$+ 0.515 \text{ in.} \times \left(\frac{10.91 \text{ in.}}{2}\right)^2 \times \frac{1}{2}$$

$$= 65.21 \text{ in.}^3$$

$I = 740 \text{ in.}^4$

$$f_v = \frac{75 \text{ k} \times 65.21 \text{ in.}^3}{740 \text{ in.}^4 \times 0.515 \text{ in.}} = 12.83 \text{ ksi}$$

A difference of about 10% occurs between an exact approach and an estimated value. The code is aware of this error which seldom is greater than 15% and compensates for it in $F_v = 0.40 F_y$.

Example 2.23. Determine if the beam shown is satisfactory. Assume full lateral support. Check both flexural and shearing stresses. Neglect the beam weight.

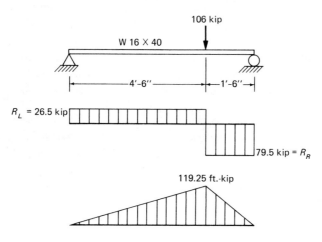

Solution.

$$d = 16.01 \text{ in.}; \quad t_w = 0.305 \text{ in.}; \quad S = 64.7 \text{ in.}^3$$

$$R_R = \frac{106 \text{ k} \times 4.5 \text{ ft}}{6.0 \text{ ft}} = 79.5 \text{ k}$$

$$R_L = 106 \text{ k} - 79.5 \text{ k} = 26.5 \text{ k}$$

$$M_{max} \text{ (from diagram)} = 119.25 \text{ ft-k}$$

The moment resisted by the section is

$$M_R = S \times F_b = 64.7 \text{ in.}^3 \times 24 \text{ ksi} \times \frac{1 \text{ ft}}{12 \text{ in.}}$$

$$= 129.4 \text{ ft-k} > 119.25 \text{ ft-k} \quad \text{ok}$$

$$V_{max} = 79.5 \text{ k}$$

$$f_v = \frac{V}{A_{web}} = \frac{79.5 \text{ k}}{16.01 \text{ in.} \times 0.305 \text{ in.}} = 16.28 \text{ ksi}$$

$$F_v = 0.40 F_y = 0.4 \times 36 \text{ ksi} = 14.4 \text{ ksi}$$

$$16.28 > 14.4 \text{ ksi} \quad \text{N.G.}$$

W 16 × 40 is not satisfactory for resisting shear.

2.6 HOLES IN BEAMS

Whenever possible, holes in beams should be avoided. If drilling or cutting holes in a beam is absolutely necessary, it is good practice to avoid holes in the web at the locations of large shear and in the flange where the moment is large (see Fig. 2.8).

As a result of numerous tests, and by comparing the yield and the tensile fracture capacities of the tensile members with holes, no hole deduction needs to be considered until A_n/A_g equals $1.2(F_y/F_u)$. This corresponds to a hole allowance of 25.5% for A36 steel, 7.7% for A572 Gr 50 steel, and 14.3% for A588 Gr 50 steel. As stated in AISCS B10, no deduction in the total area of either flange is necessary if (B10-1) is satisfied:

$$0.5F_u A_{fn} \geq 0.6F_y A_{fg} \tag{2.20}$$

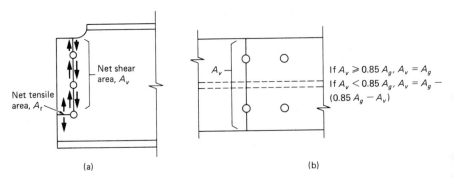

Fig. 2.8. Holes in beams. (a) Holes in beam webs should be avoided at locations of large shear. When holes are necessary, as in the end connection shown, the section must be checked for web tear-out. (b) Flange holes should be avoided at regions of large moments.

MEMBERS UNDER FLEXURE: 1   73

where A_{fg} is the gross area of the flange and A_{fn} is the net area of the flange calculated according to the provisions in AISCS B2 (see Chap. 1). If (2.20) is not satisfied (see (B10-2)), then the member flexural propeties shall be based on an effective tension flange area A_{fe} as given in (B10-3):

$$A_{fe} = \frac{5}{6} \frac{F_u}{F_y} A_{fn} \qquad (2.21)$$

Example 2.24. Calculate the design section modulus for a W 14 × 145:

a) for two 1-in.-diameter holes in each flange (A36 steel).
b) for two 2-in.-diameter holes in each flange (A572 Gr 50 Steel).

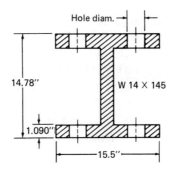

Solution.

$S_x = 232$ in.^{3}; $b_f = 15.5$ in.; $t_f = 1.090$ in.; $I_x = 1710$ in.^{4}

$A_{fg} = b_f \times t_f = 15.5$ in. $\times 1.090$ in. $= 16.90$ in.^{2}

a) Determine flange area loss due to two 1-in.-diameter holes:

$$A_{fn} = 16.90 \text{ in.}^2 - 2 \times (1.090 \text{ in.} \times 1 \text{ in.}) = 14.72 \text{ in.}^2$$

Check (2.20):

0.5×58 ksi $\times 14.72$ in.$^2 = 427$ k $> 0.6 \times 36$ ksi $\times 16.9$ in.$^2 = 365$ k

Therefore, no hole reduction is required.

$$S = S_x = 232 \text{ in.}^3$$

74 STEEL DESIGN FOR ENGINEERS AND ARCHITECTS

b) Determine flange area loss due to two 2-in.-diameter holes:

$$A_{fn} = 16.90 \text{ in.}^2 - 2 \times (1.090 \text{ in.} \times 2 \text{ in.}) = 12.54 \text{ in.}^2$$

Check (2-20):

$$0.5 \times 65 \text{ ksi} \times 12.54 \text{ in.}^2 = 408 \text{ k} < 0.6 \times 50 \text{ ksi} \times 16.9 \text{ in.}^2 = 507 \text{ k}$$

Hence, determine the effective tension flange area (2.21):

$$A_{fe} = \frac{5}{6} \frac{65 \text{ ksi}}{50 \text{ ksi}} (12.54 \text{ in.}^2) = 13.59 \text{ in.}^2$$

This is equivalent to a tension flange which is 1.090 in. thick and 13.59 in.^{2}/1.090 in. = 12.47 in. wide.

Determine effective section properties:

$$\bar{y} = \frac{(1.09)(12.47)(1.09/2) + [(0.68)(12.6)((12.6/2) + 1.09)] + (1.09)(15.5)(14.24)}{(1.09)(12.47) + (0.68)(12.6) + (1.09)(15.5)}$$

$$= \frac{311.23 \text{ in.}^3}{39.06 \text{ in.}^2} = 7.97 \text{ in.} \quad \text{(measured from the bottom of the tension flange to the centroid of the section)}$$

$$I_x = \frac{1}{12} (12.47 \times 1.09^3) + (1.09 \times 12.47)(7.97 - 1.09/2)^2$$

$$+ \frac{1}{12} (0.68 \times 12.6^3) + (0.68 \times 12.6)(7.97 - 7.39)^2$$

$$+ \frac{1}{12} (15.5 \times 1.09^3) + (15.5 \times 1.09)(14.78 - 7.97 - 1.09/2)^2$$

$$= 750.7 + 116.2 + 664.8 = 1531.7 \text{ in.}^4$$

$$S_{top} = \frac{1531.7 \text{ in.}^4}{14.78 \text{ in.} - 7.97 \text{ in.}} = 224.9 \text{ in.}^3$$

$$S_{bottom} = \frac{1531.7 \text{ in.}^4}{7.97 \text{ in.}} = 192.2 \text{ in.}^3 \text{ (governs)}$$

To eliminate the above lengthy calculations, the authors suggest deducting the same amount of area in the compression flange that is required in the tension

MEMBERS UNDER FLEXURE: 1   75

one. This results in conservative and very close values for the section properties. For example, in this case,

$$I_x = 1710 - 2 \times (16.9 - 13.59) \times (14.78/2 - 1.09/2)^2$$
$$= 1399.8 \text{ in.}^4$$

$$S = \frac{1399.8}{(14.78/2)} = 189.4 \text{ in.}^3, \text{ which is very close to the actual value of } 192.2 \text{ in.}^3 \text{ determined above (less than 1.5\% difference)}$$

Beam end connections using high-strength bolts in relatively thin webs may create a condition of web tear-out. Failure can occur by the combination of shear through the line of bolts and tension across the bolt block. This condition should be considered when designing the bolt connection and should be investigated in situations when the flange is coped or when shear resistance it at a minimum. For further discussion see Chapter 5, Bolts and Rivets (also, see Fig. 2.8a).

2.7 BEAMS WITH CONCENTRATED LOADS

For beams which are subjected to concentrated loads, AISCS K1.3 and K1.4 have established provisions governing the magnitude of the concentrated load R so as to prevent failure of the beam web by local yielding or crippling. The thickness of the beam web t_w, the length of bearing N, and the distance from the outer face of the beam flange to the web toe of fillet k all play an important role in these provisions. In a particular situation, either web yielding or web cripplings may govern.

- For the case of web yielding (AISCS K1.3):

a) If R is applied at a distance equal to or larger than the depth of the beam d from the end of the member, then (K1-2) governs and can be rewritten as follows:

$$R \leq 0.66 F_y t_w (N + 5k) \quad (2.22a)$$

or

$$R \leq 2R_1 + NR_2 \quad (2.22b)$$

b) If R is near the end of the member, then (K1-3) governs:

$$R \leq 0.66 F_y t_w (N + 2.5k) \quad (2.23a)$$

76 STEEL DESIGN FOR ENGINEERS AND ARCHITECTS

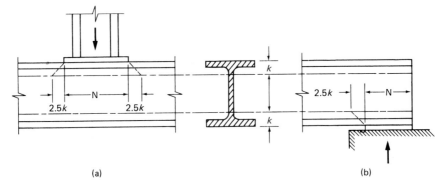

Fig. 2.9. Local web yielding. The applied load or end reaction is transferred to the web from the flange at the web-flange intersection over a length extending a distance 2.5k on either side: (a) interior concentrated load, (b) end reaction.

or

$$R \le R_1 + NR_2 \qquad (2.23b)$$

where $R_1 = 0.66(2.5)t_w k F_y$
$R_2 = 0.66 t_w F_y$

These provisions can be seen schematically in Fig. 2.9.

- For the case of web crippling (AISCS K1.4):

a) If R is applied at a distance greater than $d/2$ from the end of the member, than (K1-4) governs:

$$R \le 67.5 t_w^2 \left[1 + 3\left(\frac{N}{d}\right)\left(\frac{t_w}{t_f}\right)^{1.5} \right] \sqrt{F_{yw} t_f / t_w} \qquad (2.24a)$$

or

$$R \le 2(R_3 + NR_4) \qquad (2.24b)$$

b) If R is near the end of the member, (K1-5) governs:

$$R \le 34 t_w^2 \left[1 + 3\left(\frac{N}{d}\right)\left(\frac{t_w}{t_f}\right)^{1.5} \right] \sqrt{F_{yw} t_f / t_w} \qquad (2.25a)$$

or

$$R \le R_3 + NR_4 \qquad (2.25b)$$

MEMBERS UNDER FLEXURE: 1   77

Table 2.1 Constants for Beams with Concentrated Loads (Adapted from the AISCM)

Definition of symbols	$F_y = 36$ ksi	$F_y = 50$ ksi
V = Max. web shear, kips	14.4 dt	20 dt
R = Max. end reaction for $3\frac{1}{2}$-in. bearing, kips	$R_1 + NR_2$ or $R_3 + NR_4$	$R_1 + NR_2$ or $R_3 + NR_4$
R_1 = Constant for yielding, kips	$59.4kt_w$	$82.5kt_w$
R_2 = Constant for yielding, kips/in.	$23.8t_w$	$33.0t_w$
R_3 = Constant for crippling, kips	$204t_w^{1.5}t_f^{0.5}$	$204t_w^{1.5}t_f^{0.5}$
R_4 = Constant for crippling, kips/in.	$612t_w^3/t_f d$	$721t_w^3/t_f d$

where

F_{yw} = specified minimum yield stress of beam web, ksi

d = overall depth of the member, in.

t_f = flange thickness, in.

t_w = web thickness, in.

$R_3 = 34(F_{yw}t_f)^{0.5}t_w^{1.5}$

$R_4 = 102(F_{yw})^{0.5}t_w^3/(t_f d)$

If the governing equations above are not satisfied, bearing stiffeners extending at least one-half the web depth shall be provided.

On page 2-36 of the AISCM, the constants R_1 through R_4 have been tabulated for steels with $F_y = 36$ ksi and $F_y = 50$ ksi. These quantities are given in Table 2.1. The specific values of these constants can be obtained for a particular beam size in the Allowable Uniform Loads on Beams tables, which begin on p. 2-37. Note that the values of R given in the tables are for an end bearing length $N = 3.5$ in.

Example 2.25. Select the lighest section for the beam shown and determine the bearing length required for each support, in order not to use bearing stiffeners. Assume full lateral support and 50 ksi steel.

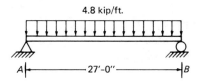

78 STEEL DESIGN FOR ENGINEERS AND ARCHITECTS

Solution.

$$W = 4.8 \text{ k/ft} \times 27 \text{ ft} = 130 \text{ k}$$
$$M = WL/8 = 437.4 \text{ ft-k}$$

Referring to the AISC beam selection table, use W 24 × 76 (M_R = 484 ft-k), t_w = 0.44 in., t_f = 0.68 in., d = 23.92 in., and k = 1.4375 in.

$$R_A = R_B = \frac{W}{2} = \frac{130.0 \text{ k}}{2} = 65.0 \text{ k}$$

Bearing length for web yielding, $N = \dfrac{R - R_1}{R_2}$ \hfill (2.23b)

Bearing length for web crippling, $N = \dfrac{R - R_3}{R_4}$ \hfill (2.25b)

Use Table 2.1 to determine the constants:

$$R_1 = 82.5 \times 1.4375 \times 0.44 = 52.2 \text{ k}$$
$$R_2 = 33.0 \times 0.44 = 14.5 \text{ k/in.}$$
$$R_3 = 240 \times (0.44)^{1.5} \times (0.68)^{0.5} = 57.8 \text{ k}$$
$$R_4 = 721 \times (0.44)^3 / (0.68 \times 23.92) = 3.78 \text{ k/in.}$$

Note that these values can be obtained from the bottom of p. 2-107 in the Allowable Loads on Beams table.

For web yielding: $N = (65 \text{ k} - 52.2 \text{ k})/14.5 \text{ k/in} = 0.88$ in.

For web crippling: $N = (65 \text{ k} - 57.8 \text{ k})/3.78 \text{ k/in} = 1.90$ in. (governs)

From a practical point of view, the actual bearing length would be longer.

Example 2.26. A column is supported by a transfer girder as shown.

a) Determine the lightest girder (neglect girder weight).
b) Determine safety against shear.
c) Determine lengths of bearing under the column and at the supports.

MEMBERS UNDER FLEXURE: 1   79

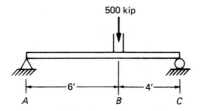

Solution.

a)
$$R_A = \frac{500 \text{ k} \times 4 \text{ ft}}{10 \text{ ft}} = 200.0 \text{ k}$$

$$R_C = 500 \text{ k} - 200 \text{ k} = 300 \text{ k}$$

$$M_{max} = 200 \text{ k} \times 6 \text{ ft} = 1200 \text{ ft-k}$$

Use W 36 × 182 at the lightest available section ($M_R = 1230$ ft-k)

b)
$$f_v = \frac{V}{A_{web}} = \frac{300 \text{ k}}{36.33 \text{ in.} \times 0.725 \text{ in.}} = 11.39 \text{ ksi}$$

$$F_v = 0.40 \, F_y = 14.4 \text{ ksi} > 11.39 \text{ ksi} \quad \text{ok}$$

W 36 × 182 is safe against shear.

c) For W 36 × 182: $R_1 = 91.5$ k, $R_2 = 17.2$ k/in.

$$R_3 = 137 \text{ k}, \quad R_4 = 5.44 \text{ k/in.}$$

(p. 2-43 AISCM)

At point A: N = larger of
$$\begin{cases} (200 \text{ k} - 91.5 \text{ k})/17.2 \text{ k/in.} \\ \quad = 6.3 \text{ in.} \quad (2.23b) \\ (200 \text{ k} - 137 \text{ k})/5.44 \text{ k/in.} \\ \quad = 11.6 \text{ in.} \quad (2.25b) \text{ (governs)} \end{cases}$$

At point C: N = larger of
$$\begin{cases} (300 \text{ k} - 91.5 \text{ k})/17.2 \text{ k/in.} \\ \quad = 12.1 \text{ in.} \quad (2.23b) \\ (300 \text{ k} - 137 \text{ k})/5.44 \text{ k/in.} \\ \quad = 30 \text{ in.} \quad (2.25b) \text{ (governs)} \end{cases}$$

80 STEEL DESIGN FOR ENGINEERS AND ARCHITECTS

For the interior load at point B, the larger N obtained from eqn. (2.22b) (web yielding) or eqn. (2.24b) (web crippling) will govern:

$$N = \frac{R - 2R_1}{R_2} = \frac{500 \text{ k} - (2 \times 91.5 \text{ k})}{17.2 \text{ k/in.}} = 18.4 \text{ in.} \quad (2.22b)$$

$$N = \frac{(R/2) - R_3}{R_4} = \frac{(500 \text{ k}/2) - 137 \text{ k}}{5.44 \text{ k/in.}} = 20.8 \text{ in.} \quad (2.24b) \text{ (governs)}$$

2.8 DESIGN OF BEARING PLATES§

When a beam is supported by a concrete or masonry base, the reaction must be distributed over an area large enough so that the average bearing pressure on the base does not exceed the allowable limit. Thus, beam bearing plates are provided, as shown in Fig. 2.10. In AISCS J9, the following allowable bearing pressures F_p are given:

On sandstone and limestone: $F_p = 0.40$ ksi

On brick in cement mortar: $F_p = 0.25$ ksi

On the full area of a concrete support: $F_p = 0.35 f'_c$

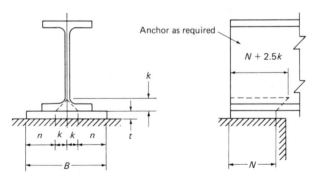

Fig. 2.10. Bearing plate for a beam supported on a concrete or masonry base (adapted from the AISCM).

---

§See AISCM pp. 2-141ff.

On less than the full area of a concrete support:

$$F_p = 0.35 f'_c \sqrt{\frac{A_2}{A_1}} \leq 0.7 f'_c$$

where f'_c = specified compressive strength of concrete, ksi
A_1 = area of steel concentrically bearing on a concrete support, in.^{2}
A_2 = maximum area of the portion of the supporting surface that is geometrically similar to and concentric with the loaded area, in.^{2}

The rules governing the length of bearing N are the same as those given in Sect. 2.7:

$$\text{For web yielding: } N = \frac{R - R_1}{R_2} \qquad (2.23\text{b})$$

$$\text{For web crippling: } N = \frac{R - R_3}{R_4} \qquad (2.25\text{b})$$

Given the allowable bearing pressure, and selecting a value of N greater than or equal to the governing one from (2.23b) or (2.25b), the width of the plate B can be calculated as follows:

$$B = \frac{R}{F_p \times N} \qquad (2.26)$$

It is common practice to use integer values for B and N.

The thickness of the plate t can be determined from the following formula, which is based on cantilever bending of the plate (about its weak axis) under uniform pressure:

$$t = \sqrt{\frac{3 f_p n^2}{F_b}} \qquad (2.27)$$

where $n = (B/2) - k$ and f_p = actual bearing pressure = $R/(B \times N)$. Since $F_b = 0.75 F_y$ (see eqn. (2.9)), the above equation can be rewritten as

$$t = 2n \sqrt{\frac{f_p}{F_y}} \qquad (2.28)$$

Example 2.27. Design a bearing plate for a W 30 × 90 with an end reaction of 85 kips which is supported by a concrete wall ($f'_c = 3000$ psi) as shown.

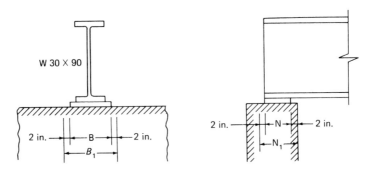

Solution. From the Allowable Uniform Load tables (p. 2-49):

$$R_1 = 36.6 \text{ k} \qquad R_3 = 51.3 \text{ k}$$
$$R_2 = 11.2 \text{ k/in.} \qquad R_4 = 3.53 \text{ k/in.}$$

$$N = \frac{85 - 36.6}{11.2} = 4.3 \text{ in.} \qquad (2.23b)$$

$$N = \frac{85 - 51.3}{3.53} = 9.5 \text{ in. (governs)} \qquad (2.25b)$$

Use $N = 10$ in.

$$\text{Area of concrete support} = A_2 = B_1 \times N_1$$
$$= (B + 4) \times (N + 4)$$

(Note that this area is geometrically similar to and concentric with the loaded area)

Try $F_p = 0.6 f'_c = 0.6 \times 3 = 1.8$ ksi

$$A_1 = B \times N = \frac{R}{F_p} \qquad (2.26)$$

$$= B \times 10 = \frac{85}{1.8} = 47.2 \text{ in.}^2 \rightarrow B = \frac{47.2}{10} = 4.72 \text{ in.}$$

Minimum B should be 10.375 in., which is the flange width of the W 30 × 90. Therefore,

MEMBERS UNDER FLEXURE: 1   83

$$f_p = \frac{85}{10 \times 10.375} = 0.819 \text{ ksi}$$

Check F_p (AISCS J9):

$$\sqrt{\frac{A_2}{A_1}} = \sqrt{\frac{(10.375 + 4)(10 + 4)}{10.375 \times 10}} = 1.4$$

$$F_p = 0.35 \times 3 \times 1.4 = 1.47 \text{ ksi} > f_p = 0.819 \text{ ksi} \quad \text{ok}$$

and 1.47 ksi < 0.7 × 3 = 2.1 ksi   ok

$$n = \left(\frac{10.375}{2}\right) - 1.3125 = 3.875 \text{ in.}$$

Determine t from (2.28):

$$t = 2 \times 3.875 \times \sqrt{\frac{0.819}{36}} = 1.17 \text{ in.}$$

Use bearing plate $1\frac{1}{4} \times 10 \times 0$ ft, $10\frac{3}{8}$ in.

Example 2.28. A S 24 × 100 is supported by a 19 × 19 in. concrete pilaster. The reaction is 200 kips, and the compressive strength of the concrete is 4000 psi. Design the required bearing plate. The smallest distance from the edge of the bearing plate to the edge of the pilaster is 2 in.

Solution. For a S 24 × 100:

$$R_1 = 77.4 \text{ k} \qquad R_3 = 122 \text{ k}$$
$$R_2 = 17.7 \text{ k/in.} \qquad R_4 = 12.1 \text{ k/in.}$$

$$N = \frac{200 - 77.4}{17.7} = 6.93 \text{ in. (governs)}$$

or

$$N = \frac{200 - 122}{12.1} = 6.45 \text{ in.}$$

Use $N = 7$ in.

84 STEEL DESIGN FOR ENGINEERS AND ARCHITECTS

Try $F_p = 0.5f'_c = 2$ ksi

$$B = \frac{200}{7 \times 2} = 14.29 \text{ in.} \quad \text{Use } B = 15 \text{ in.}$$

$$f_p = \frac{200}{7 \times 15} = 1.90 \text{ ksi}$$

$$A_2 = 19 \times (7 + 2 + 2) = 209 \text{ in.}^2$$

$$A_1 = 7 \times 15 = 105 \text{ in.}^2$$

$$F_p = \sqrt{\frac{209}{105}} \times 0.35 \times 4 = 1.98 \text{ ksi} > 1.90 \text{ ksi} \quad \text{ok}$$

and 1.98 ksi $< 0.7 \times 4 = 2.8$ ksi ok

$$n = \frac{15}{2} - 1.75 = 5.75 \text{ in.}$$

$$t = 2 \times 5.75 \times \sqrt{\frac{1.9}{36}} = 2.64 \text{ in.}$$

Use bearing plate $2\frac{3}{4} \times 7 \times 1$ ft, 3 in.

The reader can verify that the thickness of the plate can be reduced to 1.75 in. if a square bearing plate with side lengths of 12 in. is used. This plate size would also decrease the pressure on the pilaster.

2.9 DEFLECTIONS

Allowable deflections of beams are usually limited by codes and may need to be verified as part of the beam selection. The AISC sets the limit for live load deflection of beams supporting plastered ceilings at $1/360$ of the span (AISCS L3).

In addition the AISCS Commentary L3.1 suggests the following guideline: the depth of the beams for fully stressed beams should not be less than $(F_y/800) \times$ the span in inches. The deflection of a member is a function of its moment of inertia. To facilitate design, Part 2 of the AISCM has tabulated values of I_x and I_y for all W and M shapes, grouped in order of magnitude, where the shapes in bold type are the most economical in their respective groups (see pages 2-26ff).

Example 2.29. Show that beams of the same approximate depth have the same deflection when stressed to the same stress f_b.

Solution. Assume that beams are loaded with uniformly distributed loads.

$$M_1 = \frac{w_1 l^2}{8} = f_b S_1 = f_b \frac{I_1}{2d}, \quad d = \text{depth}$$

$$M_2 = \frac{w_2 l^2}{8} = f_b S_2 = f_b \frac{I_2}{2d}$$

$$\Delta_1 = \frac{5 w_1 l^4}{384 E I_1} = \frac{5}{384} \frac{w_1 l^2}{8} \frac{8 l^2}{E I_1}$$

$$= C f_b \frac{I_1}{2d} \times l^2 \times \frac{1}{E I_1} = C f_b \frac{l^2}{2dE}, \quad C = \frac{40}{384}$$

Similarly, we find that Δ_2 is equal to the same quantity. Hence, we see that the deflection is dependent on the depth and the span for the same f_b and E, and not on the load.

Example 2.30. Determine if the beam in Example 2.25 is adequate to limit deflection to $l/360$. Choose the next lightest satisfactory section if the maximum deflection is exceeded.

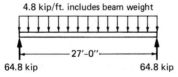

4.8 kip/ft. includes beam weight

27'-0"

64.8 kip   64.8 kip

Solution.

$$\Delta_{max} = \frac{5 \, wl^4}{384 \, EI}$$

$$E = 29{,}000 \text{ ksi}$$

For W 24 × 76, $I_x = 2100$ in.^{4}

$$\Delta_{max} = \frac{5(4.8/12) \text{k/in.} \, ((27 \times 12) \text{ in.})^4}{384 \, (29{,}000 \text{ ksi})(2100 \text{ in.}^4)} = 0.942 \text{ in.}$$

$$\frac{l}{360} = \frac{27 \times 12}{360} = 0.90 \text{ in.} < 0.942 \text{ in.} \quad \text{N.G.}$$

86 STEEL DESIGN FOR ENGINEERS AND ARCHITECTS

Determine next lightest satisfactory section

$$\frac{5\ wl^4}{384\ EI} \leq \frac{l}{360}$$

$$I_x \geq \frac{5\ wl^4}{384\ E} \times \frac{1}{l/360}$$

$$I_x \geq 2199\ \text{in.}^4$$

Choosing from the moment of inertia selection table, p. 2-27 AISCM.

$$W\ 24 \times 84, \quad I_x = 2370\ \text{in.}^4$$

$$\Delta_{\text{max}} = 0.835\ \text{in.} \quad \text{ok}$$

Use W 24 × 84.

Example 2.31. Design the beam B and girder G for the framing plan shown. Assume full lateral support, A572 Gr 50 steel, and the maximum depth of members to be 22 in. The floor is 5 in. reinforced concrete and carries a live load of 150 lb/ft². Assume there is no flooring, ceiling, or partition. Furthermore, assume that there is no restriction on deflection, but the value is requested.

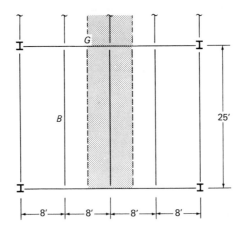

Solution. Because the weight of reinforced concrete is 150 lb/ft³, the weight of 5 in. of concrete is

$$\frac{5\ \text{in.}}{12\ \text{in.}/\text{ft}} \times 150\ \text{lb/ft}^3 = 62.5\ \text{lb/ft}^2$$

MEMBERS UNDER FLEXURE: 1   87

The tributary width of the beams is 8 ft.

$$DL_{conc} = 62.5 \text{ lb/ft}^2 \times 8 \text{ ft} = 500 \text{ lb/lin. ft of beam}$$

$$LL = 150 \text{ lb/ft}^2 \times 8 \text{ ft} = 1200 \text{ lb/lin. ft}$$

Assume weight of beam = 50 lb/lin. ft
Total distributed loading on the beams

$$w = (500 + 1200 + 50) \text{ lb/lin. ft} = 1.75 \text{ k/ft}$$

$$M_{max} = \frac{wL^2}{8} = \frac{1.75 \text{ k/ft} \times (25 \text{ ft})^2}{8} = 136.7 \text{ ft-k}$$

$$F_y = 50 \text{ ksi} \quad (\text{AISCS, Table I})$$

A W 18 × 35 may be selected as the lightest satisfactory section ($M_R = 158$ ft-k).

Because the beams frame into the girders, they are exerting their end reactions as concentrated loads on the girders. Each girder carries the end reactions of three beams on either side for a total of six concentrated loads.

The end reaction of each beam is the total load on the beam divided by 2.

$$R = \frac{1.75 \text{ lb/ft} \times 25 \text{ ft}}{2} = 21.875 \text{ k}$$

Assume weight of girder to be 150 lb/lin. ft.
End reactions of the girder

$$\frac{3 \times 2(21.875 \text{ k}) + (0.15 \text{ k/ft} \times 32 \text{ ft})}{2} = 68.0/\text{k}$$

Moment at midspan

$$(68 \text{ k} \times 16 \text{ ft}) - (43.75 \text{ k} \times 8 \text{ ft})$$
$$- (0.15 \text{ k/ft} \times 16 \text{ ft} \times 8 \text{ ft}) = 718.8 \text{ ft-k}$$

A W 30 × 99 may be selected as the lightest section. However, the depth of W 30 × 99 = 29.64 in. > 22.00 in.   N.G.

$$S_{req} = \frac{M}{F_b} = \frac{718.8 \text{ ft-k} \times 12 \text{ in./ft}}{0.66 \times 50 \text{ ksi}} = 261.4 \text{ in.}^3$$

88 STEEL DESIGN FOR ENGINEERS AND ARCHITECTS

From p. 2-10, select W 21 × 122 as the shallowest member with $S > 261.4$ in.^{3}

$$S = 273 \text{ in.}^3, \quad d = 21.68 \text{ in.} \quad \text{ok}$$

Use W 21 × 122 girder.
Check deflection:

Beam B

$$\Delta_{max} = \frac{5 \, wL^4}{384 \, EI}, \quad w = \frac{1.75 \text{ k/ft}}{12 \text{ in./ft}} = 0.15 \text{ k/in.}$$

$$I = 510 \text{ in.}^4$$

$$E = 29{,}000 \text{ ksi}$$

$$\Delta_{max} = \frac{5 \times 0.15 \text{ k/in.} \times (25 \text{ ft} \times 12 \text{ in./ft})^4}{384 \times 29{,}000 \text{ ksi} \times 510 \text{ in.}^4} = 1.07 \text{ in.}$$

Girder G

From cases 1, 7, and 9 of beam diagrams and formulas, AISCM pp. 2-296 ff

$$\Delta_{max} = \frac{5 \, wl^4}{384 \, EI} + \frac{Pl^3}{48 \, EI} + \frac{Pa}{24 \, EI}(3l^2 - 4a^2)$$

$$a = 8 \text{ ft} \times 12 \text{ in./ft} = 96 \text{ in.}$$

$$I = 2960 \text{ in.}^4$$

$$\Delta_{max} = \frac{5 \times 0.122 \times (32 \times 12)^4}{384 \times 29000 \times 2960} + \frac{43.75 \times (32 \times 12)^3}{48 \times 29000 \times 2960}$$

$$+ \frac{43.75 \times 96}{24 \times 29000 \times 2960}[3 \times (32 \times 12)^2 - 4 \times (96)^2]$$

$$= 0.402 + 0.601 + 0.827 = 1.83 \text{ in.}$$

2.10 ALLOWABLE LOADS ON BEAMS TABLES

The Allowable Loads on Beams tables starting on page 2-37 of the AISCM provide the allowable uniformly distributed loads in kips for W, M, S, and channels, which are simply supported with adequate lateral support. Separate tables are provided for $F_y = 36$ ksi and $F_y = 50$ ksi (gray border). In addition, L_c; L_u; the maximum deflection for the allowable uniform load; S_x; the maximum shear force V; the concentrated load coefficients R_1 through R_4; and the maximum end reaction R for a $3\frac{1}{2}$-in. bearing (see Sect. 2.7) are provided. The many important features of these tables are clearly explained starting on p. 2-30, which eliminates the need to repeat that material here. The examples on pages 2-34 and 2-35 illustrate the use of these tables.

PROBLEMS TO BE SOLVED

2.1. Calculate the bending stress due to a bending moment of 105 ft-k (foot-kips) about the strong axis on the following wide-flange beams:

a) W 10 × 49
b) W 12 × 45
c) W 14 × 38

2.2 Calculate the allowable bending moment for each of the beams in problem 2.1 given a) A36 steel, b) $F_y = 50$ ksi. Assume full lateral support.

2.3. Select the most economical W section for a simply supported beam carrying a 25-k concentrated load at the center of a 30-ft span. Neglect the weight of the beam, and assume full lateral support.

2.4. Select the most economical W section for a simply supported beam loaded with a uniformly distributed load of 3.0 k/ft over a span of 35 ft. Neglect the weight of the beam, and assume full lateral support.

2.5 through 2.7. Select the most economical W section for the beams shown. Assume full lateral support, and neglect the weight of the beam.

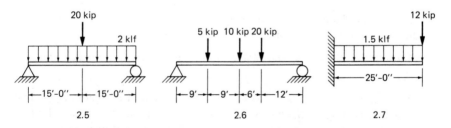

2.8 through 2.10. Redesign the beams in Problems 2.5–2.7 for $F_y = 50$ ksi.

2.11. Assuming that a W 18 × 50 wide-flange beam was mistakenly installed in the situation described in Problem 2.3, determine the load that the new beam can carry.

2.12. Determine the distributed load that can be carried by a W 21 × 57 beam over a simply supported span of 25 ft with lateral supports every 5 ft.

2.13. Determine the concentrated load that a W 18 × 40 beam can carry if the span is 42 ft with the load and a lateral support located at midspan. Assume F_y = 50 ksi.

2.14. Compute the maximum shearing stress on the following sections due to an external shear force of 60 kips.

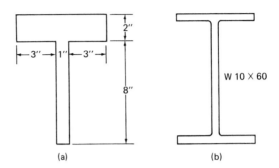

2.15. Select the most economical W 12 section that can carry a concentrated load of 85 kips at the third point of a 4-ft 6-in. simple span. Check both flexural and shear stresses.

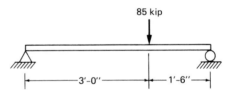

2.16. Calculate the design section modulus for a W 16 × 89 for (A572 Gr 50):

a) two $\frac{1}{2}$-in. diameter holes in each flange
b) two $1\frac{1}{2}$-in. diameter holes in each flange.

2.17 and 2.18. Solve Problem 2.16 for A36 and A588 Gr 50 steels, respectively.

2.19. Assuming that the 85-k load on the W 12 beam in Problem 2.15 is due to a column bearing on the beam, calculate the bearing lengths required under the column and for each support, such that stiffeners are not necessary.

2.20. Calculate the allowable bending stress of a W 12 × 53 with a maximum unsupported length of 18 ft using a) F_y = 36 ksi and b) F_y = 50 ksi.

2.21. Calculate the allowable bending stress of a W 10 × 39 with a maximum unsupported length of 22 ft using a) F_y = 36 ksi and b) F_y = 50 ksi.

2.22. Calculate the allowable bending stress of a W 14 × 43 with a maximum unsupported length of 24 ft using a) F_y = 36 ksi and b) F_y = 50 ksi.

2.23. Determine the allowable bending stress for a W 36 × 245 with a maximum unsupported length of a) 36 ft, b) 27 ft, c) 18 ft, d) 9 ft, and e) 6 ft. Do not use any charts.

2.24. Solve Problem 2.23 using the appropriate charts.

2.25. Design a bearing plate for a S 20 × 86 carrying a 120-k reaction. Assume that one edge of the plate must be at the edge of a concrete pilaster ($f'_c = 4000$ psi).

2.26. For the beam and reaction given in Problem 2.25, design a bearing plate resting $2\frac{1}{2}$ in. behind the face of a very large concrete wall ($f'_c = 4000$ psi).

2.27. A beam spanning continuously, as shown, supports a live load of 1.50 k/ft and a dead load of 0.50 k/ft. Assuming full lateral support, design the beam for

    a) calculated positive and negative moments
    b) positive and negative moments using redistribution allowed by AISCS

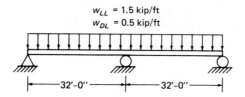

2.28. Design a continuous beam for the condition shown. Assume full lateral support.

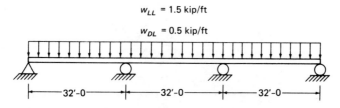

2.29. Assuming full lateral support for the loading shown, design a beam first for calculated moments with simple spans, then for continuous beam moments, and finally for redistributed moments.

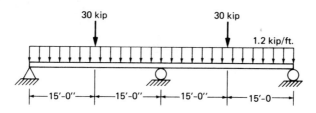

2.30. The wide-flange beam shown is loaded through the shear center, subjecting the member to a moment of 90 ft-k. Design the lightest W 12 section by breaking down the moment into components along the principal axes and considering biaxial bending. Assume full lateral support.

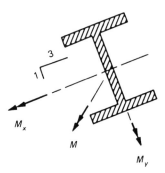

2.31. Design a C 10 channel purlin with full lateral support spanning 16 ft and subject to biaxial bending. Assume the live load of 90 lb/ft, dead load of 40 lb/ft, and wind load of 60 lb/ft to act through the shear center.

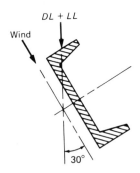

3
Members under Flexure: 2

3.1 COVER-PLATED BEAMS

The moment of inertia and section modulus of a structural steel member can be increased by the use of cover plates. In new steel work, height restrictions may limit the use of deep sections, making a cover-plated shallow beam necessary. Occasionally, the availability of rolled sections suggests the economical use of cover plates. Cover plates can also be added to increase the flexural capacity of existing beams of renovated or modified structures. Plates, in many cases, can be attached to beam flanges and cut off where they are no longer necessary.

The normal procedure for design of cover plates, where depth restrictions govern, is to select a wide-flange beam that has a depth less than the allowable depth, leaving room for the cover plates. The moments of inertia of symmetrical cover plates are added to that of the wide flange, making

$$I_{req} = I_{WF} + 2A_{pl}\left(\frac{d}{2}\right)^2 \qquad (3.1)$$

where d is the distance between the centers of gravity of the two cover plates. From eqn. (3.1), the plate area required is equal to

$$A_{pl} = 2\,\frac{I_{req} - I_{WF}}{d^2} \qquad (3.2)$$

The use of an unsymmetrical flange addition requires the location of the section neutral axis and the calculation of the moment of inertia.

Example 3.1. Determine the maximum uniform load a cover-plated W 8 × 67 can support simply spanning 30 ft. The plates are $\frac{3}{4}$ × 12 in. Assume full lateral support.

Solution.

$$M = F_b \times S, \quad M = \frac{wL^2}{8}, \quad w = \frac{8 F_b S}{L^2}$$

The section modulus is equal to the moment of inertia I of the section divided by the distance c from the neutral axis to the most extreme fiber of the section.

$$I_{section} = I_{WF} + 2 I_{pl}$$

Neglecting I_0 for the cover plates,

$$I_{section} = I_{WF} + 2 A_{pl} \left(\frac{d}{2}\right)^2 = 272.0 \text{ in.}^4 + 2 \times 9.0 \text{ in.}^2 \times (4.875 \text{ in.})^2$$

$$I_{section} = 699.8 \text{ in.}^4$$

$$c = \frac{9.0 \text{ in.}}{2} + 0.75 \text{ in.} = 5.25 \text{ in.}$$

$$S = \frac{699.8 \text{ in.}^4}{5.25 \text{ in.}} = 133.3 \text{ in.}^3$$

$$L = 30.0 \text{ ft}$$

$$w = \frac{8 F_b S}{L^2} = \frac{8 \times 24.0 \text{ ksi} \times 133.3 \text{ in.}^3}{12 \text{ in./ft} \times (30.0 \text{ ft})^2} = 2.37 \text{ k/ft}$$

Subtracting the weight of the beam and cover plates,

$$w = 2.37 \text{ k/ft} - 0.067 \text{ k/ft} - 2(0.03 \text{ k/ft}) = 2.24 \text{ k/ft}$$

The maximum applied uniform load is 2.24 k/ft

Checking deflection,

$$\Delta = \frac{5 \, Wl^3}{384 \, EI} = \frac{5 \times (2.37 \times 30) \times (30 \times 12)^3}{384 \times 29,000 \times 699.8} = 2.13 \text{ in.}$$

The maximum deflection corresponds to a $l/169$ deflection. Note that in many cases this deflection is considered excessive and therefore unsatisfactory (AISCS L3.1). It is recommended that whenever practical, a minimum depth of $(F_y/800) \, l$ be provided (AISCS Commentary L3.1).

Example 3.2. An existing W 16 × 67, whose compression flange is laterally supported, needs to be reinforced to carry a maximum moment of 325 ft-k. Determine the required reinforcement if a) both flanges have symmetric cover plates and b) if the bottom flange has a cover plate only. Assume that $\tfrac{3}{4}$-in. bolts connect the plates to the beam flanges (two rows of bolts per flange).

Solution.

$$S_{req} = \frac{325 \text{ ft-k} \times 12 \text{ in./ft}}{24 \text{ ksi}} = 162.5 \text{ in.}^3$$

a) Try $\tfrac{3}{8}$-in. cover plates

$$d = 16.33 \text{ in.} + 0.375 \text{ in.} = 16.705 \text{ in.}$$
$$\text{Total depth} = 16.33 \text{ in.} + 2 \times (0.375 \text{ in.}) = 17.08 \text{ in.}$$
$$I_{req} = 162.5 \text{ in.}^3 \times (17.08 \text{ in.}/2) = 1387.8 \text{ in.}^4$$

From eqn. (3.2):

$$A_{pl} = 2 \times \frac{1387.8 \text{ in.}^4 - 954 \text{ in.}^4}{(16.705 \text{ in.})^2} = 3.11 \text{ in.}^2$$

Required plate width = $3.11 \text{ in.}^2 / 0.375 \text{ in.} = 8.3 \text{ in.}$

Use $\tfrac{3}{8} \times 8\tfrac{1}{2}$ in. cover plate at the top and bottom ($S = 163.8 \text{ in.}^3$)

Note: Investigation of the hole reduction provisions yields no reduction in the cross-sectional properties required (see AISCS B10 and Sect. 2.6). Also, the total plate area is less than 70% of the total flange area (AISCS B10).

b) Try 4 × 16 in. cover plate

$$\bar{y} = \frac{(19.7 \times 8.17) + [4 \times 16 \times (16.33 + 2)]}{19.7 + (4 \times 16)} = 15.94 \text{ in.}$$

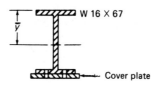

$$I = 954 + 19.7 \times (15.94 - 8.17)^2 + \frac{1}{12} \times 16 \times (4)^3$$

$$+ (16 \times 4) \times (16.33 + 2 - 15.94)^2$$

$$= 954 + 1190.9 + 85.3 + 365.6 = 2595.8 \text{ in.}^4$$

$$S = \frac{2595.8 \text{ in.}^4}{15.94 \text{ in.}} = 162.8 \text{ in.}^3 > 162.5 \text{ in.}^3 \text{ ok}$$

Use 4 × 16 in. cover plate

Note: This plate size should not actually be used, since it violates the requirement given in AISCS B10, which says that the total cross-sectional area of cover plates bolted or riveted to the member shall not exceed 70% of the total flange area. The above example shows the economic benefits that can be achieved by using symmetrical cover plates.

Example 3.3. A simply supported beam spanning 24.0 ft is to carry a uniform load of 4.2 k/ft. However, height restrictions limit the total beam depth to 12 in. Determine if a standard wide flange can be used, and design a section using cover plates if necessary. Assume full lateral support.

Solution.

$$M = \frac{wL^2}{8}, \quad S = \frac{M}{F_b}$$

$$M = \frac{4.2 \text{ k/ft} \times (24.0 \text{ ft})^2}{8} = 302.4 \text{ ft-k}$$

$$S_{req} = \frac{302.4 \text{ ft-k} \times 12 \text{ in.}/\text{ft}}{24.0 \text{ ksi}} = 151.2 \text{ in.}^3$$

Check wide flange sections that have a depth less than 12.0 in.

Largest section is W 10 × 112, $S_x = 126$ in.^{3} < 151.2 in.^{3} N.G. Cover plated section is required.

Try W 10 × 68, $d = 10.4$ in., $S = 75.7$ in.^{3}, $I = 394.0$ in.^{4}

Assume cover plate thickness = $\frac{3}{4}$ in.

$$I_{pl} = I_{req} - I_{WF} = 2A_{pl}d^2$$

$$I_{req} = S \times c = 151.2 \text{ in.}^3 \times \left(\frac{11.9 \text{ in.}}{2}\right) = 899.6 \text{ in.}^4$$

$$I_{WF} = 394.0 \text{ in.}^4$$

$$I_{pl} = 899.6 \text{ in.}^4 - 394.0 \text{ in.}^4 = 505.6 \text{ in.}^4$$

$$\frac{d}{2} = \frac{10.4 \text{ in.}}{2} + \frac{0.75 \text{ in.}}{2} = 5.575 \text{ in.}$$

$$505.6 \text{ in.}^4 = 2A_{pl}(5.575 \text{ in.})^2$$

$$A_{pl} = \frac{505.6 \text{ in.}^4}{62.2 \text{ in.}^2} = 8.13 \text{ in.}^2$$

$$A_{pl} = w \times t$$

$$w = \frac{8.13 \text{ in.}^2}{0.75 \text{ in.}} = 10.84 \text{ in.} \quad \text{say 11 in.}$$

Use W 10×68 beam with $\tfrac{3}{4} \times 11$-in. cover plates.

$$I = 906 \text{ in.}^4$$

$$d = 10.40 \text{ in.} + 2 \times 0.75 \text{ in.} = 11.90 \text{ in.} < 12 \text{ in.} \quad \text{ok}$$

Checking deflection (for full-length cover plates),

$$\Delta = \frac{5 \, Wl^3}{384 \, EI} = \frac{5 \times (4.2 \times 24) \times (24 \times 12)^3}{384 \times 29000 \times 906} = 1.19 \text{ in.}$$

$$\frac{\Delta}{l} = \frac{1.19 \text{ in.}}{24 \text{ ft} \times 12 \text{ in.}/\text{ft}} = \frac{1}{242}$$

Deflection-to-span ratio is within tolerable limits.

Cover plates to flanges and flanges to webs of built-up beams or girders are connected by bolts, rivets, or welds. The longitudinal spacing must be adequate to transfer the horizontal shear, calculated by dividing the strength of the fasteners acting across the section by the shear flow at the section. Maximum spacings for compression flange fasteners and tension flange fasteners are given

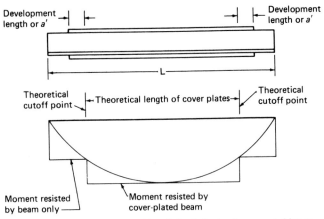

Fig. 3.1. Cover-plating a uniformly loaded, simply supported beam.

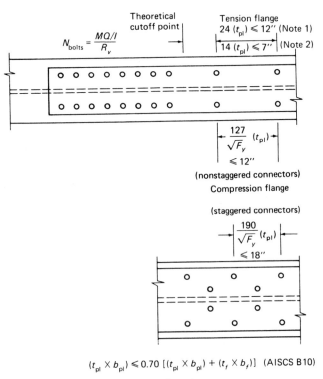

$(t_{pl} \times b_{pl}) \leq 0.70 \, [(t_{pl} \times b_{pl}) + (t_f \times b_f)]$ (AISCS B10)

Note 1: Painted member or unpainted member not subject to corrosion.

Note 2: Unpainted member of weathering steel subject to corrosion.

Fig. 3.2. Bolt and rivet requirements for cover plates. Fasteners in partial plate extensions must develop flexural stress at cutoff point. Maximum bolt or rivet spacing throughout plate must not exceed the values shown.

in Figs. 3.2 and 3.3, as limited by AISCS E4 and D2, respectively. As with beam flanges, no plate area reduction is necessary for fastener holes if (B10-1) is satisfied (see eqn. (2.20), Sect. 2.6).

Partial-length cover plates are attached to rolled shapes to increase the section modulus where increased strength is required and are then discontinued where they are unnecessary (see Fig. 3.1). The plates are extended beyond the theoretical cutoff point and adequately fastened to develop the cover plate's portion of the flexural stresses in the beam or plate girder at the theoretical cutoff point (AISCS B10). Additionally, welded connections for the plate termination must be adequate to develop the cover plate's portion of the flexural stresses in the section at the distance a' from the actual end of the cover plate. In some cases (usually under fatigue loading) the cover plate cutoff point may have to be placed at a lower bending stress than the stress at the theoretical cutoff point. To assure adequate weld strength in all cases, the cover plate must be extended the minimum length a' past the theoretical cutoff point, even if the connection can be completed in a shorter development length. The requirements for bolts or rivets are shown in Fig. 3.2, and weld requirements are shown in Fig. 3.3. For further discussion, see AISC Commentary Section B.10.

The design of connectors for attaching cover plates to members is covered in

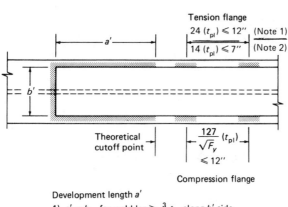

Development length a'
1) $a' = b_{pl}$ for weld leg $\geq \frac{3}{4} t_{pl}$ along b' side
2) $a' = 1\frac{1}{2} b_{pl}$ for weld leg $< \frac{3}{4} t_{pl}$ along b' side
3) $a' = 2 b_{pl}$ for no weld along end b'

Note 1: Painted member or unpainted member not subject to corrosion.

Note 2: Unpainted member of weathering steel subject to corrosion.

Fig. 3.3. Fillet weld requirements for cover plates. The length a' is used as the minimum length to ensure that all development requirements are met.

100 STEEL DESIGN FOR ENGINEERS AND ARCHITECTS

Chapters 5 and 6. Nevertheless, the connector design is included in the following examples for the sake of completeness.

Example 3.4. Design partial length cover plates for a W 27 × 94 loaded as shown. The plates will be connected to the flanges at the termination, with continuous fillet weld at the sides and across the end. The beam is fully laterally supported. Uniform load includes weight of beam.

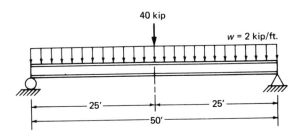

Solution.

$$V = \frac{40 \text{ k}}{2} + \left(2 \text{ k/ft} \times \frac{50 \text{ ft}}{2}\right) = 70 \text{ k}$$

$$M_{max} = \frac{40 \times 50}{4} + \frac{2 \times (50)^2}{8} = 1125 \text{ ft-k}$$

If b_c is the width and t_c the thickness of the cover plate,

$$S_{tot} = \frac{3270 + b_c t_c \times \left(\frac{26.92}{2} + \frac{t_c}{2}\right)^2 \times 2}{\left(\frac{26.92}{2} + t_c\right)}$$

$$S_{tot} \geq \frac{1125 \text{ ft-k} \times 12 \text{ in./ft}}{24.0 \text{ ksi}} = 562.5 \text{ in.}^3$$

$$2 \times 3270 + b_c t_c \times (26.92 + t_c)^2 \geq 562.5 \times (26.92 + 2 t_c)$$

Try

- $t_c = 1\frac{1}{2}$ in. → $b_c = 8.49$ in.
- $t_c = 1\frac{3}{8}$ in. → $b_c = 9.22$ in.
- $t_c = 1\frac{5}{8}$ in. → $b_c = 7.88$ in.

Select cover plates to be $1\frac{5}{8} \times 8$ in.

$$I = 3270 + 8 \times 1.625 \times \left(\frac{26.92}{2} + \frac{1.625}{2}\right)^2 \times 2 = 8566 \text{ in.}^4$$

$$S = \frac{8566}{\frac{26.92}{2} + 1.625} = 567.9 \text{ in.}^3 > 562.5 \text{ in.}^3 \quad \text{ok}$$

Moment at cutoff point is determined from wide flange capacity.

$$M = \frac{F_b \times S}{12 \text{ in./ft}} = \frac{24.0 \text{ ksi} \times 243 \text{ in.}^3}{12 \text{ in./ft}} = 486 \text{ ft-k}$$

Determine the theoretical cutoff points.

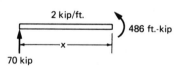

$$70 X - 2 \times \frac{X^2}{2} - 486 = 0$$

Solving for X

$$X^2 - 70 X + 486 = 0$$
$$X = 7.82 \text{ ft}$$

Determine connection of cover plates to flanges.

$$V_{\text{cutoff}} = 70 \text{ k} - (7.82 \text{ ft} \times 2 \text{ k/ft}) = 54.36 \text{ k}$$

102 STEEL DESIGN FOR ENGINEERS AND ARCHITECTS

$$Q = 1.625 \times 8 \times \left(\frac{26.92}{2} + \frac{1.625}{2}\right) = 185.5 \text{ in.}^3$$

$$q = \frac{VQ}{I} = \frac{54.36 \text{ k} \times 185.5 \text{ in.}^3}{8566 \text{ in.}^4} = 1.18 \text{ k/in.}$$

Minimum size weld is $\frac{5}{16}$ in. (AISCS Table J2.4).
Use E 60 welds. Capacity of the weld is

$$0.30 \times \frac{\sqrt{2}}{2} \times 60 \text{ ksi} \times \frac{5}{16} \text{ in.} = 3.98 \text{ k/in.}$$

Using intermittent fillet welds at 12 in. on center (AISCS D2, E4)

$$\ell_w = \frac{1.18 \text{ k/in.} \times 12 \text{ in.}}{2 \times 3.98 \text{ k/in.}} = 1.78 \text{ in.} \quad \text{Use } 1\frac{3}{4} \text{ in. weld length}$$

Termination welds (AISCS B10)

$$M_{\text{cutoff}} = 486 \text{ ft-k}$$

Force to be carried in termination length (C-B10-1)

$$H = \frac{MQ}{I} = \frac{486 \text{ ft-k} \times 12 \text{ in./ft} \times 185.5 \text{ in.}^3}{8566 \text{ in.}^4} = 126.3 \text{ k}$$

$a' = 1.5 \times 8 \text{ in.} = 12 \text{ in.}$ (see Fig. 3.3)

Total length of termination weld $= (12 \text{ in.} \times 2) + 8 \text{ in.} = 32 \text{ in.}$

$$H_w = 32 \text{ in.} \times 3.98 \text{ k/in.} = 127.4 \text{ k} > 126.3 \text{ k} \quad \text{ok}$$

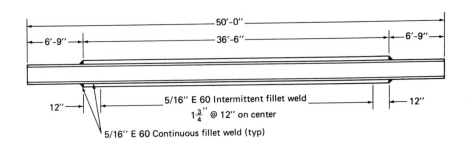

Example 3.5. A W 16 × 100 with $\frac{7}{8}$ × 10 in. cover plates spans a distance of 32 ft with a uniform load of 3.6 k/ft (including weight of beam) and point loads of 15 kips located 10 ft from each end. Determine the length of the partial length cover plates and the connections to the flanges. Use $\frac{3}{4}$-in. ϕ A325-N bolts throughout and type SC, Class A bolts for the plate termination. Assume full lateral support.

Solution.

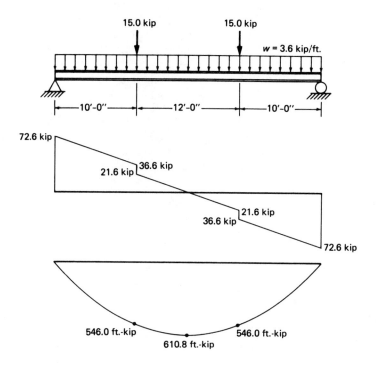

$$I = 1490 \text{ in.}^4 + 2 \times \frac{7}{8} \text{ in.} \times 10 \text{ in.} \times \left(\frac{16.97 \text{ in.}}{2} + \frac{0.875 \text{ in.}}{2}\right)^2$$

$$= 2883 \text{ in.}^4$$

$$S = \frac{I}{c} = \frac{2883 \text{ in.}^4}{\frac{16.97 \text{ in.}}{2} + 0.875 \text{ in.}} = 308.0 \text{ in.}^3$$

$$M_{\text{all}} = F_b \times S = 7392 \text{ in.-k} = 616 \text{ ft-k} > 610.8 \text{ ft-k} \quad \text{ok}$$

Moment at termination of cover plates

$$M = 175 \text{ in.}^3 \times \frac{24 \text{ ksi}}{12 \text{ in./ft}} = 350 \text{ ft-k}$$

Distance to termination point from end of beam

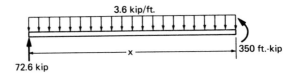

From free body diagram,

$$72.6 \, X - 3.6 \times \frac{X^2}{2} - 350 = 0$$

Solving for X,

$$X = 5.60 \text{ ft}$$

Shear at cutoff point of cover plates

$$V = 72.6 \text{ k} - 5.6 \text{ ft} \times 3.6 \text{ k/ft} = 52.44 \text{ k}$$

Shear flow at cutoff point

$$Q = \frac{7}{8} \text{ in.} \times 10 \text{ in.} \times \left(\frac{16.97 \text{ in.}}{2} + \frac{0.875 \text{ in.}}{2} \right) = 78.07 \text{ in.}^3$$

$$q = \frac{52.44 \text{ k} \times 78.07 \text{ in.}^3}{2883 \text{ in.}^4} = 1.42 \text{ k/in.}$$

Shear capacity of $\frac{3}{4}$-in. A 325-N bolts

$$R_v = 9.3 \text{ k} \quad \text{(AISCM Table I-D)}$$

Spacing of bolts throughout plated region

Minimum spacing:

- Top Plate (AISCS E4):

$$s \leq 0.875 \text{ in.} \times \frac{127}{\sqrt{F_y}} = 18.5 \text{ in.}$$

$$s \leq 12 \text{ in.}$$

Spacing of bolts for top plate must be minimum 12 in.

- Bottom Plate (AISCS D2):

$$s \leq 24 \times 0.875 \text{ in.} = 21 \text{ in.}$$

$$s \leq 12 \text{ in.}$$

Spacing of bolts for bottom plate must be a minimum of 12 in.
Spacing of bolts in termination region

$$H = \frac{MQ}{I} = \frac{350 \text{ ft-k} \times 12 \text{ in./ft} \times 78.07 \text{ in.}^3}{2883 \text{ in.}^3} = 113.7 \text{ k}$$

Capacity of termination bolts ($\frac{3}{4}$-in. A 325-SC Class A)

$$R_v = 7.51 \text{ k (AISCM Table I-D)}$$

Number of bolts needed

$$N = \frac{H}{R_v} = \frac{113.7 \text{ k}}{7.51 \text{ k}} = 15.1 \quad \text{use 16 bolts}$$

To reduce length of plate termination, stagger bolts.

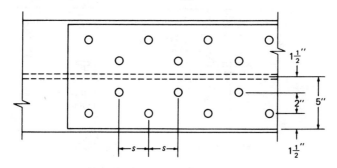

Determine s such that loss of area is less than or equal to two bolt holes.

$$10 \text{ in.} - 4 \times \left(\frac{3}{4} \text{ in.} + \frac{1}{8} \text{ in.}\right) + 2 \times \frac{s^2}{4 \times 2 \text{ in.}}$$

$$= 10 \text{ in.} - 2 \times \left(\frac{3}{4} \text{ in.} + \frac{1}{8} \text{ in.}\right)$$

$$s^2 = 7.0 \text{ in.}$$

$$s = 2.65 \text{ in.} - \text{say } 2\frac{3}{4} \text{ in.}$$

3.2 BUILT-UP MEMBERS

The use of built-up members is necessary when the flexural capacity of rolled sections becomes inadequate. Built-up members usually consist of a web with flange plates connected to the two edges of the web. Today, almost all built-up members have their flanges welded to the webs. Cover plates are sometimes used on the flanges and are cut off at the locations on the span where they are no longer required. Web stiffeners are attached to one or both sides of the web when large shear stresses occur in the web (see Fig. 3.4).

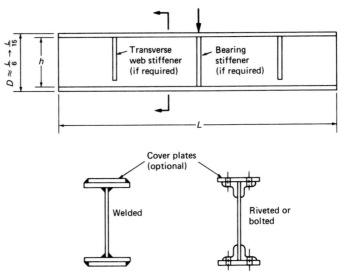

Fig. 3.4. Typical built-up members.

Typically, the depth of built-up members varies from $\frac{1}{6}$ to $\frac{1}{15}$ of the span, with $\frac{1}{10}$ to $\frac{1}{12}$ being the most common. In AISCS Commentary L3.1, it is suggested that the depth of fully stressed members in floors should not be less than $(F_y/800)$ times the span; for fully stressed roof purlins, the depth should not be less than $(F_y/1000)$ times the span. Note that these values are offered only as guides; the requirements for flexure, shear, and deflection are of primary concern.

In AISCS Chapters F and G, two types of built-up members are recognized: built-up beams and plate girders. A built-up beam is distinguished from a plate girder on the basis of the web slenderness ratio h/t_w, where h is the clear distance between flanges and t_w is the thickness of the web. When $h/t_w > 760/\sqrt{F_b}$ (where F_b is the allowable bending stress, ksi), the built-up member is referred to as a *plate girder*. In this case, the provisions of Chapter G apply for the allowable bending stress; otherwise, Chapter F is applicable. If the ratio is less than $760/\sqrt{F_b}$, the member is commonly referred to as a *built-up beam*, and the allowable bending stress is governed by Chapter F (see Chap. 2). For the allowable shear stress, Chapters F and G are applicable.

3.3 DESIGN OF BUILT-UP BEAMS

There is no unique solution for the design of a built-up beam. Although minimum cost is one of the most desired results, a beam can be designed for any number of combinations of web and flange dimensions and stiffener arrangements (when required). A number of trials may be necessary for cost optimization. Typically, assumptions are made on the sizes for the web and flanges, and these assumptions are modified as deemed necessary.

The usual design procedure is to select a trial cross section by the flange-area method and then to check by the moment of inertia method. The flange-area method assumes that the flanges will carry most of the bending moment and the web will carry all the shear forces. Required web area is then

$$A_w = \frac{V}{F_v} \tag{3.3}$$

where V is the vertical shear force at the section and F_v is the allowable shear stress on the web. The required flange area is then determined from

$$A_f = \frac{M}{F_b h} - \frac{A_w}{6} \tag{3.4}$$

where M is the bending moment at the section, F_b is the allowable bending stress, and h is the clear distance between flanges.

Verification by the moment of inertia method requires calculating the moment of inertia of the beam cross section. The allowable bending stress in the compression flange is determined from AISCS F1.

When designing a built-up beam for flexure, it is important to check both the width-thickness ratios for the compression flange and the web against the appropriate limiting values given in AISCS Table B5.1. The limiting ratios for the compression elements in a built-up beam, taken from Table B5.1, are given in Table 3.1.

The magnitudes of the width-thickness ratios of the compression elements and the maximum unsupported length of the compression flange all play a key role in determining the applicable allowable bending stress F_b (see AISCS F1 and Chap. 2). If it is found necessary to increase the section's moment of inertia, the flange area should, in general, be increased. By slightly increasing the flange thickness, the flange width-thickness ratio will remain below the appropriate limiting value given in Table 3.1 while increasing the value of I.

The allowable shear stress F_v is given in AISCS F4. As was shown in Chap. 2, F_v is given in (F4-1) when $h/t_w \leq 380/\sqrt{F_y}$:

$$F_v = 0.40 F_y \tag{3.5}$$

Note that this shear stress acts on the overall depth times the web thickness. In this case, intermediate stiffeners are not required. Such beams do not depend on tension field action (see the discussion which follows for the definition of tension field action).

For the case when $h/t_w > 380/\sqrt{F_y}$, the allowable shear stress acting on the

Table 3.1. Limiting Width-Thickness Ratios for Compression Elements in a Built-Up Beam

Description of Element	Width-Thickness Ratio	Limiting Width-Thickness Ratios	
		Compact	Noncompact
Flanges of I-shaped welded beams in flexure	$b_f/2t_f$	$65/\sqrt{F_y}$	$95/\sqrt{F_{yf}/k_c}$ [a]
Webs in flexural compression[b]	d/t_w	$640/\sqrt{F_y}$	—
	h/t_w	—	$760/\sqrt{F_b}$

[a] $k_c = \dfrac{4.05}{(h/t_w)^{0.46}}$ if $h/t_w > 70$; otherwise, $k_c = 1.0$.

F_{yf} = yield strength of the flange

[b] For hybrid beams, use F_{yf} instead of F_y.

Source: Adapted from AISCS Table B5.1.

clear distance between flanges times the web thickness is given in (F4-2):

$$F_v = \frac{F_y}{2.89} (C_v) \le 0.40 F_y \qquad (3.6)$$

where $C_v = \dfrac{45{,}000 k_v}{F_y (h/t_w)^2}$ when $C_v < 0.8$

$\phantom{\text{where } C_v } = \dfrac{190}{h/t_w} \sqrt{\dfrac{k_v}{F_y}}$ when $C_v > 0.8$

$k_v = 4.00 + \dfrac{5.34}{(a/h)^2}$ when $a/h < 1.0$

$ = 5.34 + \dfrac{4.00}{(a/h)^2}$ when $a/h > 1.0$

t_w = web thickness, in.
a = clear distance between transverse stiffeners, in.
h = clear distance between flanges at the section under investigation, in.

Tables 1-36 and 1-50 (pages 2-232 and 2-233, respectively) of AISCM give the values of the allowable shear stress in ksi according to eqn. (3.6) for steels with $F_y = 36$ ksi and $F_y = 50$ ksi, respectively. As noted above, this allowable stress does not include tension field action.

Intermediate stiffeners are not required when $h/t_w \le 260$ and the maximum web shear stress is less than or equal to F_v given in eqn. (3.6). When transverse stiffeners of adequate strength are provided at required spacings, they act as compression members. The web then behaves as a membrane that builds up diagonal tension fields consisting of shear forces greater than those associated with the theoretical buckling load of the web. The result, commonly referred to as *tension field action*, is similar to the force distribution in a Pratt truss (see Fig. 3.5). Thus, the shear capacity of the web is increased to a level where it is able to resist applied shear forces unaccounted for by the linear buckling theory. Applying thin plate theory, it is clear that the additional shear strength is dependent on the web plate dimensions and the stiffener spacing.

In lieu of eqn. (3.6), the allowable shear stress may be determined by including the contribution of tension field action (AISCS G3). If intermediate stiffeners are sized and spaced according to the provisions given in AISCS G4 and if $C_v \le 1$, the allowable shear stress, including tension field action, is given in (G3-1):

$$F_v = \frac{F_y}{2.89} \left[C_v + \frac{1 - C_v}{1.15 \sqrt{1 + (a/h)^2}} \right] \le 0.40\, F_y \qquad (3.7)$$

110 STEEL DESIGN FOR ENGINEERS AND ARCHITECTS

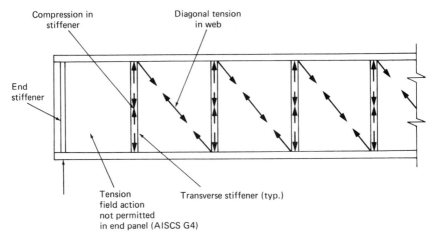

Fig. 3.5. Tension field action in a built-up member.

where all quantities have been defined previously. Values of the allowable stress given in eqn. (3.7) are listed in Tables 2-36 and 2-50 (pp. 2-234 and 2-235) for steels with $F_y = 36$ ksi and $F_y = 50$ ksi, respectively. The italic values in the tables indicate the gross area, as a percentage of the web area, required for pairs of intermediate stiffeners. It is important to note that the spacing between stiffeners at end panels for members designed on the basis of tension field action must not exceed the value given in eqn. (3.6) (AISCS G4). Also, tension field action is not considered when $0.60 F_y \sqrt{3} \leq F_u \leq 0.40 F_y$ or when the panel ratio a/h exceeds 3.0 (AISCS Commentary G3).

In order to obtain the advantages of tension field action, the provisions for minimum stiffness and area of the transverse stiffeners given in AISCS G4 must be satisfied. These requirements are depicted in Fig. 3.6. Additionally, to facilitate handling during fabrication and erection, the panel aspect ratio a/h is arbitrarily limited to the following (see (F5-1)):

$$\frac{a}{h} \leq \left[\frac{260}{(h/t_w)}\right]^2 \quad \text{and} \quad 3.0 \tag{3.8}$$

The stiffeners which are provided for tension field action must be connected for total shear transfer not less than the value given in (G4-3):

$$f_{vs} = h\sqrt{\left(\frac{F_y}{340}\right)^3} \tag{3.9}$$

MEMBERS UNDER FLEXURE: 2

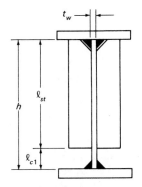

1. $A_{st} \geq \dfrac{1 - C_v}{2} \left[\dfrac{a}{h} - \dfrac{(a/h)^2}{\sqrt{1 + (a/h)^2}} \right] YDht_w$ (G4-2)

    (total area when stiffeners are furnished in pairs)

$$Y = \dfrac{F_{y,web}}{F_{y,st}}$$

D = 1.0 stiffeners in pairs
   = 1.8 single-angle stiffeners
   = 2.4 single-plate stiffeners

$$\dfrac{b_{st}}{t_{st}} \leq \dfrac{95}{\sqrt{F_y}} \quad \text{(Table B5.1)}$$

2. $I_{st} \geq \left(\dfrac{h}{50}\right)^4$ (G4-1)

3. Stiffener length, ℓ_{st}

   $\ell_{st} = h - \ell_{c1}$

    where $w + 6t_w \geq \ell_{c1} \geq w + 4t_w$

       w = size of weld connecting flange to web

If single stiffener is used, attach to compression flange.

Fig. 3.6. Requirements for Transverse Stiffeners (AISCS G4).

where f_{vs} = the shear between the web and the transverse stiffeners, kips per linear inch. Shear transfer may be reduced in the same proportion as the largest computed shear stress to allowable shear stress in adjacent panels.

When a member is designed on the basis of tension field action, a stress reduction in the web may be required due to the presence of high combined moment and shear. Consequently, the maximum bending tensile stress in the web f_b is limited to the value given in (G5-1):

$$f_b \leq \left(0.825 - 0.375 \frac{f_v}{F_v}\right) F_y \qquad (3.10)$$

but should not be taken greater than $0.60F_y$. Note that f_v is the computed average web shear stress (total shear divided by the web area), and F_v is given in eqn. (3.7).

When intermediate stiffeners must be provided for the cases when tension field action is not considered, the stiffeners should have a moment of inertia of at least $(h/50)^4$ to provide adequate lateral support of the web (AISCS Commentary G3). In addition, the limiting width-thickness ratio given in Table B5.1 and the spacing requirements given in eqn. (3.8) are applicable.

Bearing stiffeners must be placed in pairs at unframed ends and at points of concentrated loads when required (AISCS K1.8). The provisions given in AISCS K1.3 through K1.5 govern in this case (see Chap. 2). Figure 3.8 summarizes the bearing stiffener requirements as given in AISCS K1.8.

Hybrid beams are sections in which the web plate is of a different grade of steel than the flange plate. Hybrid built-up beams may be used with the following limitations: 1) the maximum allowable bending stress is $0.60F_{yf}$ for mem-

Fig. 3.7. Fascia, or edge, girder of a steel frame building with vertical web stiffeners.

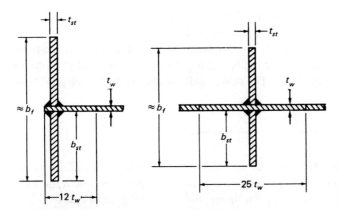

Effective area of end-bearing stiffeners. Provide at unframed ends.

Effective area of load-bearing stiffeners. Provide under concentrated loads when required.

Design requirements

1. $\dfrac{b_{st}}{t_{st}} \le \dfrac{95}{\sqrt{F_y}}$ (Table B5.1)

2. Struts designed as columns. Determine allowable axial stress

End bearing

$A_{\text{eff}} = 2\,(b_{st} \times t_{st}) + 12\,t_w^2$

Interior bearing

$A_{\text{eff}} = 2\,(b_{st} \times t_{st}) + 25\,t_w^2$

$$I_{st} = \dfrac{t_{st}\,(2b_{st} + t_w)^3}{12}$$

$$r = \sqrt{\dfrac{I_{st}}{A_{\text{eff}}}}$$

Kh, $K = 0.75$ (AISCS K1.8)

Allowable axial stress is found in AISC Tables C-36 and C-50 for value of (Kh/r). If stiffener yield stress is different from web yield stress, use lesser grade of the two.

Check compressive stress: $f_a = \dfrac{R}{A_{\text{eff}}} \le F_a$

3. Check bearing criteria:

$$f_p = \dfrac{R}{A_b} \le F_p = 0.90 F_y \quad \text{(AISCS J8)}$$

where A_b = bearing area, in^2

Fig. 3.8. Requirements for bearing stiffeners.

114 STEEL DESIGN FOR ENGINEERS AND ARCHITECTS

bers with full lateral support where F_{yf} is the yield stress of the flange (AISCS F1.2) and 2) tension field action may not be considered (AISCS G3).

To facilitate the design of built-up beams, tables of dimensions and properties are provided as a guide for selecting members of economical proportions (see p. 2-230 of the AISCM). The design aids on pp. 2-236 through 2-242 can also be used.

Example 3.6. Design a built-up beam with no intermediate stiffeners to support a uniform load of $4.8\ k/\text{ft}$ on a 65-ft span. The member will be framed between columns, and its compression flange will be laterally supported over its entire length.

Solution.

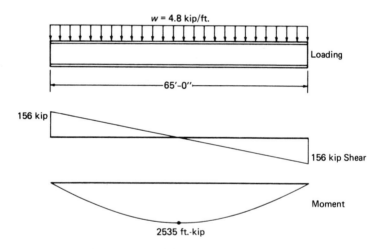

To limit the deflection, assume that $h = \dfrac{L}{10} = \dfrac{65\ \text{ft} \times 12\ \text{in.}/\text{ft}}{10} = 78\ \text{in.}$

Assuming that $F_b = 22$ ksi, for a built-up beam,

$$\frac{h}{t_w} \leq \frac{760}{\sqrt{F_b}} = \frac{760}{\sqrt{22}} = 162 \quad \text{(AISCS Chap. F)}$$

$$t_w \geq \frac{78}{162} = 0.48\ \text{in.} \quad \text{Try } t_w = \tfrac{1}{2}\ \text{in.} \quad \left(\frac{h}{t_w} = \frac{78}{0.5} = 156 < 260\right)$$

Assuming that $\frac{a}{h} > 3.0$ (no stiffeners), $F_v = 3.4$ ksi (from interpolation, Table 1-36).

$$f_v = \frac{V}{A_w} = \frac{156.0 \text{ k}}{0.5 \text{ in.} \times 78 \text{ in.}} = 4.0 \text{ ksi} > 3.4 \text{ ksi} \quad \text{N.G.}$$

Try $\frac{9}{16}$ in. web

$$\frac{h}{t_w} = 139; \quad F_v = 4.3 \text{ ksi}$$

$$f_v = \frac{156.0 \text{ k}}{0.56 \text{ in.} \times 78 \text{ in.}} = 3.57 \text{ ksi} < 4.3 \text{ ksi} \quad \text{ok}$$

Since $h/t_w = 139 > 640/\sqrt{36} = 107$, the web is noncompact. (Table B5.1).

$$S_{\text{req}} = \frac{M}{F_b} = \frac{2535 \text{ ft-k} \times 12 \text{ in./ft}}{22.0 \text{ ksi}} = 1383 \text{ in.}^3$$

Preliminary flange size

$$A_f = \frac{M}{F_b h} - \frac{A_w}{6} = \frac{2535 \text{ ft-k} \times 12 \text{ in./ft}}{22.0 \text{ ksi} \times 78 \text{ in.}} - \frac{0.56 \text{ in.} \times 78 \text{ in.}}{6}$$

$$= 10.45 \text{ in.}^2$$
$$b_f \times t_f = 10.45$$

Try $t_f = \frac{3}{4}$ in., $b_f = 14$ in.

$$k_c = \frac{4.05}{(h/t_w)^{0.46}} = \frac{4.05}{(139)^{0.46}} = 0.42$$

$$\frac{b_f}{2t_f} = \frac{14 \text{ in.}}{2 \times 0.75 \text{ in.}} = 9.33 < \frac{95}{\sqrt{F_y/k_c}} = \frac{95}{\sqrt{36/0.42}} = 10.3$$

Check by moment of inertia method:

$$I = I_{0\text{web}} + 2(Ad^2)_{\text{fl}} = \frac{9}{16} \text{ in.} \times \frac{(78 \text{ in.})^3}{12} + 2\left[10.5 \text{ in.}^2 \times \left(\frac{78.75 \text{ in.}}{2}\right)^2\right]$$

$$= 54{,}803 \text{ in.}^4$$

$$S = \frac{I}{c} = \frac{54{,}803 \text{ in.}^4}{39.75 \text{ in.}} = 1379 \text{ in.}^3 < 1383 \text{ in.}^3 \quad \text{say ok}[1]$$

Check maximum deflection:

$$\Delta = \frac{2535 \times (65)^2}{161 \times 54{,}803} = 1.21 \text{ in.} \simeq \frac{l}{643} \quad \text{ok}$$

Use $\frac{9}{16} \times 78$ in. web plate with $\frac{3}{4} \times 14$ in. flange plates.

(For the connection of the web plate to the web, see Chap. 6.)

Note: This may not be the most economic solution, as a solution with web stiffeners, much thinner web, and heavier flanges could give much less poundage, even including the extra labor cost, as its equivalent in steel weight.

Weight of web + flanges

$$\begin{array}{rl} \text{web} & 23 \text{ psf} \times 6.5 \text{ ft} = 149.5 \text{ lb/ft} \\ \text{flg} & 35.6 \text{ plf} \times 2 \phantom{\text{ ft}} = \underline{71.2 \text{ lb/ft}} \\ & \phantom{35.6 \text{ plf} \times 2 \phantom{\text{ ft}} =} 220.7 \text{ lb/ft} \times 65 \text{ ft} = 14{,}346 \text{ lb} \end{array}$$

Example 3.7. Rework Example 3.6 using intermediate stiffeners and $h = 60$ in. In this case, the beam is not framed at the supports.

Solution.

$$t_w = \frac{h}{760/\sqrt{F_b}} = \frac{60}{162} = 0.37 \text{ in.}$$

Select $\frac{3}{8}$-in. web; $\quad \dfrac{h}{t_w} = \dfrac{60 \text{ in.}}{0.375 \text{ in.}} = 160 < 260$

Assuming that $\dfrac{a}{h} > 3.0$,

$$f_v = \frac{156 \text{ k}}{0.375 \text{ in.} \times 60 \text{ in.}} = 6.93 \text{ ksi} > F_v = 3.2 \text{ ksi} \quad \text{(Table 1-36) N.G.}$$

Thus, transverse stiffeners are required when $f_v > F_v = 3.2$ ksi.

---

[1] In actuality, F_b can be determined from (F1-4), which yields $F_b = 0.62 \times 36$ ksi = 22.2 ksi; this results in $S_{\text{req}} = 1369$ in.^{3} < 1379 in.^{3}. Typically, (F1-5) can be used conservatively in these cases.

Flange design:

$$A_f = \frac{M}{F_b \times h} - \frac{A_w}{6} = \frac{2535 \text{ ft-k} \times 12 \text{ in./ft}}{22.0 \text{ ksi} \times 60 \text{ in.}} - \frac{0.375 \text{ in.} \times 60 \text{ in.}}{6}$$

$$= 23.05 - 3.75 = 19.3 \text{ in.}^2$$

Try $1\frac{1}{4} \times 16$ in. flange.

$$k_c = \frac{4.05}{(160)^{0.46}} = 0.39$$

$$\frac{b_f}{2t_f} = \frac{16 \text{ in.}}{2 \times 1.25 \text{ in.}} = 6.4 < \frac{95}{\sqrt{36/0.39}} = 9.9$$

$$I = \frac{3}{8} \text{ in.} \times \frac{(60 \text{ in.})^3}{12} + 2\left[20.0 \text{ in}^2 \times \left(\frac{61.25 \text{ in.}}{2}\right)^2\right] = 44{,}266 \text{ in.}^4$$

$$S = \frac{44{,}266 \text{ in.}^4}{31.25 \text{ in.}} = 1417 \text{ in.}^3 > 1383 \text{ in.}^3 \quad \text{ok (see Ex. 3.6)}$$

Connection of flange plates to web (see Chap. 6 for details):

$$q_v = \frac{VQ}{I}$$

$$Q = (1.25 \text{ in.} \times 16 \text{ in.}) \times \left(30 \text{ in.} + \frac{1.25 \text{ in.}}{2}\right) = 612.5 \text{ in.}^3$$

$$q_v = \frac{156 \text{ k} \times 612.5 \text{ in.}^3}{44{,}266 \text{ in.}^4} = 2.16 \text{ k/in.}$$

For minimum weld size of $\frac{5}{16}$ in. (AISCS Table J2.4):

$$F_w = 0.928 \text{ k/in.}/16\text{th} \times \tfrac{5}{16} \times 2 \text{ welds} = 9.28 \text{ k/in.}$$

Assume an intermittent fillet weld spaced 12 in. on center.

$$\text{Required weld length} = \frac{2.16 \text{ k/in.} \times 12 \text{ in.}}{9.28 \text{ k/in.}} = 2.79 \text{ in.}$$

Use $\frac{5}{16}$-in. weld, 3 in. long, spaced 12 in. on center.

118   STEEL DESIGN FOR ENGINEERS AND ARCHITECTS

Stiffener spacing:

$$A_w = 0.375 \text{ in.} \times 60 \text{ in.} = 22.5 \text{ in.}^2$$

Since $h/t_w = 160 < 260$, stiffeners are not required when $f_v \leq 3.2$ ksi.

- End Panel (Eqn. (F4-2) Governs):

    $V = 156$ k

    $$f_v = \frac{156 \text{ k}}{22.5 \text{ in.}^2} = 6.93 \text{ ksi} \longrightarrow \frac{a}{h} = 0.85 \quad \text{(Table 1-36)}$$

    $a = 0.85 \times 60 \text{ in.} = 51 \text{ in.} = 4 \text{ ft, 3 in.}$

- Second Panel (Eqn. (G3-1) Governs):

    $V = 156 \text{ k} - 4.8 \text{ k/ft} \times 4.25 \text{ ft} = 135.6 \text{ k}$

    $$f_v = \frac{135.6 \text{ k}}{22.5 \text{ in.}^2} = 6.03 \text{ ksi} \longrightarrow \frac{a}{h} = 2.50 \quad \text{(Table 2-36)}$$

    $a = 2.50 \times 60 \text{ in.} = 150 \text{ in.} = 12 \text{ ft, 6 in.}$

- Third Panel (Eqn. (G3-1) Governs):

    $V = 135.6 \text{ k} - 4.8 \text{ k/ft} \times 12.5 \text{ ft} = 75.6 \text{ k}$

    $$f_v = \frac{75.6 \text{ k}}{22.5 \text{ in.}^2} = 3.36 \text{ ksi} \longrightarrow \frac{a}{h} = 2.50 \quad \text{(Table 2-36)}$$

    $a = 2.50 \times 60 \text{ in.} = 150 \text{ in.} = 12 \text{ ft, 6 in.}$

Note that this spacing is slightly conservative; the maximum value from eqn. (F5-1) can be used:

$$\frac{a}{h} = \left(\frac{260}{160}\right)^2 = 2.64 \longrightarrow a = 2.64 \times 60 \text{ in.} = 158.4 \text{ in.}$$

- Middle Panel (Eqn. (G3-1) Governs):

    $V = 75.6 \text{ k} - 4.8 \text{ k/ft} \times 12.5 \text{ ft} = 15.6 \text{ k}$

    $$f_v = \frac{15.6 \text{ k}}{22.5 \text{ in.}^2} = 0.69 \text{ ksi} < 3.2 \text{ ksi} \longrightarrow \text{No stiffener required}$$

MEMBERS UNDER FLEXURE: 2   119

As noted above, stiffeners are no longer required when $f_v \leq 3.2$ ksi. This corresponds to $V = 3.2$ ksi $\times 22.5$ in.$^2 = 72$ k. Thus, stiffeners are no longer required at a distance of $32.5 [1 - (72/156)] = 17.5$ ft from the support. Conservatively space the stiffeners, as shown in the following figure.

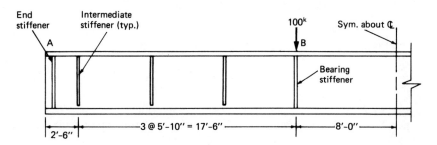

Check combined shear and tension stress at 18.25 ft from the support:

$$V = 156 \text{ k} - 4.8 \text{ k/ft} \times 18.25 \text{ ft} = 68.4 \text{ k}$$

$$f_v = \frac{68.4 \text{ k}}{22.5 \text{ in.}^2} = 3.04 \text{ ksi}$$

$$F_v = 8.5 \text{ ksi} \quad \left(\text{Table 2-36 for } \frac{a}{h} = \frac{7 \times 12}{60} = 1.4\right)$$

$$F_b = \left[0.825 - \left(0.375 \times \frac{3.04}{8.5}\right)\right] 36 \text{ ksi} = 24.9 \text{ ksi} > 22 \text{ ksi}$$

$$f_b'(\text{web}) = \frac{2048 \text{ ft-k} \times 12 \text{ in./ft} \times (31.25 \text{ in.} - 1.25 \text{ in.})}{44{,}266 \text{ in.}^4} = 16.7 \text{ ksi}$$

$$f_b < F_b = 22 \text{ ksi} \quad \text{ok}$$

Design of intermediate stiffeners: Panel 2 (first panel where tension field action occurs)

$$k_v = 5.34 + \frac{4.00}{(1.4)^2} = 7.38$$

$$C_v = \frac{45{,}000 \times 7.38}{36 \times (160)^2} = 0.36 < 1$$

$$A_{st} = \frac{1 - 0.36}{2}\left[1.4 - \frac{(1.4)^2}{\sqrt{1 + (1.4)^2}}\right] \times 1 \times 1 \times 22.5 \text{ in.}^2$$

$$= 0.083 \times 22.5 \text{ in.}^2 = 1.87 \text{ in.}^2 \text{ total area (stiffeners furnished in pairs)}$$

120 STEEL DESIGN FOR ENGINEERS AND ARCHITECTS

Note: The total stiffener area could have been obtained by multiplying the total web area by the italic value (divided by 100) given in Table 2-36 for $h/t_w = 160$ and $a/h = 1.4$:

$$A_{st} = \frac{8.3}{100} \times 22.5 \text{ in.}^2 = 1.87 \text{ in.}^2$$

Note also that this area could be reduced to a total value of

$$\left(\frac{6.03}{8.5}\right) \times 1.87 = 1.3 \text{ in.}^2$$

as per AISCS G4. However, due to the limiting width-thickness ratio given in Table B5.1, try $\frac{1}{4} \times 4$ in. stiffeners on each side:

$$\frac{4 \text{ in.}}{0.25 \text{ in.}} = 16 \approx \frac{95}{\sqrt{36}} = 15.8 \quad \text{say ok.}$$

Check the moment of inertia:

$$I_{req} = \left(\frac{60}{50}\right)^4 = 2.07 \text{ in.}^4 < \frac{1}{12} \times 0.25 \text{ in.} \times (8.375 \text{ in.})^3$$

$$= 12.2 \text{ in.}^4 \quad \text{ok}$$

Minimum length required (AISCS G4):

From Fig. 3.6, $l_{cl} = \frac{5}{16}$ in. $+ (6 \times 0.375 \text{ in.}) = 2.56$ in.

$$l_{st} = 60 \text{ in.} - 2.56 \text{ in.} = 57.44 \text{ in.}$$

Use 4 ft 10 in.-long plates on each side of the web.

Connection of stiffener plates to web:

$$f_{vs} = h\sqrt{\left(\frac{F_y}{340}\right)^3} = 60 \text{ in.} \times \sqrt{\left(\frac{36 \text{ ksi}}{340}\right)^3} = 2.07 \text{ k/in.}$$

Required weld size:

$$w = \frac{(2.07 \text{ k/in.})/(4 \text{ welds})}{0.707 \times 21 \text{ ksi}} = 0.035 \text{ in.}$$

MEMBERS UNDER FLEXURE: 2   121

Use $\frac{3}{16}$-in. continuous weld on both sides of each stiffener.

(Note: An intermittent fillet weld could be used, but it is not desirable.)

Use intermediate stiffeners $\frac{1}{4} \times 4 \times 4$ ft, 10 in.

Design of bearing stiffeners at unframed ends (AISCS K1.8):

Try two $\frac{1}{2} \times 7$ in. stiffeners:

$$\frac{b_{st}}{t_{st}} = \frac{7 \text{ in.}}{0.5 \text{ in.}} = 14 < \frac{95}{\sqrt{36}} = 15.8 \quad \text{ok}$$

$$A_{\text{eff}} = 2 \times (7 \text{ in.} \times 0.5 \text{ in.}) + 12 \times (0.375 \text{ in.})^2 = 8.69 \text{ in.}^2$$

$$I_{st} = \frac{0.5 \text{ in.} \times [(2 \times 7 \text{ in.}) + 0.375 \text{ in.}]^3}{12} = 123.8 \text{ in.}^4$$

$$r = \sqrt{\frac{128.8 \text{ in.}^4}{8.69 \text{ in.}^2}} = 3.77 \text{ in.}$$

$$\frac{Kh}{r} = \frac{0.75 \times 60 \text{ in.}}{3.77 \text{ in.}} = 11.92$$

$$F_a = 21.05 \text{ ksi} \quad \text{(Table C-36)}$$

$$f_a = \frac{156 \text{ k}}{8.69 \text{ in.}^2} = 17.95 \text{ ksi} < 21.05 \text{ ksi} \quad \text{ok}$$

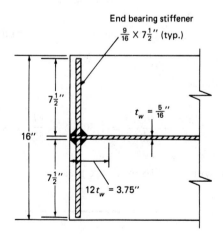

122 STEEL DESIGN FOR ENGINEERS AND ARCHITECTS

Check the bearing on the bottom flange:

Assume $1\frac{1}{2}$ in. clipped from the stiffener, as shown, to accommodate the fillet weld connecting the flange to the web.

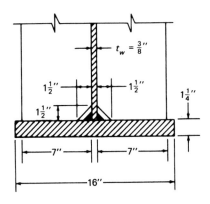

$$A_b = 0.5 \text{ in.} \times (7 \text{ in.} - 1.5 \text{ in.}) = 2.75 \text{ in.}^2$$

$$f_p = \frac{(156 \text{ k}/2)}{2.75 \text{ in.}^2} = 28.4 \text{ ksi} < 0.90 \times 36 \text{ ksi} = 32.4 \text{ ksi} \quad \text{ok}$$

Connection of bearing stiffeners to web:

$$\text{Required shear transfer} = \frac{156 \text{ k}}{4 \times 60 \text{ in.}} = 0.65 \text{ k/in.}$$

Use $\frac{3}{16}$-in. continuous fillet welds on each side of both stiffeners.

Use bearing stiffeners $\frac{1}{2}$ in. $\times$ 7 in. $\times$ 5 ft, 0 in.

Check maximum deflection:

$$\Delta = \frac{2535 \times (65)^2}{161 \times 44{,}266} = 1.50 \text{ in.} = \frac{l}{520} \quad \text{ok}$$

Use $\frac{3}{8} \times 60$ in. web with $1\frac{1}{4} \times 16$ in. flange plates.

MEMBERS UNDER FLEXURE: 2   123

3.4 DESIGN OF PLATE GIRDERS

As noted above, a plate girder is a built-up member which has a web-slenderness ratio $h/t_w > 760/\sqrt{F_b}$. The main difference between the design of a plate girder and a built-up beam is in the determination of the allowable bending stress. For plate girders, the allowable bending stress in the compression flange F_b' is given in (G2-1):

$$F_b' = F_b R_{PG} R_e \qquad (3.11)$$

where F_b is the applicable allowable bending stress given in AISCS Chap. F and

R_{PG} = plate girder factor

$$= 1 - 0.0005 \frac{A_w}{A_f}\left(\frac{h}{t_w} - \frac{760}{\sqrt{F_b}}\right) \leq 1.0$$

R_e = hybrid girder factor

$$= \frac{12 + \left(\dfrac{A_w}{A_f}\right)(3\alpha - \alpha^3)}{12 + 2\left(\dfrac{A_w}{A_f}\right)} \leq 1.0 \quad \text{(nonhybrid girders, } R_e = 1.0\text{)}$$

where A_w = area of the web at the section under investigation, in.^{2}
A_f = area of the compression flange, in.^{2}
$\alpha = 0.6 F_{yw}/F_b \leq 1.0$
F_{yw} = yield stress of the web, ksi.

Since plate girders have thin webs, a portion of the web may deflect enough laterally so that it does not provide its full share of the bending resistance. Thus, the factor R_{PG} ensures that the bending capacity is adequate for the cases when this occurs. Also, the factor R_e is included to compensate for any loss of bending resistance when portions of the web of a hybrid girder are strained beyond their yield stress limit.

In AISCS G1, limitations are given for the maximum web-slenderness ratio. For the cases when no transverse stiffeners are provided or when stiffeners are provided but spaced more than $1\frac{1}{2}h$, the maximum ratio is determined from (G1-1):

$$\frac{h}{t_w} \leq \frac{14{,}000}{\sqrt{F_{yf}(F_{yf} + 16.5)}} \qquad (3.12)$$

124 STEEL DESIGN FOR ENGINEERS AND ARCHITECTS

where F_{yf} = yield stress of the flange, ksi. When transverse stiffeners are provided and are spaced not more than $1\tfrac{1}{2}h$, (G1-2) governs:

$$\frac{h}{t_w} \leq \frac{2000}{\sqrt{F_{yf}}} \qquad (3.13)$$

The allowable shear stress is limited by (F4-2) when tension field action is not considered and by (G3-1) when it is considered. All of the provisions for the design of intermediate and bearing stiffeners required for built-up beams are also applicable for plate girders.

Example 3.8. A plate girder loaded as shown is not framed at the supports. Lateral support is provided at the points of the applied concentrated loads only. Design the girder with intermediate stiffeners.

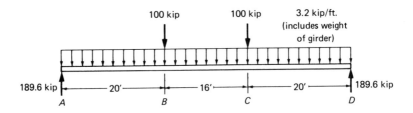

Solution.

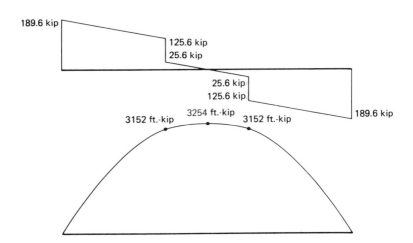

MEMBERS UNDER FLEXURE: 2   125

Assume $h = \dfrac{L}{10} = \dfrac{56 \text{ ft} \times 12 \text{ in./ft}}{10} = 67.2$ in. say 70 in.

Maximum web-slenderness ratio $\dfrac{h}{t_w} = \dfrac{2000}{\sqrt{36}} = 333$ \hfill (G1-2)

Minimum web thickness $t_w = \dfrac{70 \text{ in.}}{333} = 0.21$ in. Try $\tfrac{5}{16}$ in. web.

Considering a flange stress reduction, assume that $F_b = 21.0$ ksi.

$$\dfrac{h}{t_w} = \dfrac{70}{0.3125} = 224 > \dfrac{760}{\sqrt{21}} = 166$$

Flange design:

$$A_f = \dfrac{3254 \text{ ft-k} \times 12 \text{ in./ft}}{21.0 \text{ ksi} \times 70 \text{ in.}} - \dfrac{0.3125 \text{ in.} \times 70 \text{ in.}}{6} = 22.92 \text{ in.}^2$$

Try $1\tfrac{1}{2} \times 16$ in. flange.

$$k_c = \dfrac{4.05}{(224)^{0.46}} = 0.34$$

$$\dfrac{b_f}{2t_f} = \dfrac{16 \text{ in.}}{2 \times 1.5 \text{ in.}} = 5.33 < \dfrac{95}{\sqrt{36/0.34}} = 9.23 \quad \text{ok}$$

$$I = \dfrac{5}{16} \text{ in.} \times \dfrac{(70 \text{ in.})^3}{12} + 2\left[24.0 \text{ in.}^2 \times \left(\dfrac{71.5 \text{ in.}}{2}\right)^2\right] = 70{,}279 \text{ in.}^4$$

$$S = \dfrac{70{,}279 \text{ in.}^4}{36.5 \text{ in.}} = 1925.5 \text{ in.}^3$$

$$f_b = \dfrac{3254 \text{ ft-k} \times 12 \text{ in./ft}}{1925.5 \text{ in.}^3} = 20.28 \text{ ksi}$$

Check the bending stress in the 16-ft center panel (between B and C):
Moment of inertia of flange plus $\tfrac{1}{6}$ of the web:

$$I_T = \dfrac{1.50 \text{ in.} \times (16 \text{ in.})^3}{12} = 512.0 \text{ in.}^4$$

$$A_T = A_f + \tfrac{1}{6} A_w = (1.5 \text{ in.} \times 16 \text{ in.}) + \tfrac{1}{6}(0.3125 \text{ in.} \times 70 \text{ in.})$$
$$= 27.65 \text{ in.}^2$$
$$r_T = \sqrt{\frac{512 \text{ in.}^4}{27.65 \text{ in.}^2}} = 4.30 \text{ in.}$$

For panel BC, $C_b = 1.0$ (AISCS F1.3)

$$\frac{l}{r_T} = \frac{16 \text{ ft} \times 12 \text{ in./ft}}{4.30 \text{ in.}} = 44.65 < \sqrt{\frac{102 \times 10^3 \times 1.0}{36}} = 53.2$$

Therefore, $F_b = 22.0$ ksi in panel BC.

Reduced allowable bending stress in the compression flange (AISCS G2):

$$A_w = 0.3125 \text{ in.} \times 70 \text{ in.} = 21.88 \text{ in.}^2$$
$$A_f = 1.5 \text{ in.} \times 16 \text{ in.} = 24 \text{ in.}^2$$
$$R_{PG} = 1 - 0.0005 \left(\frac{21.88 \text{ in.}^2}{24 \text{ in.}^2}\right)\left(224 - \frac{760}{\sqrt{22}}\right) = 0.972$$
$$R_e = 1.0$$
$$F'_b = 22 \text{ ksi} \times 0.972 \times 1.0 = 21.4 \text{ ksi} > f_b = 20.28 \text{ ksi} \quad \text{ok}$$

Check the bending stress in the 20-ft end panels (between A and B, C and D):

$$f_b = \frac{3152 \text{ ft-k} \times 12 \text{ in./ft}}{1925.5 \text{ in.}^3} = 19.64 \text{ ksi}$$

For end panels, $C_b = 1.75$ ($M_1 = 0$)

$$\frac{l}{r_T} = \frac{20 \text{ ft} \times 12 \text{ in./ft}}{4.30 \text{ in.}} = 55.8 < \sqrt{\frac{102 \times 10^3 \times 1.75}{36}} = 70.4$$

$F_b = 22.0$ ksi

$F'_b = 21.4$ ksi > 19.64 ksi ok

Therefore, use $\tfrac{5}{16} \times 70$ in. plate for the web and $1\tfrac{1}{2} \times 16$ in. plates for the flanges.

Connection of flange plates to web (see Chap. 6 for details):

$$Q = (1.5 \text{ in.} \times 16 \text{ in.}) \times \left(35 \text{ in.} + \frac{1.5 \text{ in.}}{2}\right) = 858 \text{ in.}^3$$

$$q_v = \frac{189.6 \text{ k} \times 858 \text{ in.}^3}{70{,}279 \text{ in.}^4} = 2.31 \text{ k/in.}$$

Using minimum weld size of $\frac{5}{16}$ in.,

$$F_w = 0.928 \times 5 \times 2 = 9.28 \text{ k/in.}$$

For intermittent fillet welds spaced 12 in. on center,

$$\text{Required weld length} = \frac{2.31 \text{ k/in.} \times 12 \text{ in.}}{9.28 \text{ k/in.}} = 2.99 \text{ in.}$$

Use $\frac{5}{16}$-in. fillet weld, 3 in. long, spaced 12 in. on center.

Required intermediate stiffeners:

End panel stiffener (no tension field action permitted per AISCS G4):

$V = 189.6$ k

$$f_v = \frac{189.6 \text{ k}}{21.88 \text{ in.}^2} = 8.67 \text{ ksi} \longrightarrow \frac{a}{h} = 0.47 \quad \text{(this value was obtained by solving (F4-2) for } a/h\text{)}$$

$a = 0.47 \times 70$ in. $= 32.9$ in.   Use 2 ft, 6 in.

Shear at first intermediate stiffener:

$$V = 189.6 \text{ k} - 3.2 \text{ k/ft} \times 2.5 \text{ ft} = 181.6 \text{ k}$$

$$f_v = \frac{181.6 \text{ k}}{21.88 \text{ in.}^2} = 8.3 \text{ ksi} \longrightarrow \frac{a}{h} = 1.0 \quad \text{(Table 2-36)}$$

$a = 1.0 \times 70$ in. $= 70$ in.   Use 5 ft, 10 in.

Shear at second intermediate stiffener:

$$V = 181.6 \text{ k} - 3.2 \text{ k/ft} \times 5.83 \text{ ft} = 162.9 \text{ k}$$

$$f_v = \frac{162.9 \text{ k}}{21.88 \text{ in.}^2} = 7.5 \text{ ksi} \longrightarrow \frac{a}{h} = \max\left(\frac{a}{h}\right) = 1.35 \quad \text{(F5-1)}$$

$a = 1.35 \times 70 \text{ in.} = 94.5 \text{ in.}$ Use 7 ft, 6 in.

Shear at third intermediate stiffener:

$$V = 162.9 \text{ k} - 3.2 \text{ k/ft} \times 7.5 \text{ ft} = 138.9 \text{ k}$$

$$f_v = \frac{138.9 \text{ k}}{21.88 \text{ in.}^2} = 6.3 \text{ ksi} \longrightarrow \frac{a}{h} = 1.35$$

$a = 1.35 \times 70 \text{ in.} = 94.5 \text{ in.}$ Use 7 ft, 6 in.

Check the 16-ft center panel:

$V = 25.6 \text{ k}$

$$f_v = \frac{25.6 \text{ k}}{21.88 \text{ in.}^2} = 1.17 \text{ ksi} < F_v = 1.66 \text{ ksi} \quad \left(\text{(F4-2) with } \frac{a}{h} > 3\right)$$

No intermediate stiffeners are required in the center panel.
Conservatively provide stiffeners as shown.

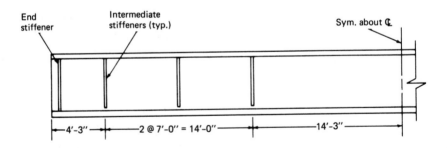

Check combined shear and tension stress at point B:

$V = 125.6 \text{ k}$

$$f_v = \frac{125.6 \text{ k}}{21.88 \text{ in.}^2} = 5.74 \text{ ksi}$$

For $\dfrac{a}{h} = \dfrac{5.83 \text{ ft} \times 12 \text{ in./ft}}{70 \text{ in.}} = 1.0$, $F_v = 8.8 \text{ ksi}$ (Table 2-36)

$$F_b = \left[0.825 - \left(0.375 \times \frac{5.74}{8.8}\right)\right] 36 \text{ ksi} = 20.9 \text{ ksi} < 22 \text{ ksi}$$

$$f_b(\text{web}) = 19.64 \text{ ksi} \left(\frac{35 \text{ in.}}{36.5 \text{ in.}}\right) = 18.8 \text{ ksi} < 20.9 \text{ ksi} \quad \text{ok}$$

Design of intermediate stiffeners:

$$A_{st} = \% \text{ web area (Table 2-36)} \times (f_v/F_v)$$

For $h/t_w = 224$ and $a/h = 1.0$, % web area = 11.2

(from interpolation, Table 2-36)

$$A_{st} = 0.112 \times 21.88 \text{ in.}^2 \times (8.3 \text{ ksi}/8.8 \text{ ksi}) = 2.31 \text{ in.}^2$$

Try $\frac{5}{16} \times 4$ in. stiffeners ($A_{st} = 2 \times 0.3125$ in. $\times 4$ in. $= 2.5$ in.^{2}).

$$\frac{4.0 \text{ in.}}{0.3125 \text{ in.}} = 12.8 < \frac{95}{\sqrt{36}} = 15.8 \quad \text{ok}$$

Check moment of inertia:

$$I_{req} = \left(\frac{70}{50}\right)^4 = 3.84 \text{ in.}^4 < \frac{1}{12} \times 0.3125 \text{ in.} \times (8.3125 \text{ in.})^3$$

$$= 14.96 \text{ in.}^4 \quad \text{ok}$$

Minimum stiffener length (AISCS G4):

$$l_{cl} = \tfrac{5}{16} \text{ in.} + (6 \times 0.3125 \text{ in.}) = 21.19 \text{ in.}$$

$$l_{st} = 70 \text{ in.} - 2.19 \text{ in.} = 67.8 \text{ in.}$$

Use 5 ft 8 in. long plates on each side of the web.

Connection of stiffener plates to web:

$$f_{vs} = 70 \text{ in.} \times \sqrt{\left(\frac{36 \text{ ksi}}{340}\right)^3} = 2.41 \text{ k/in.}$$

Use $\tfrac{3}{16}$-in. continuous welds on both sides of each stiffener.

Use intermediate stiffeners $\tfrac{5}{16} \times 4 \times 5$ ft, 8 in.

130 STEEL DESIGN FOR ENGINEERS AND ARCHITECTS

Design of bearing stiffeners at unframed ends (AISCS K1.8):

Try two $\tfrac{9}{16} \times 7\tfrac{1}{2}$ in. stiffeners.

$$\frac{b_{st}}{t_{st}} = \frac{7.5 \text{ in.}}{0.5625 \text{ in.}} = 13.3 < \frac{95}{\sqrt{36}} = 15.8 \quad \text{ok}$$

$$A_{\text{eff}} = 2 \times (7.5 \text{ in.} \times 0.5625 \text{ in.}) + 12 \times (0.3125 \text{ in.})^2 = 9.61 \text{ in.}^2$$

$$I_{st} = \frac{0.5625 \text{ in.} \times [(2 \times 7.5 \text{ in.}) + 0.3125 \text{ in.}]^3}{12} = 168.3 \text{ in.}^4$$

$$r = \sqrt{\frac{168.3 \text{ in.}^4}{9.61 \text{ in.}^2}} = 4.18 \text{ in.}$$

$$\frac{Kh}{r} = \frac{0.75 \times 70 \text{ in.}}{4.18 \text{ in.}} = 12.6$$

$$F_a = 21.02 \text{ ksi} \quad \text{(Table C-36)}$$

$$f_a = \frac{189.6 \text{ k}}{9.61 \text{ in.}^2} = 19.73 \text{ ksi} < 21.02 \text{ ksi} \quad \text{ok}$$

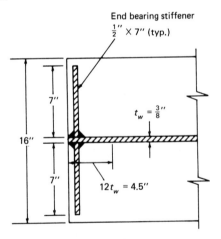

Check bearing on bottom flange:

Assume that $1\tfrac{1}{2}$ in. is clipped from the end stiffener at the bottom to accommodate the fillet weld connecting the flange to the web.

$$A_b = 0.5625 \text{ in.} \times (7.5 \text{ in.} - 1.5 \text{ in.}) = 3.375 \text{ in.}^2$$

$$f_p = \frac{(189.6 \text{ k}/2)}{3.375 \text{ in.}^2} = 28.1 \text{ ksi} < 0.90 \times 36 \text{ ksi} = 32.4 \text{ ksi} \quad \text{ok}$$

MEMBERS UNDER FLEXURE: 2   131

Connection of end bearing stiffener to web:

$$\text{Required shear transfer} = \frac{189.6 \text{ k}}{4 \times 70 \text{ in.}} = 0.68 \text{ k/in.}$$

Use $\frac{3}{16}$-in. continuous fillet welds on each side of both stiffeners.
Use end bearing stiffeners $\frac{9}{16} \times 7 \times 5$ ft 10, in.
Design of bearing stiffeners under concentrated loads:

Check local web yielding (AISCS K1.3), assuming point bearing (i.e., $N = 0$):

$$k = t_f + w(\text{flange to web})$$
$$= 1.5 \text{ in.} + 0.3125 \text{ in.} = 1.81 \text{ in.}$$

$$\frac{R}{t_w(N + 5k)} < 0.66 F_y \quad (\text{K1-2})$$

$$\frac{100 \text{ k}}{0.3125 \text{ in.} [0 + (5 \times 1.81 \text{ in.})]} = 35.4 \text{ ksi} > 0.66 \times 36 = 24 \text{ ksi} \quad \text{N.G.}$$

(Note: If the local web yielding criterion is satisfied, the criteria for web crippling must then be checked.)

Try two $\frac{3}{8} \times 4$ in. stiffeners.

$$\frac{b_{st}}{t_{st}} = \frac{4.0 \text{ in.}}{0.375 \text{ in.}} = 10.7 < 15.8 \quad \text{ok}$$

$$A_{\text{eff}} = 2 \times (4.0 \text{ in.} \times 0.375 \text{ in.}) + 25 \times (0.3125 \text{ in.})^2 = 5.44 \text{ in.}^2$$

$$I_{st} = \frac{0.375 \text{ in.} \times [(2 \times 4.0 \text{ in.}) + 0.3125 \text{ in.}]^3}{12} = 17.95 \text{ in.}^4$$

$$r = \sqrt{\frac{17.95 \text{ in.}^4}{5.44 \text{ in.}^2}} = 1.82 \text{ in.}$$

$$\frac{Kh}{r} = \frac{0.75 \times 70 \text{ in.}}{1.82 \text{ in.}} = 28.8$$

$$F_a = 20.02 \text{ ksi} \quad (\text{Table C-36})$$

$$f_a = \frac{100 \text{ k}}{5.44 \text{ in.}^2} = 18.38 \text{ ksi} < 20.02 \text{ ksi} \quad \text{ok}$$

Check bearing:

Again, assume $1\frac{1}{2}$ in. clipped from the stiffener to accommodate the fillet weld connecting the flange to the web.

$$A_b = 0.375 \text{ in.} \times (4.0 \text{ in.} - 1.5 \text{ in.}) = 0.94 \text{ in.}^2$$

$$f_p = \frac{(100 \text{ k}/2)}{0.94 \text{ in.}^2} = 53.3 \text{ ksi} > 32.4 \text{ ksi} \quad \text{N.G.}$$

$$\text{Required } A_b = \frac{(100 \text{ k}/2)}{32.4 \text{ ksi}} = 1.54 \text{ in.}^2$$

Conservatively use $\frac{9}{16} \times 7\frac{1}{2}$ in. stiffeners under concentrated loads (same as for end stiffeners).

3.5 OPEN-WEB STEEL JOISTS

In common usage are shop-fabricated, lightweight truss members referred to as *open-web steel joists* (see Fig. 3.9). Charts are readily available that provide standard fabricated sections for desired span and loading conditions. When the use of open-web joists is desired, the designer should refer to any one of the numerous design tables available from joist suppliers and the Steel Joist Institute.

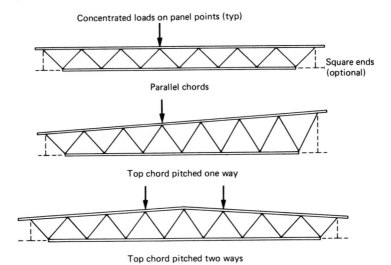

Fig. 3.9. Types of open-web steel joists. Generally, chords are double-angles, and the web members are round bars. Parallel chord joists are most commonly used.

Fig. 3.10. Long-span joist roof system of the American Royal Arena in Kansas City, Missouri (courtesy of U.S. Steel Corp.).

Fig. 3.11. View of a steel truss made from wide flange sections during fabrication.

PROBLEMS TO BE SOLVED

3.1. A W 16 × 100 beam has a simple span of 45 ft. If the member is fully laterally supported, determine

a) the maximum distributed load the beam can carry and
b) the maximum distributed load that can be carried by the beam if both flanges are cover-plated with $1\frac{1}{2}$ × 12 in. plates.

3.2. Determine the width of $1\frac{1}{2}$-in. cover plates necessary for a W 12 × 65 beam to resist a moment of 320 ft-k. Design plates for top and bottom flanges. Use $F_b = 22.0$ ksi.

3.3. The loading of an existing W 12 × 65 beam spanning 25 ft is to be increased to 2.75 k/ft. Because of existing conditions, only the bottom flange can be reinforced. Determine the thickness of a 12-in.-wide cover plate for the bottom flange necessary to carry the increased load.

3.4.[2] A simple beam spanning of 30 ft is to carry a uniform distributed load of 1.8 k/ft. However, height restrictions limit the total depth of $10\frac{1}{2}$ in. Determine the width of $\frac{3}{4}$-in. partial length cover plates necessary to increase a W 8 × 67 section to carry the load. Assuming bolted plates with two rows of $\frac{3}{4}$-in. A 325 N or SC bolts as applicable, calculate the required spacing of bolts.

3.5.[2] Design partial-length cover plates for Problem 3,4, assuming $\frac{3}{4}$ × 11-in. plates welded continuously along the sides and across the ends with $\frac{5}{16}$-in. welds.

3.6.[2] A W 10 × 49 beam with $\frac{3}{4}$ × 12-in. cover plates spans a distance of 35 ft, with loads as shown. Assuming $F_b = 22.0$ ksi, design partial-length cover plates bolted to the flanges with two rows of $\frac{3}{4}$-in. A 325-N bolts, using an allowable load of 9 kips per bolt; and determine how many bolts are required in the development length.

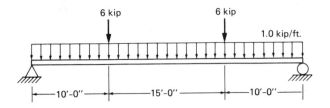

3.7. Show that the required plate girder flange area equation $A_f = M/(F_b h) - A_w/6$ is derived from the gross moment of inertia equation. Hint: Assume flange thickness to be small compared with the web plate height.

---

[2]Knowledge of connection design (Chaps. 5, 6, and 7) is required.

MEMBERS UNDER FLEXURE: 2 135

3.8.[3] A built-up beam framed between two columns 70 ft apart supports a uniform load 4.5 k/ft. Design trial web and flange plates that will not require transverse stiffeners. Assume the beam depth to be $L/12$. Check using the moment of inertia method. Calculate the welds.

3.9.[3] Design bearing and intermediate stiffeners, in pairs, and their connections for a member with loading as shown. The section consists of a $56 \times \frac{5}{16}$-in. web plate and $1\frac{1}{4} \times 20$-in. flange plates and is unframed at the ends. Assume that the member is fully supported laterally.

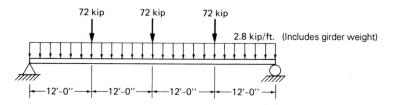

3.10.[3] Completely design the built-up beam shown. The beam is laterally supported at points A, B, C, and D, the compressive flange is not prevented from rotating between the points of support; and the ends are not framed. Design such that intermediate stiffeners are not required.

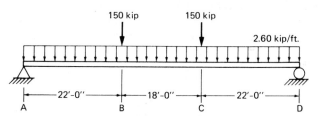

3.11.[3] Redesign Problem 3.10 as a plate girder using minimum thickness web plate and intermediate web stiffeners.

---

[3]Knowledge of connection design (Chaps. 5, 6, and 7) is required.

4
Columns

4.1 COLUMNS

Straight members that are subject to compression by axial forces are known as *columns*. The strength of a column is governed by the yielding of the material for short ones, by elastic buckling for long ones, and by inelastic (plastic) buckling for ones of intermediate lengths. A "perfect column," that is, one made of isotropic material, free of residual stresses, loaded at its centroid, and perfectly straight, will shorten uniformly due to uniform compressive strain on its cross section. If the load on the column is gradually increased, it will eventually cause the column to deflect laterally and fail in a bending mode. This load, called the *critical load*, is considered the maximum load that can be safely carried by the column.

Example 4.1. For a column with pin-connected ends and subjected to its critical load P_{cr}, show that

a) $y'' + k^2 y = 0$ for $k^2 = \dfrac{P}{EI}$

b) $kl = \pi$ and that $P_{cr} = \dfrac{\pi^2 EI}{l^2}$.

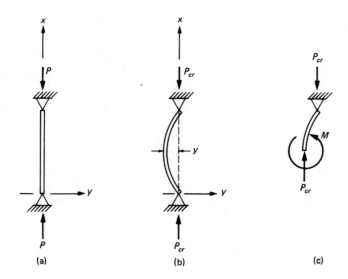

(a) (b) (c)

Solution.

a) From mechanics of materials, the moment in a beam is given as

$$M = EIy''$$

for the axes indicated in figure b. From figure c

$$M = -Py$$

Equating the two

$$EIy'' = -Py$$
$$EIy'' + Py = 0$$
$$y'' + (P/EI)y = 0$$

Using $P/EI = k^2$

$$y'' + k^2 y = 0$$

b) For the differential equation $y'' + k^2 y = 0$, the end conditions are

$$\text{at } x = 0, \ y = 0$$
$$\text{at } x = \frac{\ell}{2}, \ y' = 0 \text{ (slope of beam)}$$

The solution of the differential equation $y'' + k^2 y = 0$ is

$$y = A \cos kx + B \sin kx$$

$$\text{at } x = 0, \; y = A \cos 0 + B \sin 0 = 0$$

$$\cos 0 = 1, \sin 0 = 0. \text{ Hence, } A = 0$$

$$y = B \sin kx$$

$$y' = kB \cos kx$$

At $x = \ell/2$, $y' = 0$. Hence the second condition can be satisfied only if $\cos kl/2 = 0$ or $kl/2 = \pi/2$, which gives

$$k^2 l^2 = \pi^2$$

or

$$P = P_{cr} = \frac{\pi^2 EI}{l^2}$$

As developed above for a long column with pinned ends

$$P_{cr} = \frac{\pi^2 EI}{l^2} \quad (4.1)$$

where P_{cr} is the critical load, called the *Euler load*. Because the axial stress $f_a = P/A$

$$F_{acr} = \frac{\pi^2 EI}{l^2 A} = F_e \quad (4.2)$$

and using $\sqrt{I/A} = r$ (radius of gyration), we find

$$F_e = \frac{\pi^2 E}{(l/r)^2} \quad (4.3)$$

where F_e is Euler's stress.

When we graphically represent the ideal failure stresses (F_F) of a column

Fig. 4.1. Detail of a wide-flange beam to pipe column connection.

versus the ratio l/r (called slenderness) we obtain a curve made of the branches AB and BCD (see Fig. 4.2).

The branch AB is where a column can be expected to fail by yielding ($F_F = F_y$) and

$$\frac{l}{r} \leq \left(\frac{l}{r}\right)^* = \pi\sqrt{\frac{E}{F_y}}$$

where $(l/r)^*$ is the slenderness for which the Euler stress is equal to the yield stress. On branch BCD, the column can be expected to fail by elastic buckling with a failure stress

$$F_F = F_e = \frac{\pi^2 E}{(l/r)^2} < F_y$$

where $l/r > (l/r)^* = \pi\sqrt{E/F_y}$.

However, tests have shown that columns fail in the zone shaded in Fig. 4.2. This variation in failure stresses of test samples was originally attributed to imperfections in the columns, but it has been proven that residual stresses created this condition. The residual stresses are created by uneven cooling of rolled sections such as in wide flange sections where the tips of the flanges and the middle portion of the web cools much faster than the juncture of the web and

140 STEEL DESIGN FOR ENGINEERS AND ARCHITECTS

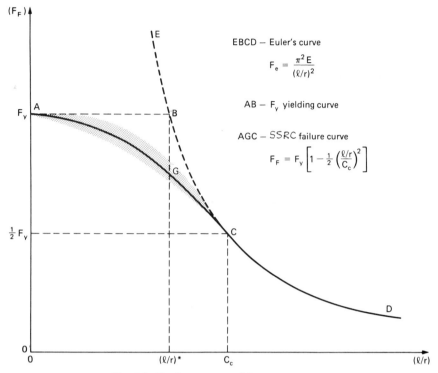

Fig. 4.2. Slenderness versus failure stress curve.

flanges. As a result of these observations, the Structural Stability Research Council (SSRC) has adopted the slenderness C_c to separate the elastic from the nonelastic buckling. C_c is the slenderness corresponding to $F_F = F_y/2$ and is equal to $\pi\sqrt{2E/F_y}$. The branch AGC is a quadratic curve which fits the test results, has a value of $F_F = F_y$ and a horizontal tangent at $l/r = 0$, and matches Euler's curve at point C by having the same ordinate and the same tangent for $l/r = C_c$.

The branch CD is Euler's curve. Hence the failure stresses can be given by

$$F_F = F_y\left[1 - \frac{1}{2}\left(\frac{l/r}{C_c}\right)^2\right] \tag{4.4}$$

for $l/r \leq C_c$ where $C_c = \sqrt{2\pi^2 E/F_y}$ and by

$$F_F = \frac{\pi^2 E}{(l/r)^2} \tag{4.5}$$

for $l/r \geq C_c$.

COLUMNS 141

Example 4.2. Determine the Euler stress and critical load for a pin-connected W 8 × 31 column with a length of 16 ft. $E = 29{,}000$ ksi.

Solution. For a W 8 × 31

$$A = 9.13 \text{ in.}^2, \quad I_x = 110 \text{ in.}^4, \quad I_y = 37.1 \text{ in.}^4,$$
$$r_x = 3.47 \text{ in.}, \quad r_y = 2.02 \text{ in.}$$

I_y and r_y govern.

$$F_e = \frac{\pi^2 E}{(l/r)^2} = \frac{\pi^2 \times 29{,}000}{\left(\dfrac{16 \times 12}{2.02}\right)^2} = 31.68 \text{ ksi}$$

$$P_{cr} = A \times F_e = 9.13 \times 31.68 = 289.25 \text{ k}$$

Example 4.3. Determine the Euler stress and critical load for the pin-connected column as shown, with a length of 30 ft. $E = 29{,}000$ ksi.

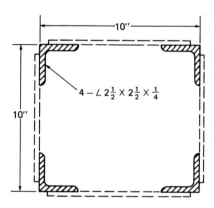

Solution. For a $2\frac{1}{2} \times 2\frac{1}{2} \times \frac{1}{4}$ in. angle

$$A = 1.19 \text{ in.}^2, \quad I = 0.703 \text{ in.}^4, \quad x = y = 0.717 \text{ in.}$$

Due to symmetry, $\bar{y}$ is 5 in. from the top or bottom.

$$I = 4 \times (I_0 + Ad^2) = 4(0.703 + 1.19(5 - .717)^2)$$
$$I = 90.1 \text{ in.}^4$$

142 STEEL DESIGN FOR ENGINEERS AND ARCHITECTS

$$r = \sqrt{\frac{I}{A}} = \sqrt{\frac{90.1}{4 \times 1.19}} = 4.35 \text{ in.}$$

$$F_e = \frac{\pi^2 E}{(l/r)^2} = \frac{\pi^2 \times 29{,}000}{\left(\dfrac{30 \times 12}{4.35}\right)^2} = 41.80 \text{ ksi}$$

$$P_{cr} = A \times F_e = (4 \times 1.19) \times 41.80 = 199.0 \text{ k}$$

The critical load equation can also be obtained by transforming the Euler stress equation

$$P_{cr} = A \times F_e = A \times \frac{\pi^2 E}{(l/\sqrt{I/A})^2} = \frac{\pi^2 EI}{l^2}$$

$$P_{cr} = \frac{\pi^2 \times 29{,}000 \times 90.1}{(30 \times 12)^2} = 199.0 \text{ k}$$

Note: Euler stress and critical load are theoretical values. For actual allowable stresses of columns, see examples beginning with 4.6.

Example 4.4. Determine the Euler stress and critical load for the pin-connected column as shown with a length of 30 ft. $E = 29{,}000$ ksi.

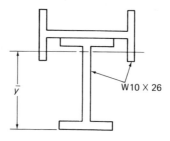

W 10 × 26

Solution. For a W 10 × 26

$$A = 7.61 \text{ in.}^2, \quad I_x = 144 \text{ in.}^4, \quad I_y = 14.1 \text{ in.}^4, \quad d = 10.33 \text{ in.},$$

$$t_w = 0.26 \text{ in.}$$

$$\bar{y} = \frac{\left(7.61 \times \dfrac{10.33}{2}\right) + 7.61 \times \left(10.33 + \dfrac{0.26}{2}\right)}{2 \times 7.61} = 7.81 \text{ in.}$$

$$I_x = \Sigma I_0 + Ad^2 = 1.44 + 7.61(7.81 - 5.17)^2 + 14.1$$
$$+ 7.61(10.46 - 5.17)^2 = 424.1 \text{ in.}^4$$

$$I_y = \Sigma I_0 = 144 + 14.1 = 158.1 \text{ in.}^4 \text{ (governs)}$$

$$\text{least } r = \sqrt{\frac{I}{A}} = \sqrt{\frac{158.1}{2(7.61)}} = 3.22 \text{ in.}$$

$$F_e = \frac{\pi^2 E}{(l/r)^2} = \frac{\pi^2 \times 29{,}000}{\left(\dfrac{30 \times 12}{3.22}\right)^2} = 22.9 \text{ ksi}$$

$$P_{cr} = A \times F_e = (2 \times 7.61) \times 22.9 \text{ ksi} = 349 \text{ k}$$

or

$$P_{cr} = \frac{\pi^2 EI}{l^2} = \frac{\pi^2 \times 29{,}000 \times 158.1}{(30 \times 12)^2} = 349 \text{ k}$$

4.2 EFFECTIVE LENGTHS

Columns with supporting conditions other than pinned at both ends have critical loads different from Euler columns.

Example 4.5. For the column conditions shown, prove that

a) P_{cr} is four times the P_{cr} for the same pin-supported column.
b) P_{cr} is one-fourth that for the same pin-supported column.
c) P_{cr} is two times the P_{cr} of the same pin-supported column.

Note that the following problems can be solved exactly using analytical methods. What follows, however, is a heuristic approach to the solutions.

Solution.

a)

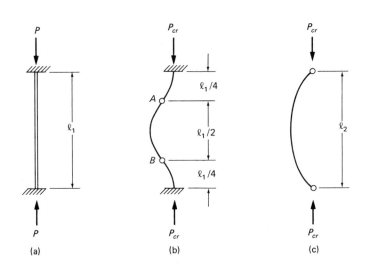

Considering the deflected shape of the column fixed at both ends, we see that contraflexure points (i.e., points of zero moment which are equivalent to a hinge) occur at A and B. That portion of the column between A and B is the same as for a pin-ended column. Hence, from figure b:

$$P_{cr} = \frac{\pi^2 EI}{(l_1/2)^2} = \frac{\pi^2 EI}{(0.5\, l_1)^2}$$

Defining K as the effective length factor, the above equation can be rewritten as

$$P_{cr} = \frac{\pi^2 EI}{(Kl_1)^2} = \frac{\pi^2 EI}{\frac{1}{4}(l_1)^2} = \frac{4\pi^2 EI}{(l_1)^2}$$

with $K = 0.5$. Due to the term in the numerator, the capacity is four times the critical load of the pin-supported column.

b)

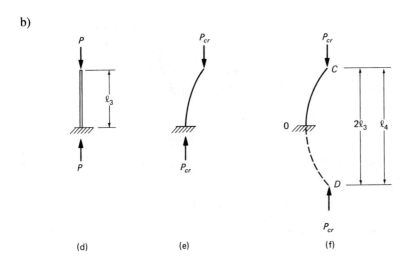

(d)　　　　　(e)　　　　　(f)

By taking the symmetric portion of the deflected column OC in figure f, the column CD behaves as an Euler column of length $2l_3$.

The column of length C to D is the same as for a pin-ended column. Hence

$$P_{cr} = \frac{\pi^2 EI}{(2\,l_3)^2} = \frac{\pi^2 EI}{(Kl_3)^2} = \frac{\pi^2 EI}{4\,(l_3)^2}$$

with 2 for the effective length factor. Due to the term in the denominator, the load capacity is one-fourth the critical load of the pin-supported column.

c)

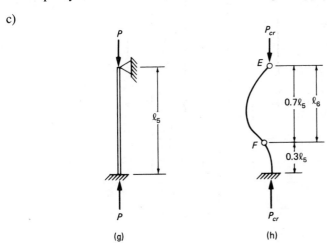

(g)　　　　　(h)

146 STEEL DESIGN FOR ENGINEERS AND ARCHITECTS

The distance EF is approximately $0.7\, l_5$, where F is the point of contraflexure. We then have

$$P_{cr} = \frac{\pi^2 EI}{(0.7\, l_5)^2} = \frac{2\, \pi^2 EI}{(l_5)^2}$$

with the effective length factor $K = 0.7$. The load capacity is approximately 2 times the critical load of the pin-supported column.

The values of K (effective length factor) above agree with the values found in Table 4.1.

As seen in the above example, for a column with both ends fixed the critical load is

$$F_{cr} = \frac{\pi^2 E}{(0.5\, l/r)^2} \tag{4.6}$$

Table 4.1. Effective Length Factor K Based on AISCS C2 (AISCS Commentary Table C-C2.1, reprinted with permission).

	(a)	(b)	(c)	(d)	(e)	(f)
Buckled shape of column is shown by dashed line						
Theoretical K value	0.5	0.7	1.0	1.0	2.0	2.0
Recommended design value when ideal conditions are approximated	0.65	0.80	1.2	1.0	2.10	2.0
End condition code	Rotation fixed and translation fixed Rotation free and translation fixed Rotation fixed and translation free Rotation free and translation free					

and with one end fixed and the other pinned

$$F_{cr} = \frac{\pi^2 E}{(0.7\, l/r)^2} \tag{4.7}$$

This leads to the development of the expression

$$F_{cr} = \frac{\pi^2 E}{(Kl/r)^2} \tag{4.8}$$

where K, called the *effective length factor*, depends on the end connections of the column.

Because the allowable axial stress for a column depends upon the slenderness ratio (l/r), the AISC allows the slenderness ratio to be modified by a factor K,

Fig. 4.3. Erection of prefabricated, two-story column-and-beam system for the Sears Tower, Chicago. Note reinforced openings in the beam webs for heating and cooling ducts. (Courtesy of U.S. Steel Corp.)

148 STEEL DESIGN FOR ENGINEERS AND ARCHITECTS

so that the allowable axial stress also depends upon the end conditions of the column. The effective length factor K can be obtained from Table C-C2.1 of the AISC Commentary (reproduced as Table 4.1 here), and the value r for rolled sections can be found in the properties for designing tables.

The AISC code in Specifications C2.1 and C2.2 allows for laterally braced frames the use of $K = 1$ as a conservative simplification, while for unbraced frames the use of analysis is suggested. This can be done by interpolation between appropriate cases of Table 4.1 or by a more precise approach, (see AISCS Commentary C2 pages 5-134 through 5-138 and AISCS Fig. C-C2.2). Determination of K for laterally unbraced frames is beyond the scope of this book and will not be discussed here. Examples are given on pages 3-4 through 3-7 of the AISCM.

Whether a column connection is pinned or fully fixed can not always be readily determined. Quite often, connections meant to be fully fixed do not have full fixity due to improper detailing. In general, pinned column base connections have the base plate anchored to the foundation with bolts located at or near the centerline of the base plate, and pinned column-beam connections have the column connected to the beam web only (see Fig. 4.4, left). Column base plates for fixed columns usually have the anchor bolts near the column flanges, as far away from the centerline as practical and in fixed column-beam connec-

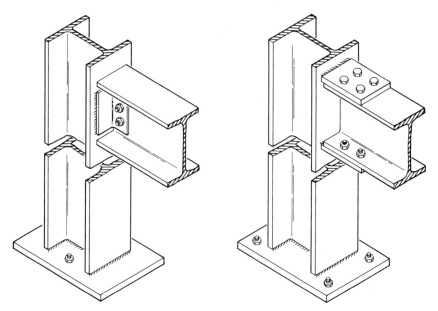

Fig. 4.4. Column connections. Left, pinned column base and pinned column-to-beam connection. Right, fixed column base and fixed column-to-beam connection.

tions the column flange is rigidly connected to the beam flanges (see Fig. 4.4, right).

4.3 ALLOWABLE AXIAL STRESS

AISCS E2 specifies the allowable axial compressive stress for members whose cross sections meet those of AISCS Table B5.1 (width-thickness ratios). When Kl/r, the largest effective slenderness ratio of any unbraced segment is less than C_c, (E2-1) governs:

$$F_a = \frac{\left[1 - \frac{(Kl/r)^2}{2 C_c^2}\right] F_y}{\frac{5}{3} + \frac{3(Kl/r)}{8 C_c} - \frac{(Kl/r)^3}{8 C_c^3}} \qquad (4.9)$$

where $C_c = \sqrt{2\pi^2 E/F_y}$. Referring to Section 4.1, the value C_c is the slenderness corresponding in Euler's curve to the stress $F_y/2$, as at a stress close to this one, Euler's curve and the SSRC curve match. When Kl/r exceeds C_c, (E2-2) governs:

$$F_a = \frac{12 \pi^2 E}{23 (Kl/r)^2} \qquad (4.10)$$

It should be noted that the denominator in eqn. (4.9) and the ratio $\frac{23}{12}$ in eqn. (4.10) are the safety factors for the allowable compressive stresses, and the numerators are the failure stresses for the particular slenderness. For very short columns ($Kl/r \cong 0$), the safety factor is $\frac{5}{3}$ (the same as for tension members) and is increased by 15% to $\frac{23}{12}$ as slenderness increases to when Kl/r equals C_c and remains at $\frac{23}{12}$ for Kl/r greater than C_c. The safety factor is increased as slenderness increases because slender columns are more sensitive to eccentricities in loading and flaws in the steel itself than short columns.

To determine the allowable axial compressive stress F_a without lengthy calculations, the AISCS has established tables that give the allowable stress F_a as functions of Kl/r values and yield stress of the steel. AISCM Tables C-36 and C-50 are provided for steels with $F_y = 36$ ksi and $F_y = 50$ ksi, respectively (see pages 3-16, 3-17). For other grades of steels, Tables 3 and 4 in the Numerical Values Section (pages 5-119, 5-120) are needed. Figure 4.5 shows a graphical representation of the failure stresses and the allowable axial stress for $F_y = 36$ ksi and $F_y = 50$ ksi as a function of slenderness of the member.

In AISCS B7, it is recommended that the slenderness ratio be limited to a value of 200.

150 STEEL DESIGN FOR ENGINEERS AND ARCHITECTS

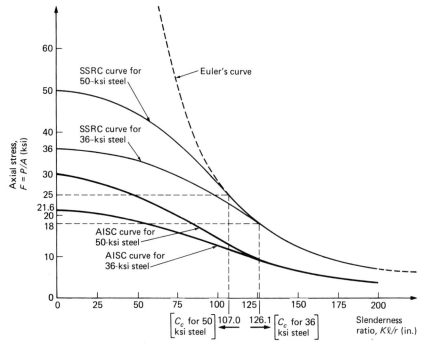

Fig. 4.5. Slenderness curves for the most commonly used steels. Euler's curve for theoretical stress and SSRC curve for failure stress are also shown for comparison with the AISC specified curves for allowable stress.

Example 4.6. Calculate $F_{\text{allowable}}$ and $P_{\text{allowable}}$ for a W 8 × 48 column with fixed ends and a length of 16 ft, 6 in.

Solution. The effective length concept is used to equate framed compression member length to that of an equivalent pin-ended member. For this purpose, theoretical K values and suggested design values are tabulated in the AISC Commentary (see Table 4.1)

$K = 0.65$ (AISCS Table C-C2.1)

$r_y = 2.08$ in. (minimum r)

$$\frac{Kl}{r} = \frac{0.65 \times (16.5 \times 12)}{2.08} = 61.88$$

COLUMNS 151

$$C_c = \sqrt{\frac{2\pi^2 E}{F_y}} = \sqrt{\frac{2\pi^2 29{,}000}{36}} = 126.1 \quad \text{(AISCS E2 or Numerical Values Table 4, p. 5-120)}$$

$61.88 < 126.1$

Because the slenderness ratio is less than C_c, AISC eqn. (E2-1) is to be used:

$$F_a = \frac{\left[1 - \frac{(Kl/r)^2}{2C_c^2}\right]F_y}{\frac{5}{3} + \frac{3(Kl/r)}{8C_c} - \frac{(Kl/r)^3}{8C_c^3}}$$

$$F_a = \frac{\left[1 - \frac{(61.88)^2}{2(126.1)^2}\right] \times 36.0}{\frac{5}{3} + \frac{3(61.88)}{8(126.1)} - \frac{(61.88)^3}{8(126.1)^3}} = 17.25 \text{ ksi} \quad \text{(or from Table C-36 by interpolation)}$$

$A = 14.1 \text{ in.}^2$

$P_{\text{all}} = F_a \times A = 17.25 \text{ ksi} \times 14.1 \text{ in.}^2 = 243.2 \text{ k}$

Example 4.7. Calculate the allowable stress F_a and allowable load for a W 8 × 48 column with one end fixed and one end pinned with a length of 35 ft.

Solution.

$$K = 0.80$$

$$r_y = 2.08 \text{ in.} \quad (\text{minimum } r)$$

$$C_c = 126.1 \quad (\text{from Example 4.6})$$

$$\frac{Kl}{r} = \frac{0.80 \times (35 \times 12)}{2.08} = 161.54$$

$$161.54 > 126.1$$

For slenderness ratios greater than C_c, AISC eqn. (E2-2) is to be used (AISCS E2).

152 STEEL DESIGN FOR ENGINEERS AND ARCHITECTS

$$F_a = \frac{12\pi^2 E}{23\left(\frac{Kl}{r}\right)^2}$$

$$= \frac{12 \times \pi^2 \times 29{,}000}{23 \times (161.54)^2} = 5.72 \text{ ksi} \quad \text{(or from Table C-36 by interpolation)}$$

$A = 14.1 \text{ in.}^2$

$P = F_a \times A = 5.72 \text{ ksi} \times 14.1 \text{ in.}^2 = 80.7 \text{ k}$

Example 4.8. Using the AISC table for allowable stress for compression members of 36-ksi yield stress, determine the allowable load a W 10 × 77 can support with a length of 16 ft and pinned ends.

Solution.

$K = 1.0$

$r_y = 2.60$ in. (minimum r)

$\dfrac{Kl}{r} = \dfrac{1.0 \times 16.0 \times 12}{2.60} = 73.85$

$F_a = 16.01 \text{ ksi} \left(\dfrac{Kl}{r} = 74\right)$ (AISC Table C-36)

$P_{\text{all}} = F_a \times A = 16.01 \times 22.6 = 361.8 \text{ k}$

Example 4.9. Determine the allowable load a W 12 × 65 can support with a length of 20 ft and pinned ends.

Solution.

$K = 1.0$

$r_y = 3.02$ in.

$\dfrac{Kl}{r} = \dfrac{1.0 \times 20 \text{ ft} \times 12 \text{ in.}/\text{ft}}{3.02 \text{ in.}} = 79.47$

$F_a = 15.42 \text{ ksi}$ (AISC Table C-36)

$P_{\text{all}} = F_a \times A = 15.42 \text{ ksi} \times 19.1 \text{ in.}^2 = 294.5 \text{ k}$

Example 4.10. Determine the allowable load a W 10 × 49 can support with a length of 24 ft, 6 in. One end is fixed and the other pinned. Steel yield stress $F_y = 50$ ksi.

Solution.

$$K = 0.80$$

$$r_y = 2.54 \text{ in.}$$

$$\frac{Kl}{r} = \frac{0.80 \times 24.5 \times 12}{2.54} = 92.6$$

$$F_a = 16.37 \text{ ksi} \quad \text{(AISC Table C-50)}$$

$$P_{\text{all}} = F_a \times A = 16.37 \text{ ksi} \times 14.4 \text{ in.}^2 = 235.73 \text{ k}$$

4.4 DESIGN OF AXIALLY LOADED COLUMNS

The following procedure can be used to design an axially loaded column without design-aid tables:

1. Determine the effective length factor K from Table 4.1.
2. Selecting an arbitrary r, obtain Kl/r and determine the corresponding allowable axial stress F_a from AISCM Tables C-36 or C-50 or from AISCS eqn. (E2-1) or (E2-2).
3. Determine the required area ($A = P/F_a$), and select an appropriate rolled section.
4. Recalculate the actual slenderness ratios of the selected section with respect to the x and y axes and the corresponding F_a values.
5. Repeat the above process until satisfactory convergence has been obtained.

When determining the slenderness ratio of a column, the radii of gyration for the x and y directions, as well as the respective unsupported lengths, must be obtained. The value Kl/r must be calculated for both x and y directions, and the larger of the two values shall be used for determining the allowable compressive stress F_a. In addition, the AISC has allowable concentric load tables, starting on p. 3-19, which can be used to determine the capacity of a column for a given effective length based on the provisions in AISCS E2.

Example 4.11. Design a W 14 column pinned at both ends to support an axial load of 610 kips. $L_x = L_y = 21$ ft, 6 in.

Solution.

$$K = 1.0$$

$$\text{Assume } \frac{Kl}{r} = 70$$

$$F_a = 16.43 \text{ ksi}$$

$$A_{req} = \frac{P}{F_a} = \frac{610 \text{ k}}{16.43 \text{ ksi}} = 37.13 \text{ in.}^2$$

Try W 14 × 120

$$A = 35.3 \text{ in.}^2, \quad r_x = 6.24 \text{ in.}, \quad r_y = 3.74 \text{ in.}$$

Since $L_x = L_y$, r_y governs

$$\frac{Kl_x}{r_x} = \frac{1.0 \times 21.5 \times 12}{6.24 \text{ in.}} = 41.35$$

$$\frac{Kl_y}{r_y} = \frac{1.0 \times 21.5 \times 12}{3.74 \text{ in.}} = 68.98 \text{ (governs)}$$

$$F_a = 16.53 \text{ ksi}$$

$$A_{req} = \frac{P}{F_a} = \frac{610 \text{ k}}{16.53 \text{ ksi}} = 36.90 \text{ in.}^2$$

36.90 in.^{2} > 35.2 in.^{2} Try next heaviest section

Try W 14 × 132

$$A = 38.8 \text{ in.}^2, \quad r_x = 6.28 \text{ in.}, \quad r_y = 3.76 \text{ in.}$$

$$\frac{Kl_y}{r_y} = \frac{1.0 \times 21.5 \times 12}{3.76 \text{ in.}} = 68.6 \text{ (critical)}$$

$$F_a = 16.59 \text{ ksi}$$

$$f_a = \frac{610 \text{ k}}{38.8 \text{ in.}^2} = 15.72 \text{ ksi} < 16.59 \text{ ksi}$$

Use W 14 × 132.

Using the allowable load tables for a W 14, for $KL_y = 21$ ft, 6 in., use a W 14 × 132 which has a capacity of 643 k (by interpolation).

Example 4.12. Select the lightest W section for a column supporting a 300-kip load, fixed at the bottom and pinned at the top. $L_x = L_y = 17$ ft, 6 in. Refer to AISC columns tables.

Solution.

$$K = 0.80 \quad \text{(AISC Table C-C2.1)}$$
$$KL = 0.80 \times 17.5 \text{ ft} = 14.0$$

W shapes capable of supporting the load:

$$\text{W } 8 \times 67, \quad P = 304 \text{ k}$$
$$\text{W } 10 \times 68, \quad P = 339 \text{ k}$$
$$\text{W } 12 \times 65, \quad P = 341 \text{ k}$$
$$\text{W } 14 \times 68, \quad P = 332 \text{ k}$$

Note: W 10 × 60 and W 14 × 62 may be considered satisfactory, as they are only 1% overstressed with $P = 297$ k.

Use W 12 × 65 (lightest section) or W 10 × 60 if a slight overstress is permissible.

Example 4.13. Select the lightest pipe or square structural tube for $P = 300$ kips and $L_x = L_y = 14$ ft, 0 in.

Solution.

$10 \times 10 \times 3/8 \times 47.90$ lb/ft square tube ($P = 332$ k)

$8 \times 8 \times 1/2 \times 48.85$ lb/ft square tube ($P = 315$ k)

$10 \phi \times 54.74$ lb/ft extra strong steel pipe ($P = 301$ k)

$8 \phi \times 72.42$ lb/ft double-extra strong steel pipe ($P = 369$ k)

Use 10 × 10 × 3/8 square tube, 47.90 lb/ft.

Example 4.14. Design the lightest W 12 column for an axial load of 350 kips. The column is pinned at the bottom and fixed at the top, but subject to sway in the x and y directions. $L_x = 22$ ft, 0 in., $L_y = 11$ ft, 0 in.

Solution. Assume $Kl/r = 70$, $K = 2.0$

$$F_a = 16.43 \text{ ksi}$$

$$A_{req} = \frac{P}{F_a} = \frac{350 \text{ k}}{16.43 \text{ ksi}} = 21.30 \text{ in.}^2$$

Try W 12 × 72

$$A = 21.1 \text{ in.}^2, \quad r_x = 5.31 \text{ in.}, \quad r_y = 3.04 \text{ in.}$$

$$\frac{K_x l_x}{r_x} = \frac{2.0 \times (22.0 \times 12)}{5.31} = 99.4 \text{ (governs)}$$

$$\frac{K_y l_y}{r_y} = \frac{2.0 \times (11.0 \times 12)}{3.04} = 86.8$$

$$F_a = 13.05 \text{ ksi}$$

$$A_{req} = \frac{350 \text{ k}}{13.05 \text{ ksi}} = 26.82 \text{ in.}^2 > 21.1 \text{ in.}^2 \quad \text{N.G.}$$

Try W 12 × 87

$$A = 25.6 \text{ in.}^2 \quad r_x = 5.38 \text{ in.}, \quad r_y = 3.07 \text{ in.}$$

$$\frac{K_x l_x}{r_x} = \frac{2.0 \times (22.0 \times 12)}{5.38} = 98.1 \text{ (governs)}$$

$$\frac{K_y l_y}{r_y} = \frac{2.0 \times (11.0 \times 12)}{3.07} = 86.0$$

$$F_a = 13.22 \text{ ksi}$$

$$A_{req} = \frac{350 \text{ k}}{13.22 \text{ ksi}} = 26.48 \text{ in.}^2 > 25.6 \text{ in.}^2 \quad \text{N.G.}$$

By inspection, a W 12 × 96 will be adequate because of the small overstress of the W 12 × 87, as shown above.

As an alternate approach, use the column load tables as provided in the AISCM. For a W 12 with $P = 350$ k and $K_y L_y = 2 \times 11 = 22$ ft, choose a W 12 × 87 which can adequately support 376 k for the weak direction. However, since $K_x L_x = 2 \times 22 = 44$ ft is greater than $K_y L_y$, the strong axis may govern. By dividing $K_x L_x$ by the ratio r_x/r_y (given in the column tables), we obtain the equivalent length in the y direction of $K_x L_x$:

$$\frac{K_x L_x}{(r_x/r_y)} = \frac{44}{1.75} = 25.1 \text{ ft.} \quad \text{say } 25 \text{ ft.}$$

For this length, a W 12 × 87 can support a load of only 340 k, which is less than 350 k, which is no good. Therefore, a W 12 × 96 should be used, since it can carry 376 k for a length of 25 ft.

Note that the allowable concentric load tables can be used for steels with $F_y = 36$ ksi and $F_y = 50$ ksi (shaded). In general, to use these tables, the following procedure must be used:

1. Determine the effective length factor K from AISCM Table C-C2.1, and calculate the effective length KL in feet, for the y direction.
2. From the tables, select an appropriate section based on the effective length and the axial load P.
3. Dividing the effective length $K_x L_x$ by the value r_x/r_y of the selected section, obtain the effective length with respect to the minor axis equivalent in load-carrying capacity to the actual effective length about the major axis. The column then must be designed for the larger of the two effective lengths $K_y L_y$ or $K_x L_x/(r_x/r_y)$.

Example 4.15. Design the column shown using the AISC column tables.

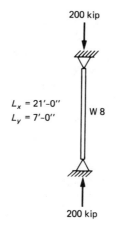

Solution.

$$K = 1.0$$

$$KL_y = 1.0 \times 7.0 \text{ ft} = 7.0 \text{ ft (weak direction)}$$

For $(KL) = 7$ ft and loading of 200 kips, select W 8 × 40, which can carry 223 kips and has $r_x/r_y = 1.73$.

$$\frac{L_x}{r_x/r_y} = \frac{21 \text{ ft}}{1.73} = 12.14 \text{ ft}$$

From the same table, for $(KL) = 12.14$ ft and $P = 200$ kips, a W 8 × 48 is needed. Use W 8 × 48.

Example 4.16. Design a W 10 column pinned at top and bottom to support an axial load of 200 kips. $L_x = 14$ ft, 0 in., $L_y = 7$ ft, 0 in.

Solution.

$$K = K_x = K_y = 1.0$$

$$\text{Assume } \frac{Kl}{r} = 70$$

$$F_a = 16.4 \text{ ksi}$$

$$A_{req} = \frac{P}{F_a} = \frac{200 \text{ k}}{16.4 \text{ ksi}} = 12.2 \text{ in.}^2$$

Try W 10 × 45

$$A = 13.3 \text{ in.}^2, \quad r_x = 4.32 \text{ in.}, \quad r_y = 2.01 \text{ in.}$$

$$\frac{Kl_x}{r_x} = \frac{1.0 \times 14 \times 12}{4.32} = 38.9$$

$$\frac{Kl_y}{r_y} = \frac{1.0 \times 7 \times 12}{2.01} = 41.8 \text{ (governs)}$$

$$F_a = 19.05 \text{ ksi}$$

$$A_{req} = \frac{P}{F_a} = \frac{200 \text{ k}}{19.05 \text{ ksi}} = 10.50 \text{ in.}^2$$

10.50 in.^{2} < 13.3 in.^{2} Try the next lighter section.

Try W 10 × 39

$$A = 11.5 \text{ in.}^2, \quad r_x = 4.27 \text{ in.}, \quad r_y = 1.98 \text{ in.}$$

$$\frac{Kl_x}{r_x} = \frac{1.0 \times 14 \times 12}{4.27} = 39.3$$

$$\frac{Kl_y}{r_y} = \frac{1.0 \times 7 \times 12}{1.98} = 42.4 \text{ (critical)}$$

$$F_a = 19.0 \text{ ksi}$$

$$f_a = \frac{200 \text{ k}}{11.5 \text{ in.}^2} = 17.4 \text{ ksi} < 19.0 \text{ ksi}$$

We note that if W 10 × 33, the next lighter section, is used

$$A = 9.71 \text{ in.}^2$$

$$f_a = \frac{200 \text{ k}}{9.71 \text{ in.}^2} = 20.60 \text{ ksi} > 19.0 \text{ ksi} \quad \text{N.G.}$$

Use W 10 × 39.

Example 4.17. Select the lightest column from AISCM columns tables for $K_x = 2.0$, $K_y = 1.2$, $L_x = 18.0$ ft, $L_y = 12.5$ ft, $F_y = 50$ ksi, and $P = 470$ kips.

Solution.

$$K_y L_y = 1.2 \times 12.5 \text{ ft} = 15.0 \text{ ft}$$

Try W 10 × 77; $r_x/r_y = 1.73$

$$\frac{K_x L_x}{r_x/r_y} = \frac{2.0 \times 18.0 \text{ ft}}{1.73} = 20.8 \text{ ft}$$

Redesigning for $(KL) = 20.8$ ft, select W 10 × 100.
Try W 12 × 72, $r_x/r_y = 1.75$

$$\frac{K_x L_x}{r_x/r_y} = \frac{2.0 \times 18.0 \text{ ft}}{1.75} = 20.6 \text{ ft}$$

Redesigning for $(KL) = 20.6$ ft, select W 12 × 87.

160 STEEL DESIGN FOR ENGINEERS AND ARCHITECTS

Try W 14 × 82, $r_x/r_y = 2.44$

$$\frac{K_x L_x}{r_x/r_y} = \frac{2.0 \times 18.0 \text{ ft}}{2.44} = 14.8 \text{ ft} < 15.0 \text{ ft}$$

We see then that W 14 × 82 is satisfactory.

W 14 × 82 is the lightest satisfactory section.

Example 4.18. Find the allowable load a 10-in. standard steel pipe can support with a length of 20 ft and pinned ends.

Solution.

$$K = 1.0$$
$$r = 3.67 \text{ in.}$$
$$A = 11.9 \text{ in.}^2$$

$$\frac{Kl}{r} = \frac{1.0 \times 20 \times 12}{3.67} = 65.4$$

$$F_a = 16.9 \text{ ksi}$$

$$P_{\text{all}} = F_a \times A = 16.9 \times 11.9 = 201 \text{ k}$$

Alternatively, use AISCM design tables:

From page 3-36, for 10-in. standard pipe with $KL = 20$ ft, read $P = 201$ k.

Example 4.19. Determine what compressive load a bracing member made up of two angles 4 × 3 × $\frac{3}{8}$ (long legs back to back) separated by a $\frac{3}{8}$-in. plate can carry when $L = 20$ ft, 0 in. Also, determine the location of the intermediate connectors. Assume pinned ends.

Solution. For double angle and WT compressive members, allowable concentric loads are tabulated in Part 3 of the AISCM (see pp. 3-59ff). The allowable compressive loads about the X-X axis are determined in accordance with AISCS E2; about the Y-Y axis, the allowable loads are based on flexural-torsional buckling (p. 3-53). The derivation of these critical loads is beyond the scope of this text.

For the 2L4 × 3 × $\frac{3}{8}$:

$$KL_x = 1 \times 20 = 20 \text{ ft}, \quad P_{cr} = 20 \text{ k (governs)}$$

$$KL_y = 1 \times 20 = 20 \text{ ft}, \quad P_{cr} = 22 \text{ k}$$

The corresponding slenderness ratio $= \dfrac{KL_x}{r_x}$

$$= \dfrac{20 \times 12}{1.26} = 190.5$$

Along the length of the built-up compression member, intermediate connectors should be provided at a spacing a, such that the slenderness ratio Ka/r of each component does not exceed $\frac{3}{4}$ times the governing slenderness ratio of the built-up member (AISCS E4). The smallest radius of gyration r should be used when computing the slenderness ratio of each component.

For a single L4 × 3 × $\frac{3}{8}$, $r_{min} = r_z = 0.644$ in.

$$\dfrac{Ka}{r_z} \leq 0.75 \dfrac{KL_x}{r_x} = 0.75 \times 190.5 = 142.9$$

$a =$ spacing between fully tensioned, high-strength bolts or welds (in this case, high-strength bolts)

$= 142.9 \times 0.644$ in. $= 92$ in. $= 7.67$ ft

Therefore, provide two connectors, one at each of the $\frac{1}{3}$ points along the length of the brace. Note that this satisfies the requirement for the minimum number of connectors that must be provided.

Note: The design of single-angle compression members will not be discussed here, as the construction of such struts is for all practical purposes infeasible. See p. 3-55 of the AISCM and Sect. 4 of the Specification for Single-Angle Members (p. 5-310) for further discussion.

Example 4.20. Determine the allowable load a WT 4 × 14 can carry if it is pinned at both ends and is 8 ft, 0 in.

Solution. From page 3-105 of the AISCM:

$$KL_x = 1 \times 8 = 8 \text{ ft}, \quad P_{cr} = 56 \text{ k (governs)}$$

$$KL_y = 1 \times 8 = 8 \text{ ft}, \quad P_{cr} = 72 \text{ k}$$

4.5. COMBINED STRESSES

Example 4.21. The top chord of a bridge truss is shown in the accompanying figure. Determine the maximum stresses at the extreme fibers of this member if it is subjected to a moment of 15 ft-k about the x axis and an axial compression of 100 kips.

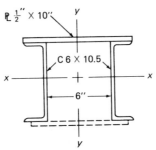

Solution.

$$A_{channel} = 3.09 \text{ in.}^2$$

$$A_{plate} = 5.0 \text{ in.}^2$$

$$I_c = 15.2 \text{ in.}^4$$

$$d_c = 6 \text{ in.}$$

$$\bar{y} = \frac{(0.5 \text{ in.} \times 10 \text{ in.}) \times 6.25 \text{ in.} + 2(3.09 \text{ in.}^2 \times 3.0 \text{ in.})}{5.0 \text{ in.}^2 + 2(3.09 \text{ in.}^2)}$$

$$= 4.45 \text{ in. from bottom}$$

$$I_x = \Sigma I_0 + Ad^2 = \frac{10(0.5)^3}{12} + 5(6.25 - 4.45)^2 + 2(15.2)$$

$$+ 2(3.09)(3 - 4.45)^2 = 59.7 \text{ in.}^4$$

Combined stress at top $= -P/A - Mc/I$ ($-$ stress indicates compression)

$$\frac{-100}{5.0 + 2(3.09)} - \frac{(15 \times 12)(6.5 - 4.45)}{59.7} = -15.13 \text{ ksi}$$

Combined stress at bottom $= -P/A + Mc/I$

$$\frac{-100}{5.0 + 2(3.09)} + \frac{(15 \times 12)(4.45)}{59.7} = 4.47 \text{ ksi}$$

From the preceding example, we can see that the actual calculated stresses $f_t = -15.13$ ksi and $f_b = +4.47$ ksi can be obtained from the strength of materials. However, this approach cannot be used to design the member, since F_b can vary from a minimum of $0.6 F_y$ or less to a maximum of $0.75 F_y$. Similarly, F_a can vary from almost a nominal zero value to a maximum of $0.6 F_y$. In order to take care of these different stresses and to determine the capacity of the member (i.e., design), the following approach has been devised.

When a straight member is subjected to a significant bending stress as well as a compressive stress, the member is sometimes referred to as a *beam-column*. The bending stress may be caused by 1) eccentric application of an axial load, 2) applied end moments, or 3) lateral loads such as wind forces.

A beam-column carrying some given moments deflects laterally more than a beam carrying the same moments because of the presence of the axial load. As the axial load is gradually increased, the deflection increases at a much faster rate, since the increase in moment depends not only on the increased axial load but also on the increased eccentricity (moment arm) caused by the increased load. Thus, a beam-column fails due to instability caused by excessive bending.

The increase in deflection due to the axial load increases the bending stress f_b, which, in effect, requires it to be amplified by the factor

$$\frac{1}{\left(1 - \dfrac{f_a}{F_e}\right)} \tag{4.11}$$

where f_a = computed axial stress and F_e = Euler stress for a prismatic member. For safety reasons, the Euler stress F_e is divided by the same factor of safety for the axial load and is commonly represented by F'_e.

Considering the effects of the moments on the beam-column, as mentioned previously, the moments can usually be applied in two different ways: 1) at the ends of the column (frame action) or 2) via transverse loading. The moments applied at the ends of the column will bend the member in either a single or a double curvature. Transverse loadings can be applied to the member in many different ways: distributed loads, concentrated loads, etc. To compensate for these effects, the constant C_m is introduced, which modifies the amplification factor given in eqn. (4.11) above. The appropriate values for C_m are given in AISCS H1; further discussion can be seen in Commentary section H1.

If the allowable axial stress F_a were equal to the allowable bending stress F_b, the capacity of the column could be represented by

$$f_a + f_b \leq F \tag{4.12a}$$

where $F = F_a = F_b$. Dividing through by F, we can write

$$\frac{f_a}{F} + \frac{f_b}{F} \leq 1 \qquad (4.12b)$$

Since F_a and F_b are not necessarily equal, these allowable values should be introduced in the proper places in the above inequality:

$$\frac{f_a}{F_a} + \frac{f_b}{F_b} \leq 1 \qquad (4.13)$$

which shows that the fraction of the capacity used by the axial load (f_a/F_a) plus the same for bending (f_b/F_b) must be ≤ 1. If biaxial loading is considered, then

$$\frac{f_a}{F_a} + \frac{f_{bx}}{F_{bx}} + \frac{f_{by}}{F_{by}} \leq 1.0 \qquad (4.14)$$

which is (H1-3) in AISCS H1. This equation is valid in the case of small axial loads, i.e., $f_a/F_a \leq 0.15$.

When the axial load produces a stress such that $f_a/F_a > 0.15$, the bending stresses have to be modified to account for the secondary bending moments, as discussed previously. This can be seen in (H1-1):

$$\frac{f_a}{F_a} + \frac{C_{mx}f_{bx}}{\left(1 - \frac{f_a}{F'_{ex}}\right)F_{bx}} + \frac{C_{my}f_{by}}{\left(1 - \frac{f_a}{F'_{ey}}\right)F_{by}} \leq 1.0 \qquad (4.15)$$

For the case when significant axial loads with moments applied at the supports are considered, the code requires that (H1-2) be satisfied:

$$\frac{f_a}{0.60 F_y} + \frac{f_{bx}}{F_{bx}} + \frac{f_{by}}{F_{by}} \leq 1.0 \qquad (4.16)$$

Thus, it can be seen from this last equation that the effects of slenderness and the second-order moments are omitted for this case.

For the benefit of the reader, all of the quantities used in eqns. (4.14) through (4.16) will now be defined:

F_a = allowable axial stress if axial force alone existed, ksi.
F_b = allowable compressive bending stress if bending moment alone existed, ksi. The reader is advised to determine the values of F_b carefully,

COLUMNS 165

as they may vary from a value close to zero up to $0.75 F_y$, depending on the direction of bending, the locations of lateral support, etc. (see Chapter 2).

$F'_e = \dfrac{12\pi^2 E}{23(Kl_b/r_b)^2}$ = Euler stress divided by a factor of safety, ksi. The quantities l_b and r_b are the actual unbraced length and corresponding radius of gyration in the plane of bending, respectively. K is the effective

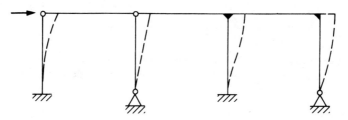

a. $C_m = 0.85$ for all the columns (subparagraph (a)).

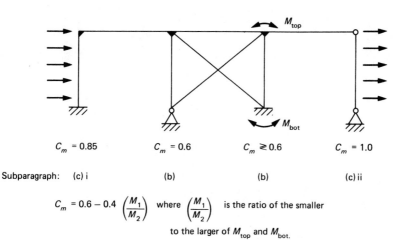

b.

Fig. 4.6. Values of C_m for compression members in frames (AISCS H1): (a) not braced against sidesway, (b) braced against sidesway.

166 STEEL DESIGN FOR ENGINEERS AND ARCHITECTS

length factor in the plane of bending. Note that the increases of the allowable stresses for wind and seismic loads (Sect. 0.2) apply here to F_a, F_b, 0.6 F_y, and F'_e.

f_a = computed axial stress, ksi.
f_b = computed compressive bending stress at the point under consideration, ksi.
C_m = coefficient as defined in AISCS H1 under subparagraphs (a), (b), and (c) and represented in Fig. 4.6.

Example 4.22. For the column shown, determine if this member is satisfactory. The column is from part of a building where sidesway is prevented, and the member is bending about the major axis.

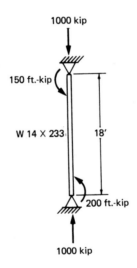

Solution. For a W 14 × 233

$A = 68.5$ in.^{2}, $S_x = 375$ in.^{3}, $r_x = 6.63$ in., $r_y = 4.10$ in.

$L_c = 16.8$ ft, $L_u = 78.5$ ft, $K = 1.0$ (AISCS C2.1)

$$\frac{Kl}{r_y} = \frac{1.0 \times 18.0 \text{ ft} \times 12 \text{ in.}/\text{ft}}{4.10} = 52.68$$

$F_a = 18.11$ ksi   (AISC Table C-36)

COLUMNS 167

$$f_a = \frac{P}{A} = \frac{1000 \text{ k}}{68.5 \text{ in.}^2} = 14.60 \text{ ksi}$$

$$f_a/F_a = \frac{14.60 \text{ ksi}}{18.11 \text{ ksi}} = 0.806 > 0.15$$

AISC formulas (H1-1) and (H1-2) must be satisfied:

$$L_c < L < L_u \quad F_{bx} = 0.6 \times F_y = 22.0 \text{ ksi}$$

$$f_{b\max} = \frac{M_{\max}}{S_x} = \frac{200 \text{ ft-k} \times 12 \text{ in.}/\text{ft}}{375 \text{ in.}^3} = 6.4 \text{ ksi}$$

$$C_{mx} = 0.6 - 0.4 \left(\frac{150 \text{ ft-k}}{200 \text{ ft-k}} \right) = 0.30 \quad \text{(AISCS H1, subparagraph (b))}$$

$$F'_{ex} = \frac{12\pi^2 E}{23 (Kl_b/r_b)^2} = \frac{12\pi^2 (29,000)}{23 \left(\frac{1.0(18 \times 12)}{6.63} \right)^2} = 140.5 \text{ ksi}$$

Note that this value could have also been obtained in Table 8 of the Numerical Values section, p. 5-122. Also, the value of F'_{ex} can be obtained from the bottom of the column load tables. For the W 14 × 233 (p. 3-21):

$$\frac{F'_{ex}(K_x L_x)^2}{10^2} = 456 \text{ k}$$

$$F'_{ex} = \frac{456 \times 100}{(1.0 \times 18)^2} = 140.7 \text{ ksi}$$

$$\frac{f_a}{F_a} + \frac{C_{mx} f_{bx}}{(1 - (f_a/F'_{ex})) F_{bx}} \leq 1.0 \quad \text{(AISCS H1-1)}$$

$$\frac{14.6}{18.11} + \frac{0.3 \times 6.4}{(1 - (14.6/140.5)) \times 22.0} = 0.9 < 1.0 \quad \text{ok}$$

$$\frac{f_a}{0.6 F_y} + \frac{f_{bx}}{F_{bx}} \leq 1.0 \quad \text{(AISCS H1-2)}$$

$$\frac{14.6}{22.0} + \frac{6.4}{22.0} = 0.95 < 1.0 \quad \text{ok}$$

The W 4 × 223 column is satisfactory.

168 STEEL DESIGN FOR ENGINEERS AND ARCHITECTS

Example 4.23. Check the adequacy of the column shown below. The column is braced against sidesway.

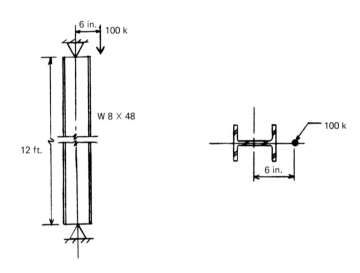

Solution. For a W 8 × 48

$A = 14.1$ in.^{2}, $S_x = 43.3$ in.^{3}, $r_x = 3.61$ in., $r_y = 2.08$ in.

$L_c = 8.6$ ft, $L_u = 30.3$ ft, $K = 1.0$

$$\frac{Kl}{r_y} = \frac{1.0 \times 12 \text{ ft} \times 12 \text{ in./ft}}{2.08 \text{ in.}} = 69.2$$

$F_a = 16.51$ ksi

$$f_a = \frac{100 \text{ k}}{14.1 \text{ in.}^2} = 7.09 \text{ ksi}$$

$$\frac{f_a}{F_a} = \frac{7.09}{16.51} = 0.43 > 0.15 \therefore \text{(H1-1) and (H1-2) applicable}$$

$F_{bx} = 22$ ksi

$M_x = 100 \text{ k} \times 6 \text{ in.} = 600$ in.-k

$$f_{bx} = \frac{600}{43.3} = 13.86 \text{ ksi}$$

$C_{mx} = 0.6 - 0.4(\frac{0}{50}) = 0.6$

$$\frac{Kl}{r_x} = \frac{1.0 \times 12 \times 12}{3.61} = 39.9, \quad F'_{ex} = 93.75 \text{ ksi}$$

(H1-1): $\dfrac{7.09}{16.51} + \dfrac{0.6 \times 13.86}{\left[1 - \left(\dfrac{7.09}{93.75}\right)\right]22} = 0.43 + 0.41 = 0.84 < 1$ ok

(H1-2): $\dfrac{7.09}{22} + \dfrac{13.86}{22} = 0.32 + 0.63 = 0.95 < 1$ ok

The W 8 × 48 is adequate.

Example 4.24. Design a W 14 column for an axial load of 300 kips and the moments shown below.

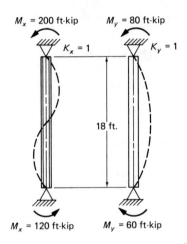

Solution.

To begin the design, the values of C_{mx} and C_{my} are determined:

$$C_m = 0.60 - 0.40 \frac{M_1}{M_2}$$

$$C_{mx} = 0.60 - 0.40 \frac{120}{200} = 0.36$$

$$C_{my} = 0.60 - 0.40 \frac{-60}{80} = 0.90$$

170 STEEL DESIGN FOR ENGINEERS AND ARCHITECTS

The design of a column subjected to axial and flexural loads is an iterative process which begins by choosing an initial column size and then checking it. To facilitate the design of these members, the modified versions of eqns. (H1-1), (H1-2), and (H1-3) on p. 3-9 of the AISCM can be used. These modified formulas convert the applied moments into equivalent axial loads P' by using coefficients B_x, B_y, a_x, and a_y found in the column tables starting on p. 3-19. The equivalent axial load P', when added to the actual load P, will give the total load which the column must resist; the column size can then be selected from the column load tables and checked. This process must repeated until a satisfactory column size is obtained.

The initial selection of the column size for this problem will be done by using Table B on p. 3-10 as follows:

- First Trial:

In the first trial, always take $U = 3.0$. The values of m are given for $C_m = 0.85$; when C_m is other than 0.85, multiply the tabular values of m by $C_m/0.85$. From Table B, the value of m for $C_m = 0.85$ is 2.1 for the first approximation.

$$P_{eff} = P_0 + M_x m + M_y mU \quad \text{(p. 3-10)}$$

$$P_{eff} = 300 + \left[200 \times 2.1 \times \frac{0.36}{0.85} \right] + \left[80 \times 2.1 \times 3 \times \frac{0.90}{0.85} \right] = 1012 \text{ k}$$

Note that in the above equation that M_x and M_y are given in ft-k, and the loads in kips.

From the column load tables, select W 14 × 193 ($P = 1025$ k for $KL = 18$ ft).

- Second Trial (W 14 × 193)

$m = 1.7$ (p. 3-10)

$U = 2.29$ (p. 3-22)

$$P_{eff} = 300 + \left[200 \times 1.7 \times \frac{0.36}{0.85} \right] + \left[80 \times 1.7 \times 2.29 \times \frac{0.90}{0.85} \right]$$

$$= 774 \text{ k} < 1025 \text{ k} \quad \text{ok}$$

W 14 × 193 is adequate; however, try to obtain a more economical section.

- Third Trial (W 14 × 159, $P = 840$ k)

Refine by using modified equation (H1-1), (H1-2), or (H1-3) (AISCM, p. 3-9).

$L_c = 16.4$ ft $< L = 18.0$ ft $< L_u = 57.2$ ft

$F_{bx} = 0.60 F_y = 22.0$ ksi

$F_{by} = 0.75 F_y = 27.0$ ksi

$A = 46.7$ in.^{2}, $S_x = 254$ in.^{3}, $r_x = 6.38$ in., $S_y = 96.2$ in.^{3},

$r_y = 4.0$ in.

$B_x = 0.184$, $B_y = 0.485$, $a_x = 283 \times 10^6$, $a_y = 111 \times 10^6$

$$\frac{Kl_x}{r_x} = \frac{1.0 \times (18 \times 12)}{6.38} = 33.8, \quad F'_{ex} = 130.80 \text{ ksi}$$

$$\frac{Kl_y}{r_y} = \frac{1.0 \times (18 \times 12)}{4.00} = 54.0, \quad F_a = 17.99 \text{ ksi}$$

$$F'_{ey} = 51.21 \text{ ksi}$$

$$f_a = \frac{300 \text{ k}}{46.7 \text{ in.}^2} = 6.42 \text{ ksi}$$

$$\frac{f_a}{F_a} = \frac{6.42 \text{ ksi}}{17.99 \text{ ksi}} = 0.36 > 0.15$$

Use modified formula (H1-1) (AISCM p. 3-9):

$$P + \left[B_x M_x C_{mx} \left(\frac{F_a}{F_{bx}} \right) \left(\frac{a_x}{a_x - P(Kl)^2} \right) \right] + \left[B_y M_y C_{my} \left(\frac{F_a}{F_{by}} \right) \left(\frac{a_y}{a_y - P(Kl)^2} \right) \right]$$

$$= 300 + \left(0.184 \times 200 \times 12 \times 0.36 \right.$$

$$\left. \times \frac{17.99}{22} \times \frac{283 \times 10^6}{283 \times 10^6 - 300(18 \times 12)^2} \right)$$

$$+ \left(0.485 \times 80 \times 12 \times 0.90 \times \frac{17.99}{27} \right.$$

$$\left. \times \frac{111 \times 10^6}{111 \times 10^6 - 300(18 \times 12)^2} \right) = 756 \text{ k}$$

Equivalent load $= 756$ k < 840 k ok

Verify the modified formula (H1-2):

172 STEEL DESIGN FOR ENGINEERS AND ARCHITECTS

$$P\left(\frac{F_a}{0.60 F_y}\right) + \left[B_x M_x \left(\frac{F_a}{F_{bx}}\right)\right] + \left[B_y M_y \left(\frac{F_a}{F_{by}}\right)\right]$$

$$= 300\left(\frac{17.99}{22}\right) + \left[0.184 \times 200 \times 12 \times \frac{17.99}{22}\right]$$

$$+ \left[0.485 \times 80 \times 12 \times \frac{17.99}{27}\right] = 916.7 \text{ k}$$

Equivalent load = 916.7 k > 840 k N.G.

Try W 14 × 176

$$A = 51.8 \text{ in.}^2, \quad S_x = 281 \text{ in.}^3, \quad r_x = 6.43 \text{ in.},$$
$$S_y = 107 \text{ in.}^3, \quad r_y = 4.02 \text{ in.}$$

Because the lighter section satisfies all conditions except formula (H1-2), we need verify only the one equation

$$\frac{f_a}{0.60 F_y} + \frac{f_{bx}}{F_{bx}} + \frac{f_{by}}{F_{by}} \leq 1.0$$

Since $L_c = 16.5$ ft $< L = 18.0$ ft $< L_u = 62.6$ ft

$$F_{bx} = 0.60 F_y = 22 \text{ ksi}$$

$$F_{by} = 0.75 F_y = 27 \text{ ksi}$$

$$f_a = \frac{300 \text{ k}}{51.8 \text{ in.}^2} = 5.79 \text{ ksi}$$

$$f_{bx} = \frac{200 \text{ ft-k} \times 12 \text{ in.}/\text{ft}}{281 \text{ in.}^3} = 8.54 \text{ ksi}$$

$$f_{by} = \frac{80 \text{ ft-k} \times 12 \text{ in.}/\text{ft}}{107 \text{ in.}^3} = 8.97 \text{ ksi}$$

$$\frac{5.79}{22.0} + \frac{8.54}{22.0} + \frac{8.97}{27.0} = 0.98 < 1.0 \quad \text{ok}$$

Use W 14 × 176.

Example 4.25. Solve Example 4.24 if the axial loads and moments are due to gravity and wind loads.

In Example 4.24, the first trial P_{eff} was calculated to be 1012 k. Here, instead of increasing the allowable stresses by $\frac{1}{3}$, we will divide the load by $\frac{4}{3}$, which yields $0.75 \times 1012 \text{ k} = 759 \text{ k}$. From the column load tables, try W 14 × 145 ($P = 767$ k):

$$m = 1.7, \quad U = 2.34$$

$$P_{\text{eff}} = 0.75 \times \{300 + [200 \times 1.7 \times (0.36/0.85)]$$

$$+ [80 \times 1.7 \times 2.34 \times (0.9/0.85)]\}$$

$$= 0.75 \times 781 = 586 \text{ k}$$

Try W 14 × 109 ($P = 564$ k)

Refine by use of modified equations (H1-1), (H1-2), and (H1-3):

$$L_c = 15.4 \text{ ft}, \quad L_u = 40.6 \text{ ft}$$

$$F_{bx} = 0.6 \times 36 = 22 \text{ ksi}, \quad F_{by} = 0.75 \times 36 = 27 \text{ ksi}$$

$$A = 32.0 \text{ in.}^2, \quad B_x = 0.185, \quad B_y = 0.523$$

$$a_x = 184.5 \times 10^6, \quad a_y = 66.3 \times 10^6$$

$$F'_{ex} = [401/(18)^2] \times 100 = 123.8 \text{ ksi}$$

$$F'_{ey} = [144/(18)^2] \times 100 = 44.4 \text{ ksi}$$

$$KL_y/r_y = 18 \times 12/3.73 = 57.9$$

$$F_a = 17.63 \text{ ksi}$$

$$f_a = 300 \text{ k}/32 \text{ in.}^2 = 9.4 \text{ ksi}$$

$$f_a/F_a = 0.53 > 0.15$$

Use modified formula (H1-1):

$$0.75 \left\{ 300 + \left[0.185 \times 200 \times 12 \times 0.36 \times \frac{17.63}{22} \right.\right.$$

$$\times \frac{184.5 \times 10^6}{184.5 \times 10^6 - 0.75 \times 300 \times (18 \times 12)^2} \right]$$

$$+ \left[0.523 \times 80 \times 12 \times 0.9 \times \frac{17.63}{27} \right.$$

$$\left.\left. \times \frac{66.3 \times 10^6}{66.3 \times 10^6 - 0.75 \times 300 \times (18 \times 12)^2} \right] \right\}$$

$$= 0.75 \times (300 + 135.8 + 350.6) = 589.8 \text{ k} > 564 \text{ k} \quad \text{N.G.}$$

Try W 14 × 120 ($P = 623$ k)

We will check this column by using eqns. (4.15) and (4.16):

$A = 35.3$ in.^{2}, $S_x = 190$ in.^{3}, $r_x = 6.24$ in., $S_y = 67.5$ in.^{3},
$r_y = 3.74$ in., $L_c = 15.5$ ft, $L_u = 44.1$ ft

$$(KL/r)_x = 18 \times 12/6.24 = 34.6$$

$$(KL/r)_y = 18 \times 12/3.74 = 57.8$$

$$F_a = 17.64 \text{ ksi}$$

$$F'_{ex} = 124.8 \text{ ksi}$$

$$F'_{ey} = 44.7 \text{ ksi}$$

$$f_a = 300 \text{ k}/35.3 \text{ in.}^2 = 8.5 \text{ ksi}$$

$$f_a/F_a = 8.5/17.64 = 0.48 > 0.15$$

$$f_{bx} = 200 \times 12/190 = 12.6 \text{ ksi}$$

$$f_{by} = 80 \times 12/67.5 = 14.2 \text{ ksi}$$

$$F_{bx} = 22 \text{ ksi}, \quad F_{by} = 27 \text{ ksi}$$

(4.15): $\dfrac{8.5}{1.33 \times 17.64} + \dfrac{0.36 \times 12.6}{(1.33 \times 22)\left[1 - \dfrac{8.5}{1.33 \times 124.8}\right]}$

$+ \dfrac{0.9 \times 14.2}{(1.33 \times 27)\left[1 - \dfrac{8.5}{1.33 \times 44.7}\right]}$

$= 0.36 + 0.16 + 0.42 = 0.94 < 1 \quad \text{ok}$

(4.16): $\dfrac{8.5}{1.33 \times 22} + \dfrac{12.6}{1.33 \times 22} + \dfrac{14.2}{1.33 \times 27}$

$= 0.29 + 0.43 + 0.40 = 1.12 \quad \text{N.G.}$

The reader can verify that a W 14 × 132 is adequate.

Example 4.26. Assuming that sidesway is prevented, select a W 10 column for the situation shown below, when the bending occurs about the strong axis.

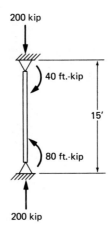

Solution.

$$C_m = 0.6 - 0.4(M_1/M_2) = 0.6 - 0.4\left(\frac{-40}{80}\right) = 0.8$$

$$K = 1.0$$

$$KL = 15.0 \text{ ft}$$

$$m = 2.2 \times \frac{0.8}{0.85} = 2.1$$

$$P_{\text{eff}} = 200 + (80 \times 2.1) = 368 \text{ k}$$

Try W 10×77 ($P = 373$ k for $KL = 15$ ft)

$$m = 2.45 \times \frac{0.8}{0.85} = 2.3$$

$$P_{\text{eff}} = 200 + (80 \times 2.3) = 384 \text{ k}$$

Check the W 10×77 with the modified equations:

$$A = 22.6 \text{ in.}^2, \quad B_x = 0.263, \quad a_x = 67.9 \times 10^6, \quad r_y = 2.6 \text{ in.}$$

$$\frac{Kl}{r} = \frac{1.0 \times 180 \text{ in.}}{2.6 \text{ in.}} = 69.23$$

$$F_a = 16.51 \text{ ksi}$$

$$f_a = \frac{200}{22.6} = 8.85 \text{ ksi}$$

$$\frac{f_a}{F_a} = 0.54 > 0.15$$

176 STEEL DESIGN FOR ENGINEERS AND ARCHITECTS

$$\text{Required load} = 200 + 0.263 \times (80 \times 12) \times 0.8 \times \frac{16.51}{22.0}$$

$$\times \frac{67.9 \times 10^6}{(67.9 \times 10^6) - 200 \times (1.0 \times 180)^2}$$

$$= 367.6 < 373 \text{ for W } 10 \times 77 \quad \text{ok}$$

$$P + P'_x = P\frac{F_a}{0.6 F_y} + B_x M_x \frac{F_a}{F_{bx}}$$

$$\text{Required load} = 200\left(\frac{16.51}{22.0}\right) + 0.263 \times (80 \times 12)\left(\frac{16.51}{22.0}\right)$$

$$= 339.6 \text{ k} < 373 \text{ k for W } 10 \times 77 \quad \text{ok}$$

Use W 10 × 77.

Example 4.27. Check the adequacy of the $\frac{1}{2} \times 6$ in. plate subjected to the loads shown below. The 30-kip tensile load acts through the centroid of the member.

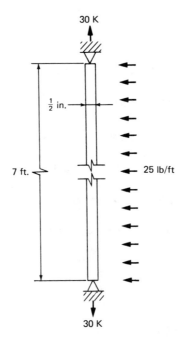

Solution. When a member is subjected to both an axial tensile stress and a flexural stress, the axial tension tends to reduce the bending compressive stress;

the secondary moment caused by the axial load is in the opposite direction to the applied moment. Since the primary moment is not amplified, AISCS H2 requires that (H2-1) be satisfied:

$$\frac{f_a}{F_t} + \frac{f_{bx}}{F_{bx}} + \frac{f_{by}}{F_{by}} \le 1.0 \qquad (4.17)$$

where f_a = computed axial tensile stress, ksi
f_b = computed bending tensile stress, ksi
F_t = allowable tensile stress, ksi (see Chapter 1)
F_b = allowable bending stress, ksi (see Chapter 2)

It is important to note that the computed bending compressive stress, relative to the axial tensile stress, should not exceed the appropriate allowable stress given in Chapter 2.

For the $\frac{1}{2} \times 6$ in. plate: $A = 3$ in.^{2}; $S = \frac{1}{6} \times 6 \times (\frac{1}{2})^2 = 0.25$ in.^{3}

$$f_a = \frac{30}{3} = 10 \text{ ksi}$$

$$F_t = 0.6 \times 36 = 22 \text{ ksi}$$

$$M = \frac{0.025 \times (7)^2}{8} = 0.153 \text{ ft-k}$$

$$f_b = \frac{0.153 \times 12}{0.25} = 7.32 \text{ ksi} \quad \text{(tensile and compressive)}$$

$$F_b = 0.75 \times 36 = 27 \text{ ksi}$$

Check (4.17):

$$\frac{10}{22} + \frac{7.34}{27} = 0.46 + 0.27 = 0.73 < 1 \quad \text{ok}$$

Note that in this case, since the computed tensile stress is larger than the computed bending compressive stress, the net stress on the entire cross section is tensile.

4.6 COLUMN BASE PLATES

Steel base plates are generally required under columns for distributing the load over a sufficient area of bearing support. The design procedure discussed here is different from the one presented in the ninth edition of the AISCM (p. 3-106). After the appearance of the ninth edition, it was shown that some inconsistencies existed in the method which made it inapplicable.[1] A new method was subsequently developed[2] similar to the one in the 8th edition but with a small modification which will be discussed below.[3]

In general, it is assumed that the pressure between the base plate and the concrete is uniform. For base plates which overhang a large amount beyond the column dimensions (i.e., m and n dimensions are large; see Fig. 4.7) this assumption is conservative, since the actual pressure near the plate edges will be smaller than the pressure directly under the column. For cases when the base plate overhangs only a small amount (m and n dimensions are small), the bending stress in the base plate is more critical at the column web, halfway between the two flanges, than at the locations m or n distances away from the plate edges. The new method requires that $n' = \frac{1}{4}\sqrt{db_f}$ be determined along with the previously required values of m and n. The value n' determines the required

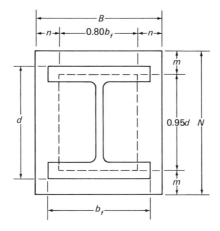

Fig. 4.7. Column base plate. These are generally used under columns to distribute the column load over a sufficient area of concrete pier or foundation.

---

[1] S. Ahmed and R. P. Kreps, "Inconsistencies in Column Base Plate Design in the New AISC ASD Manual," *Engineering Journal*, American Institute of Steel Construction, Third Quarter, 1990, pp. 106-107.
[2] W. A. Thorton, "Design of Small Base Plates for Wide Flange Columns," *Engineering Journal*, American Institute of Steel Construction, Third Quarter, 1990, pp. 108-110.
[3] The eighth edition is based on the elastic approach of the bending stress in the plate supported on three sides and free at the fourth side, while the new method is based on the yield line theory.

thickness at the column web for the portion of the base plate bounded by the column web, the two flanges, and the free edge, while m and n are used to determine the base plate thickness that projects beyond the column flanges as inverted cantilevers.

The following steps can be used for designing base plates:

1. Establish the bearing values of the concrete according to AISCS J9.
2. Determine the required area of the base plate ($A = P/F_p$).
3. Establish B and N, preferably rounded to full inches, such that m and n are approximately equal and $B \times N \geq A$. When either value is limited by other considerations, that dimension shall be established first and the other from $B \times N \geq A$.
4. Determine $m = (N - 0.95\, d)/2$, $n = (B - 0.80\, b_f)/2$, and $n' = \frac{1}{4}\sqrt{db_f}$.
5. Determine the actual bearing pressure on the support ($f_p = P/(B \times N)$).
6. Use the largest values of m, n, or n' to solve for the thickness t_p by the formula

$$t_p = 2(m, n, \text{ or } n')\sqrt{\frac{f_p}{F_y}}$$

Fig. 4.8. Pin-connected bases of trusses for wall-roof structural system for the atrium of an office building.

180 STEEL DESIGN FOR ENGINEERS AND ARCHITECTS

Note that the procedure for determining the minimum size base plate given on p. 3-107 of the AISCM will not be discussed here.

Example 4.28. A W 8 × 58 column supports an axial load of 300 kips. Design a base plate for the column if the supporting reinforced concrete footing has a very large pedestal and $f'_c = 3000$ psi.

Solution.

$$F_p = 0.35 f'_c \sqrt{A_2/A_1} \le 0.7 f'_c \quad \text{(AISCS J9)}$$

Use $F_p = 0.7 f'_c$, since $A_2 \gg A_1$

$$A_{req} = \frac{P}{F_p} = \frac{300 \text{ k}}{2.1 \text{ ksi}} = 142.86 \text{ in.}^2$$

Choose base plate 11 × 13 in. so that $A = 11$ in. × 13 in. = 143 in.^{2}, and m and n are almost equal.

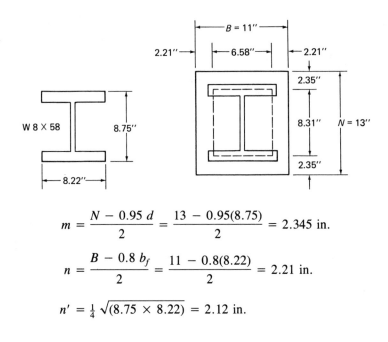

$$m = \frac{N - 0.95 d}{2} = \frac{13 - 0.95(8.75)}{2} = 2.345 \text{ in.}$$

$$n = \frac{B - 0.8 b_f}{2} = \frac{11 - 0.8(8.22)}{2} = 2.21 \text{ in.}$$

$$n' = \tfrac{1}{4} \sqrt{(8.75 \times 8.22)} = 2.12 \text{ in.}$$

$$f_p = \frac{P}{B \times N} = \frac{300 \text{ k}}{11 \text{ in.} \times 13 \text{ in.}} = 2.1 \text{ ksi}$$

$$t_p = 2m\sqrt{\frac{f_p}{F_y}} = (2 \times 2.345)\sqrt{\frac{2.1}{36}} = 1.13 \text{ in.}$$

Use $t_p = 1\frac{1}{4}$ in.
Use $11 \times 1\frac{1}{4} \times 1$ ft, 1 in. base plate.

Example 4.29. Redesign the base plate of the previous problem for pedestal size of the same area as the base plate.

Solution.

$$F_p = 0.35 f'_c = 0.35 \times 3000 \text{ psi} = 1.05 \text{ ksi}$$

$$A_{req} = \frac{P}{F_p} = \frac{300 \text{ k}}{1.05 \text{ ksi}} = 285.7 \text{ in.}^2$$

Choose base plate 16 in. × 18 in., so that $A = 16 \text{ in.} \times 18 \text{ in.} = 288 \text{ in.}^2$

$$m = \frac{N - 0.95 d}{2} = \frac{18 - 0.95(8.75)}{2} = 4.84 \text{ in.}$$

$$n = \frac{B - 0.8 b}{2} = \frac{16 - 0.8(8.22)}{2} = 4.71 \text{ in.}$$

$$n' = 2.12 \text{ in.} \quad \text{(see Example 4.28)}$$

$$f_p = \frac{P}{B \times N} = \frac{300 \text{ k}}{16 \text{ in.} \times 18 \text{ in.}} = 1.04 \text{ ksi}$$

$$t_p = (2 \times 4.84)\sqrt{\frac{1.04}{36}} = 1.65 \text{ in.}$$

Use $t_p = 1\frac{3}{4}$ in.
Use $16 \times 1\frac{3}{4} \times 1$ ft, 6 in. base plate.

PROBLEMS TO BE SOLVED

4.1. Determine the Euler stress and the critical load for the pin-connected column as shown, with a length of 24 ft and $E = 29{,}000$ ksi. (For an isometric view of the column, see Fig. 1.13.) Note: Assume that the lacing is spaced such that the stability of each channel by itself can be disregarded.

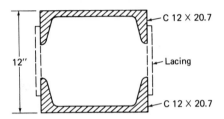

4.2. For the pin-connected column shown, determine the Euler stress and critical load if $L = 18$ ft and $E = 29{,}000$ ksi.

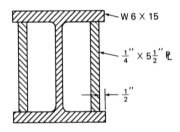

4.3. Rework Problems 4.1 and 4.2, assuming that the columns are fully fixed at both ends.

4.4. For the cantilever (flagpole type) columns shown, determine the Euler stress and critical load for each column if the unsupported length is 20 ft and $E = 29{,}000$ ksi. Consider the note in Problem 4.1.

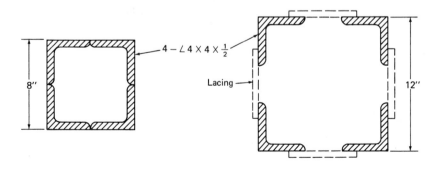

4.5. Calculate the allowable compressive axial stress F_a for (1) a W 14 × 145 column with one end fixed and the other pinned, and (2) a W 8 × 35 column with both ends pinned. Use $L = 30$ ft.

4.6. For the columns shown in Problems 4.1 and 4.2, calculate the allowable stress F_a for each column for both 36- and 50-ksi steels.

4.7. Using the AISC table for allowable stress for compression members of 36 ksi, determine the allowable load P a W 8 × 67 can support with a length of 12 ft and pinned ends.

4.8. Rework Problem 4.7, assuming 50-ksi steel and pinned ends.

4.9. Determine the allowable load a structural tube TS 8 × 8 × .375 can support with a length of 18 ft and pinned ends. $F_y = 46$ ksi.

4.10. Determine the allowable axial load a truss member can support if the member is made from two 6 × 6 × $\frac{1}{2}$-in. angles separated by a $\frac{3}{4}$-in. gap back to back. Assume that the member is the truss top chord, subject to compressive stress, and the unsupported length is 10 ft.

4.11. Design a WT 7 section for a truss top chord if it will be subject to an axial load of 120 kips. Assume that the member will be laterally supported every 12 ft.

4.12. Design a W 12 column, pinned at both ends, to support an axial load of 450 kips: $L_x = L_y = 14$ ft; $F_y = 36$ and 50 ksi.

4.13. Rework Problem 4.12 if $L_x = 16$ ft and $L_y = 8$ ft.

4.14. Design the lightest wide-flange column for an axial load of 225 kips. The column is fixed at the top, pinned at the bottom, and subject to sway. Assume that $L_x = 20$ ft and $L_y = 10$ ft.

4.15. Determine if a W 12 × 58 with axial load $P = 200$ kips, weak axis bending $M_y = 40$ ft-kips at the top and $M_y = 20$ ft-kips at the bottom is adequate for unbraced lengths of $L_x = L_y = 18$ ft. The frame is secured against sway. Ends of columns are restrained. Check column for

    a) single curvature
    b) double curvature.

4.16. A W 14 × 211 column with $P = 500$ kips, $M_x = 200$ ft-kips and $M_y = 70$ ft-kips has unbraced lengths of $L_x = L_y = 14$ ft. The frame is braced against sidesway, and ends of columns are restrained. a) Is the column safe? b) If the moments are due to transverse wind loads, is the column safe? In both cases, assume that the member is bent in single curvature, and that the moments given act at both the top and bottom of the member.

4.17. Redesign the column of Problem 4.14, without sidesway, but with a moment of 55 ft-kips applied to the top of the column about the major (x-x) axis. Use $L_x = L_y = 20$ ft. Try a W 12 member.

4.18. A W 12 × 170 is subject to an axial load of 700 kips and a moment of 85 ft-kips, applied about the major axis. Determine the maximum stresses at the extreme fibers of the member due to the loads.

4.19. Design the lightest W 8 wide-flange column for the loading shown. Assume that the 20-kip load is applied at the face of the column flange. The column is braced against sidesway.

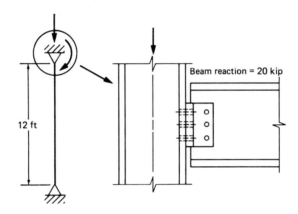

4.20 and 4.21. For the columns shown, design the lightest W 14 sections if 65-kip loads are applied at the locations shown by dots. $L = 22$ ft, $K_x = K_y = 1.0$. The column is braced against sidesway.

4.22. A W 12 × 65 column supports an axial load of 360 kips. Design a base plate for the column if the supporting concrete foundation is very large. $F_y = 36$ ksi and $f'_c = 3000$ psi.

4.23. Design a base plate for a W 14 × 120 column supporting an axial load of 650 kips. Assume that the pedestal is the same size as the base plate and f'_c = 3500 psi.

4.24. Design a W 14 column for the following:

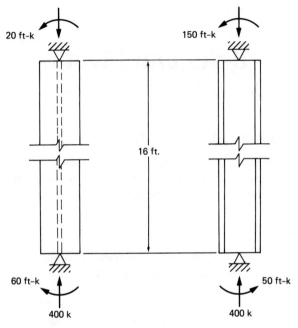

$L_x = L_y$ = 16 ft. Braced against sidesway.
$K_x = K_y$ = 1.0
F_y = 50 ksi

5
Bolts and Rivets

5.1 RIVETED AND BOLTED CONNECTIONS

Connections in structural steel are made with mechanical fasteners (rivets or bolts) or welds. Riveted and bolted connections will be covered in this chapter, and welds will be discussed in Chapter 6.

5.2 RIVETS

For many years, riveting was the sole method of connecting structural steel. In recent years, due to the ease and economy of welding and high-strength bolting, the use of rivets has declined to the point where they are completely obsolete today. Nevertheless, the Code still recognizes them mostly for existing structures. The rivets used in construction work are usually made from a soft grade of steel that does not become brittle when heated and hammered with a riveting gun. Rivets are manufactured with one formed head and are installed in holes that are punched or drilled $\frac{1}{16}$ in. larger in diameter than the nominal diameter of the rivet. The rivet is usually heated to a cherry-red color (approximately 1800°F) before placing it in the hole, and a second head is formed on the other end by a riveting hammer or a pressure type riveter. While the second head is being formed, the heat-softened shank is forced to fill the hole completely. As the rivet cools, it shrinks and squeezes the connected parts together, causing friction between them. However, this friction is usually neglected in calculations.

According to the AISC, rivets shall conform to the provisions of the "Specifications for Structural Rivets" ASTM A502, Grades 1, 2, or 3. The size of rivets used in general steel construction ranges from $\frac{5}{8}$ to $1\frac{1}{2}$ in. in diameter by $\frac{1}{8}$-in. increments. The allowable stresses are tabulated in the AISCM in accordance with AISCS Table J3.2.

Rivets usually require a crew of three to four people to be put in place. Due

BOLTS AND RIVETS   187

to high labor costs and extreme noise, rivets are no longer commonly used in modern construction. Even though bolt material is more expensive than rivet steel, the advantages of bolt placement far outweigh the disadvantage of using rivets. Hence, rivets have completely been replaced by bolts.

5.3 BOLTS

Bolting of steel structures is a very rapid field-erection process. It requires less skilled labor than riveting or welding and therefore has a distinct advantage over the other connection methods. The joints obtained using high-strength bolts are superior to riveted joints in performance and economy, and bolting has become the leading method of fastening structural steel in the field.

There are two types of bolts that are most commonly used in steel construction. The first type, *unfinished bolts*, also called *common* or *ordinary* bolts, is primarily used in light structures subjected to static loads or for connecting secondary members; it is the cheapest type of connection available. Unfinished bolts must conform to the "Specifications for Low Carbon Steel Externally and Internally Threaded Fasteners," ASTM A307. AISCS J1.12 lists specific connection types for which A307 bolts may not be used.

The second type, *high-strength bolts*, is made from medium carbon, heat-treated, or alloy steel and has tensile strengths much greater than those of ordinary bolts. High-strength bolts shall conform to the "Specifications for Structural Joints Using ASTM A325 or A490 Bolts," AISCS J3 (and to the Specification starting on p. 5-263).

Allowable tensile and shear loads for all types of bolts are given in AISCM Tables I-A, I-B, I-C, and I-D, starting on p. 4-3, in accordance with the allowable stresses given in Table J3.2 (reproduced here as Table 5.1).

Connections made of high-strength bolts are categorized into three types: (1) slip-critical (SC), (2) bearing type connection with threads included in shear plane (N) (see Fig. 5.1a), and (3) bearing type connection with threads excluded from shear plane (X), (see Fig. 5.1b). The two latter types of connections are similar to riveted connections and are designed by the same method, the only difference being in their respective shearing and bearing capacities (see AISCM Table J3.2). Bolts or rivets are said to be working in single or double shear if one or two shear planes are acting on the shank of the connector. For example, the bolts shown in Fig. 5.1 are in single shear. Table I-D gives allowable shear values for both single (S) and double (D) shear cases.

Shear connections subjected to stress reversals or severe stress fluctuations or where slippage is undesirable are required by the AISC to be of the SC type. When high-strength bolts are used in combination with welds, only those installed as SC type connections prior to welding may be considered as sharing the stresses with the weld (AISCS J.10). In SC connections, since the load is

Table 5.1. Allowable Stress on Fasteners, ksi (AISCS Table J3.2, reprinted with permission)

Description of Fasteners	Allowable Tension[g] (F_t)	Allowable Shear[g] (F_v)				Bearing-type Connections[i]
		Slip-critical Connections[e,i]				
		Standard size Holes	Oversized and Short-slotted Holes	Long-slotted holes		
				Transverse[j] Load	Parallel[j] Load	
A502, Gr. 1, hot-driven rivets	23.0[a]					17.5[f]
A502, Gr. 2 and 3, hot-driven rivets	29.0[a]					22.0[f]
A307 bolts	20.0[a]					10.0[b,f]
Threaded parts meeting the requirements of Sects. A3.1 and A3.4 and A449 bolts meeting the requirements of Sect. A3.4, when threads are not excluded from shear planes	0.33 $F_u^{a,c,h}$					0.17 F_u^h
Threaded parts meeting the requirements of Sects. A3.1 and A3.4, and A449 bolts meeting the requirements of Sect. A3.4, when threads are excluded from shear planes	0.33 $F_u^{a,h}$					0.22 F_u^h
A325 bolts, when threads are not excluded from shear planes	44.0[d]	17.0	15.0	12.0	10.0	21.0[f]
A325 bolts, when threads are excluded from shear planes	44.0[d]	17.0	15.0	12.0	10.0	30.0[f]
A490 bolts, when threads are not excluded from shear planes	54.0[d]	21.0	18.0	15.0	13.0	28.0[f]
A490 bolts, when threads are excluded from shear planes	54.0[d]	21.0	18.0	15.0	13.0	40.0[f]

[a]Static loading only.
[b]Threads permitted in shear planes.
[c]The tensile capacity of the threaded portion of an upset rod, based upon the cross-sectional area at its major thread diameter A_b shall be larger than the nominal body area of the rod before upsetting times 0.60 F_y.
[d]For A325 and A490 bolts subject to tensile fatigue loading, see Appendix K4.3.
[e]Class A (slip coefficient 0.33). Clean mill scale and blast-cleaned surfaces with Class A coatings. When specified by the designer, the allowable shear stress, F_v, for slip-critical connections having special faying surface conditions may be increased to the applicable value given in the RCSC Specification.
[f]When bearing-type connections used to splice tension members have a fastener pattern whose length, measured parallel to the line of force, extends 50 in., tabulated values shall be reduced by 20%.
[g]See Sect. A5.2
[h]See Table 2, Numerical Values Section for values for specific ASTM steel specifications.
[i]For limitations on use of oversized and slotted holes, see Sect. J3.2.
[j]Direction of load application relative to long axis of slot.

BOLTS AND RIVETS 189

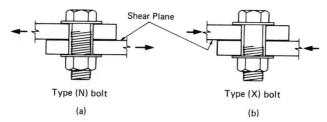

Type (N) bolt
(a)

Type (X) bolt
(b)

Fig. 5.1. Bearing-type connections of high-strength bolts: (a) for bolts with threads included in shear plane and (b) for bolts with threads excluded from shear plane.

primarily transferred by friction between the connected parts, the design is based on the assumption that if and when the connection fails, the bolts will fail in shear alone, and therefore the bearing stress of the fasteners on the members need not be checked.

5.4 DESIGN OF RIVETED OR BOLTED CONNECTIONS

A connection is said to be concentrically loaded if the resultant of the applied forces passes through the centroid of the fastener group. For such cases, tests have shown that just before yielding, all fasteners in the group carry equal portions of the load. When the resultant force does not pass through the centroid, the force may be replaced by an equal force at the centroid accompanied by a moment equal in magnitude to the force times its eccentricity. In situations such as this, each fastener in the connection group is assumed to resist the axial forces uniformly and to resist the moment in proportion to its respective distance to the centroid of the connection; nevertheless, this eccentricity may be neglected in many cases (AISCS J1.9).

A bolted (N or X type) or riveted connection can fail four different ways. First, one of the connected members may fail in tension through one or more of the fastener holes (Fig. 5.2a). Second, if the holes are drilled too close to the edge of the tension member, the steel behind the fasteners may shear off (Fig. 5.2b). Third, the fasteners may fail in shear (Fig. 5.2c). And fourth, one or more of the tension members may fail in bearing of the fasteners on the member(s) (Fig. 5.2d). To prevent failure, a connection and the connected parts must be designed to resist failure in all four of the ways mentioned.

First, to assure nonfailure of the connected parts, the members shall be designed such that the tensile stress is less than $0.60 \, F_y$ on the gross area and less than $0.50 \, F_u$ on the effective net area (AISCS D1). Therefore, the effective net area of any tensile member with bolt holes must be greater than or equal to $P/(0.50 \, F_u)$; also see Chapter 1.

Second, to prevent the steel behind the fasteners from tearing off, the minimum edge distance from the center of a fastener hole to the edge, in the direc-

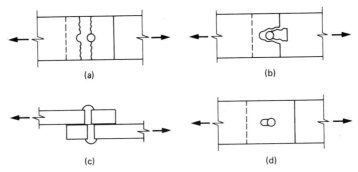

Fig. 5.2. Failure modes of connections made with bolts or rivets. (a) Tension failure of a connected part. (b) Shear failure of a connected part behind a fastener. (c) Shear failure of the fasteners. (d) Bearing failure of a connected part.

tion of the force, shall be not less than $2\,P/F_u t$ (AISCS (J3-6)) or the values given in Table J3.5, as applicable. In (J3-6), P is the force carried by one fastener, t is the thickness of the critical connected part, and F_u is the specified minimum tensile strength of the critical connected part.

Third, to assure that the fasteners will not fail in shear, the number of fasteners should be determined to limit the maximum shear stress in the critical fastener to that listed in AISCS Table J3.2 for the particular type. To determine the minimum number of fasteners required, divide the load on the entire connection by the allowable shear stress of one fastener, $(n = P/F_v)$.

Last, to prevent a connected part from crushing due to the bearing force of the fasteners on the material, the minimum number of fasteners is determined to prevent such crushing. It has been experimentally shown that the lower bound for the allowable bearing stress F_p can be given by (J3-1) for the case when the deformation around the standard hole is a design consideration and by (J3-4) if the deformation is not. In addition, special criteria are given in AISCS J3.8 for minimum spacing requirements along the line of transmitted forces. All of the above requirements have been tabulated in Tables I-E and I-F, pp. 4-6 and 4-7, respectively. The reader is urged to study the notes at the bottom of these tables.

The analysis of SC type connections is quite complex. The connection is designed similarly to N or X type connections, with an appropriate allowable stress for the shear capacity of the bolts. High-strength bolts are tightened until they acquire very high tensile stresses, and the connected parts are clamped tightly together, permitting loads to be transferred primarily by friction. Though bearing should not be encountered in this type of connection, the bearing of the bolts against the connected parts must be considered in the event that the friction bolts slip into bearing ("Specifications for Structural Joints Using ASTM A325 or A490 Bolts," Commentary). Allowable stresses for shear are given in AISCS Table J3.2. For cases of special treatment of the connected parts, Table 3 in

BOLTS AND RIVETS 191

Fig. 5.3. A beam-to-column shear connection with web angles.

"Specifications for Structural Joints Using A325 or A490 Bolts" recognizes three classes: A, B, and C. Class A, which has a slip coefficient of 0.33, is part of Table J3.2. Classes B and C have larger slip coefficients and allowable stresses, and should be used only if the need arises.

Example 5.1. Determine the maximum tensile load P that can be resisted by the existing connection shown. The $\frac{3}{4}$ in. × 14 in. plates are connected with $\frac{7}{8}$ in. diameter A502 Grade 1 rivets.

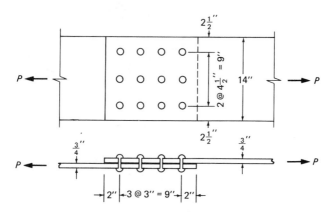

192 STEEL DESIGN FOR ENGINEERS AND ARCHITECTS

Solution. The rivets may fail in shear or bearing, or the plate in tension. Each condition must be investigated to determine the critical condition.

Shear failure in the rivet

$$P_{all} = F_v \times A_{riv} \times N_{rivets}$$

For A502 Grade 1 rivets

$$F_v = 17.5 \text{ ksi} \quad (\text{AISCS Table I-D, p. 4-5})$$

$$A_{riv} = \frac{\pi d^2}{4} = \frac{\pi \left(\frac{7}{8} \text{ in.}\right)^2}{4} = 0.601 \text{ in.}^2$$

Number of rivets $N = 12$

$$P_{all} = 17.5 \text{ ksi} \times 0.601 \text{ in.}^2 \times 12 = 126.2 \text{ k}$$

For failure in bearing Table I-E shows that for a $\frac{3}{4}$-in. plate with $F_u = 58$ ksi (A36 steel) and $\frac{7}{8}$-in.-diameter connectors, the capacity is 45.7 k per rivet. Therefore, for 12 rivets,

$$P_{all} = 12 \times 45.7 \text{ k} = 548.4 \text{ k}$$

Note that the edge distance and spacing requirements conform to the notes given at the bottom of this table.

Tensile capacity of member on gross area

$$F_t = 0.6 F_y = 22.0 \text{ ksi} \quad (\text{AISCS D1})$$

$$A_g = \tfrac{3}{4} \text{ in.} \times 14 \text{ in.} = 10.5 \text{ in.}^2$$

$$P_{all} = F_t \times A_g = 22.0 \text{ ksi} \times 10.5 \text{ in.}^2 = 231.0 \text{ k}$$

Tensile capacity of member on effective net area

$$U = 1.0 \quad (\text{AISCS B3})$$

$$F_t = 0.5 F_u = 29.0 \text{ ksi} \quad (\text{AISCS D1})$$

$$A_n = (14 \text{ in.} - 3(\tfrac{7}{8} \text{ in.} + \tfrac{1}{8} \text{ in.})) \times \tfrac{3}{4} \text{ in.} = 8.25 \text{ in.}^2$$

$$A_e = U \times A_n = 1.0 \times 8.25 \text{ in.}^2 = 8.25 \text{ in.}^2$$

$$P_{all} = F_t \times A_e = 29.0 \text{ ksi} \times 8.25 \text{ in.}^2 = 239.3 \text{ k}$$

Maximum tensile load is least value of cases calculated:

$$P_{max} = P_{all\,shear} = 126.2 \text{ k}$$

Example 5.2. Determine the maximum tensile load P that can be resisted by the connection in Example 5.1 using 1-in.-diameter A325 high-strength bolts (HSB), assuming standard holes and

a) a bearing connection with the threads included in the shear plane (A325-N).
b) a bearing connection with the threads excluded from the shear plane (A325-X).
c) an SC connection (A325-SC)

Solution. Determine the capacity for bearing, tension, and shear.

$$A_b = 0.785 \text{ in.}^2$$
$$N_b = 12$$

Bolt bearing will be same value for either friction type or bearing type connections. From Table I-E, allowable load/bolt = 52.2 k.

$$P_{all} = 12 \times 52.2 \text{ k} = 626.4 \text{ k}$$

As in Example 5.1, both edge distance and spacing requirements conform to the notes at the bottom of Table I-E.

Tension on member will be the same for all cases.

Tension on gross area of member

$$P_{all} = 231.0 \text{ k} \quad \text{(Example 5.1)}$$

Tension on effective net area of member

$$F_t = 29.0 \text{ ksi}$$
$$A_n = (14 \text{ in.} - 3(1 \text{ in.} + \tfrac{1}{8} \text{ in.})) \times \tfrac{3}{4} \text{ in.} = 7.97 \text{ in.}^2$$
$$A_e = 1.0 \times 7.97 \text{ in.}^2 = 7.97 \text{ in.}^2$$
$$P_{all} = F_t \times A_e = 29.0 \text{ ksi} \times 7.97 \text{ in.}^2 = 231.1 \text{ k}$$

194   STEEL DESIGN FOR ENGINEERS AND ARCHITECTS

Shear capacity of connectors   (Table I-D)

$$a)\ 16.5\ k \times 12 = 198\ k$$
$$b)\ 23.6\ k \times 12 = 283.2\ k$$
$$c)\ 13.4\ k \times 12 = 160.8\ k$$

Note: only SC Class A values (the lowest) are given in Tables I-D and J3.2. Other values for other classes must be determined according to the Specification for A325 and A490 high-strength bolts, p. 5-263.

The bolt shear capacity is critical in cases (a) and (c), and the tension on the gross area is critical in case (b). Maximum tensile load P:

$$a)\ P_{max} = 198\ k$$
$$b)\ P_{max} = 231\ k$$
$$c)\ P_{max} = 160.8\ k$$

Example 5.3. A truss member in tension consists of a single angle $4 \times 4 \times \frac{1}{2}$ in. The member is connected to a $\frac{1}{2}$-in. gusset plate with $\frac{3}{4}$-in.-diameter A325-N bolts as shown. Determine the maximum load the connection can carry if bolts are spaced $2\frac{1}{2}$ in. on center.

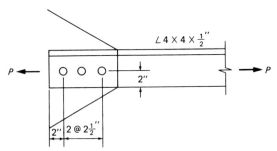

Solution. Eccentricity between the gravity axes of such members and the gage lines for their riveted or bolted end connections may be neglected (AISCS J1.9).

$$F_v = 21.0\ \text{ksi},\quad F_t = 22.0\ \text{ksi}$$
$$A_b = 0.442\ \text{in.}^2$$
$$A_g = 3.75\ \text{in.}^2$$
$$A_e = 2.81\ \text{in.}^2\quad (\text{Example 1.5})$$
$$P = 22.0\ \text{ksi} \times 3.75\ \text{in.}^2 = 82.5\ k$$
$$P = 29.0\ \text{ksi} \times 2.81\ \text{in.}^2 = 81.5\ k$$

BOLTS AND RIVETS 195

Bolt shear

$$P = 3 \times 9.3 \text{ k} = 27.9 \text{ k}$$

Bolt bearing

$$P = 3 \times 26.1 \text{ k} = 78.3 \text{ k}$$

Use least value

$$P_{all} = 27.9 \text{ k} \quad \text{(bolt shear)}$$

Example 5.4. For the plates shown, determine the allowable value of P using 1-in.-diameter A307 bolts.

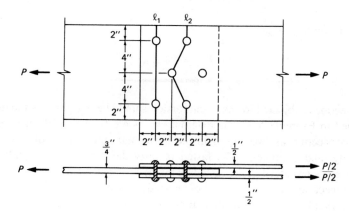

Solution.

$$l = l_{tot} - l_{holes} + \frac{s^2}{4g}$$

$$l_1 = 12 \text{ in.} - 2\left(1 \text{ in.} + \frac{1}{8} \text{ in.}\right) + 0 = 9.75 \text{ in.}$$

$$l_2 = 12 \text{ in.} - 3\left(1 \text{ in.} + \frac{1}{8} \text{ in.}\right) + \frac{2(2 \text{ in.})^2}{4 \times 4 \text{ in.}} = 9.125 \text{ in.}$$

a) P_{all} of plate

Gross area

$$P = F_y \times A_g = 22.0 \text{ ksi} \times (\tfrac{3}{4} \text{ in.} \times 12 \text{ in.}) = 198.0 \text{ k}$$

Effective net area

$$P = F_u \times A_e = 29.0 \text{ ksi} \times 1.0(\tfrac{3}{4} \text{ in.} \times 9.125 \text{ in.}) = 198.5 \text{ k}$$

b) P_{all} by shear

Bolts are in double shear.

$$P = 6 \times 15.7 \text{ k} = 94.2 \text{ k}$$

c) P_{all} by bearing

Bearing on $\tfrac{3}{4}$-in. plates is more critical than two $\tfrac{1}{2}$-in. plates.

$$P = 6 \times 52.2 \text{ k} = 313.2 \text{ k}$$

Case b) governs.

$$P_{\text{all}} = 94.2 \text{ k}$$

Connections subjected to shear and torsion should be designed by the method explained in Example 5.5. The external effects are transferred to the centroid of the connection as two forces in the horizontal and vertical directions, respectively, and a moment. The fasteners carry the direct forces equally. The shear stresses produced in the fasteners by the moment are perpendicular to the lines connecting the fastener to the centroid, and the magnitude of the shear stress carried by each connector is proportional to the distance of the fastener from the centroid (see Fig. 5.4).

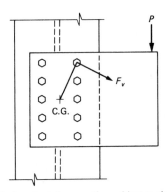

Fig. 5.4. Bolted or riveted connection subject to shear and torsion.

Example 5.5. Find the load on a fastener in the general situation shown in the sketch.

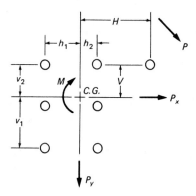

Solution. The centroid of the fastener group is first located, and the force P is broken into a vertical component P_y, a horizontal component P_x, and an applied moment M at that point.

It is assumed that the two components are equally divided among all the fasteners as resisting forces. The magnitude of the resisting force in each connector due to the moment is proportional to the distance of that connector to the centroid, and its direction is perpendicular to that line.

If we designate for the ith connector, its resisting force by R_{mi}, its distance to the centroid d_i, its resisting moment is

$$M_r = d_i \times R_{mi}$$

The applied moment then becomes

$$\Sigma(d_i \times R_{mi}) = M$$

Also referring to the farthest connector for d_{max} and $R_{m\,max}$,

$$R_{mi} = d_i \times \frac{R_{m\,max}}{d_{max}}$$

and

$$M = \frac{\Sigma d_i^2 \times R_{m\,max}}{d_{max}}$$

From here we obtain

$$R_{m\,max} = \frac{M\,d_{max}}{\Sigma d_i^2}$$

and for any fastener,

$$R_{mp} = \frac{M\,d_p}{\Sigma d_i^2}$$

Also breaking up R_{mp} into its x and y components, we obtain

$$R_{mx} = \frac{Mv}{\Sigma d_i^2} \quad R_{my} = \frac{Mh}{\Sigma d_i^2}$$

Here v and h are the ordinate and the abscissa of the connector under consideration with respect to the centroid.

Also,

$$\Sigma d_i^2 = \Sigma(v_i^i + h_i^2)$$

With n being the number of fasteners, we can then calculate the components of the force in the connector as

$$R_x = \frac{P_x}{n} + \frac{Mv}{\Sigma v_i^2 + \Sigma h_i^2}$$

and

$$R_y = \frac{P_y}{n} + \frac{Mh}{\Sigma v_i^2 + \Sigma h_i^2}$$

The total resisting force in the connector as

$$R = \sqrt{R_x^2 + R_y^2}$$

Because the most critically loaded connector may not be determined by inspection prior to analysis, it may be necessary to verify all "candidates" for the most critically loaded connector.

(Note: This problem has been solved by the elastic method in AISCM, "Eccentric Loads on Fastener Groups" (Alternate Method 1, p. 4-59). The results from this method are somewhat conservative; however, this method can be easily applied to any bolt configuration with an eccentric load inclined at any angle from the vertical.)

Ultimate Strength Method

A method based on the ultimate strength capacity, somewhat more liberal than the method described above, but still safe, is discussed in the ninth edition of the AISCM (1989). The tables starting on p. 4-62 have been developed based on this method. Note that the values given in these tables are applicable only to connector groups following a definite pattern and an eccentric vertical load. To overcome this limitation, Alternate Method 2 on p. 4-59 has been developed, which is an approximate method for inclined loads on connector patterns covered by the AISCM tables for vertical loads.

Example 5.6. Determine the resultant load on the most stressed fastener in the eccentrically loaded connection shown. Use the elastic method.

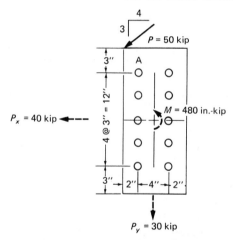

Solution. Components of the 50-kip load are 30 kips down and 40 kips to the left.

Centroid of fasteners by inspection is 9 in. from the bottom and 4 in. from either side of the plate.

$$M = 40 \text{ k} \times 9 \text{ in.} + 30 \text{ k} \times 4 \text{ in.} = 480 \text{ in.-k}$$

$\Sigma d^2 = \Sigma(h^2 + v^2)$ where h is the horizontal distance from center of gravity to each fastener, and v is the vertical distance from center of gravity to each fastener

$$= \Sigma h^2 + \Sigma v^2 = 10(2 \text{ in.})^2 + 4(6 \text{ in.})^2 + 4(3 \text{ in.})^2$$
$$= 220 \text{ in.}^2$$

At corner (critical) fasteners,

$$V_M = \text{vertical load due to moment} = \frac{Mh}{\Sigma d^2}$$

$$= \frac{480 \text{ in.-k} \times 2 \text{ in.}}{220 \text{ in.}^2} = 4.36 \text{ k}$$

$$H_M = \text{horizontal load due to moment} = \frac{Mv}{\Sigma d^2}$$

$$= \frac{480 \text{ in.-k} \times 6 \text{ in.}}{220 \text{ in.}^2} = 13.09 \text{ k}$$

$V_S = $ vertical load due to direct component

$$\frac{P_y}{N} = \frac{30 \text{ k}}{10} = 3 \text{ k}$$

$H_S = $ horizontal load due to direct component

$$\frac{P_x}{N} = \frac{40 \text{ k}}{10} = 4 \text{ k}$$

Fastener A is the critical fastener, as there the loads added to each other

$V_M = 4.36 \text{ k} \downarrow \quad H_M = 13.09 \text{ k} \leftarrow \quad H_S = 4.0 \text{ k} \leftarrow \quad 18.61 \text{ k}$

$V_S = 3.0 \text{ k} \downarrow$

$$R = \sqrt{V^2 + H^2} = \sqrt{(4.36 + 3)^2 + (13.09 + 4)^2}$$

$$= 18.61 \text{ k}$$

Maximum resultant is on fastener A

$$R = 18.61 \text{ k}$$

Example 5.7. Determine the load P that can be carried by the bracket when $\tfrac{3}{4}$-in. diameter A325-SC bolts are used. Use the elastic method.

BOLTS AND RIVETS

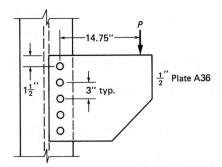

Solution.

$$l = 14.75 \text{ in.}$$
$$\bar{y} = 6 \text{ in. from bottom bolt (by inspection)}$$
$$\Sigma d^2 = \Sigma v^2 = 2(3 \text{ in.})^2 + 2(6 \text{ in.})^2 = 90 \text{ in.}^2$$
$$M = Pl = 14.75\,P$$

Maximum load is on the top and bottom bolts.

$$V_S = \text{vertical load due to shear} = \frac{P}{N} = \frac{P}{5} = 0.2\,P$$

H_M = horizontal load due to moment

$$= \frac{Mv}{\Sigma d^2} = \frac{14.75\,P \times 6 \text{ in.}}{90 \text{ in.}^2} = 0.983\,P$$

$$R_{\text{tot}} = \sqrt{V^2 + H^2} = \sqrt{(.2\,P)^2 + (.983\,P)^2} = 1.003\,P$$

Shear capacity of $\frac{3}{4}$-in. A325-SC bolts

$$R_v = 7.51 \text{ k} \quad \text{(AISC Table I-D)}$$
$$1.003\,P = 7.51 \text{ k}$$
$$P = \frac{7.51 \text{ k}}{1.003} = 7.49 \text{ k},$$

As an alternative, calculate P using Table XI, p. 4-62:

$C = 1.15$ for $l = 14.75$ in., $b = 3$ in., $n = 5$ (by interpolation)
$$P = C \times R_v = 1.15 \times 7.51 = 8.64 \text{ k}$$

202 STEEL DESIGN FOR ENGINEERS AND ARCHITECTS

As expected, this result is more liberal than the one obtained using the elastic method.

The maximum load that can be carried by the bracket is $P = 7.49$ k

Check bracket plate (AISCM, page 4-88).

$$S_{net} = 14 \text{ in.}^3 \text{ for } \tfrac{1}{2} \text{ in.} \times 15 \text{ in. plate}$$

$$M = 14.75 \text{ in.} \times P$$

$$F_b = 22.0 \text{ ksi} \quad (\text{AISCS F2.2})$$

$$M_{all} = 22.0 \text{ ksi} \times 14 \text{ in.}^3 = 308 \text{ in.-k}$$

$$P = \frac{308 \text{ in.-k}}{14.75 \text{ in.}} = 20.9 \text{ k} > 7.49 \text{ k} \quad \text{ok}$$

Example 5.8. Determine the maximum load on the fastener group shown, and determine the plate size.

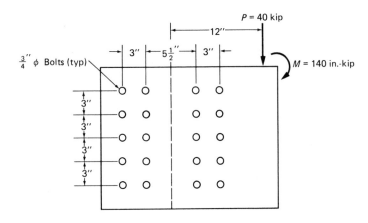

Solution.

$$M_{tot} = (40 \text{ k} \times 12 \text{ in.}) + 140 \text{ in.-k} = 620 \text{ in.-k}$$

$$l = M/P = \frac{620 \text{ in.-k}}{40 \text{ k}} = 15.5 \text{ in.}$$

$\bar{y} = 6$ in. from bottom row of bolts (by inspection)

$\bar{x} = 5.75$ in. from left vertical row of bolts (by inspection)

$\Sigma v^2 = 8((6 \text{ in.})^2 + (3 \text{ in.})^2) = 360 \text{ in.}^2$

$\Sigma h^2 = 10((5.75 \text{ in.})^2 + (2.75 \text{ in.})^2) = 406.25 \text{ in.}^2$

$\Sigma d = \Sigma v^2 + \Sigma h^2 = 360 + 406.25 = 766.25 \text{ in.}^2$

$V_S = P/N = \dfrac{40 \text{ k}}{20} = 2 \text{ k}$

$V_M = \dfrac{Mh}{\Sigma d^2} = \dfrac{620 \text{ in.-k} \times 5.75 \text{ in.}}{766.25 \text{ in.}^2} = 4.65 \text{ k}$ (for a corner fastener)

$V_{tot} = 2.0 \text{ k} + 4.65 \text{ k} = 6.65 \text{ k}$

$H_S = 0$

$H_M = \dfrac{Mv}{\Sigma d^2} = \dfrac{620 \text{ in.-k} \times 6 \text{ in.}}{766.25 \text{ in.}^2} = 4.85 \text{ k}$

$H_{tot} = 0 + 4.85 \text{ k} = 4.85 \text{ k}$

$R_{tot} = \sqrt{V^2 + H^2} = \sqrt{(6.65)^2 + (4.85)^2} = 8.23 \text{ k}$

The maximum load on the critical fastener is 8.23 k

Plate dimensions

$M = 140 \text{ in.-k} + 40 \text{ k} \times (12 \text{ in.} - 5.75 \text{ in.}) = 390 \text{ in.-k}$

$S = \dfrac{390 \text{ in.-k}}{22 \text{ ksi}} = 17.73 \text{ in.}^3$

Referring to bracket plates table (AISCM, p. 4-88), a plate of $\frac{11}{16}$-in. thickness and 15-in. depth is needed (by interpolation).

Use 15 in. × $\frac{11}{16}$ in.

Example 5.9. Using the AISC tables, determine the maximum load on the fastener group shown.

204 STEEL DESIGN FOR ENGINEERS AND ARCHITECTS

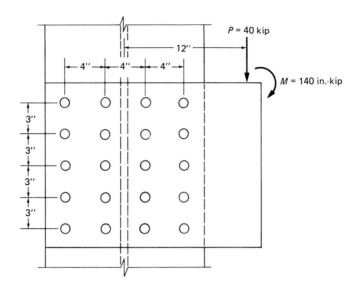

Solution. The table to be used is Table XVIII (AISCM, p. 4-69)

$$M_{tot} = 40 \text{ k} \times 12 \text{ in.} + 140 \text{ in.-k} = 620 \text{ in.-k}$$

$$\text{Equivalent } l = \frac{M}{P} = \frac{620 \text{ in.-k}}{40 \text{ k}} = 15.5 \text{ in.}$$

For $l = 14$ in. and $n = 5$,

$$C = 6.86$$

For $l = 16$ in. and $n = 5$,

$$C = 6.15$$

For $l = 15.5$ in. and $n = 5$,

$$C = 6.86 - \frac{1.5}{2}(6.86 - 6.15) = 6.33$$

$$R_v = \frac{P}{C} = \frac{40}{6.33} = 6.32 \text{ k}$$

Maximum load on critical fastener = 6.32 k.

Example 5.10. Determine P for the existing connection shown. The rivets are $\frac{7}{8}$-in. diameter A502 Grade 1.

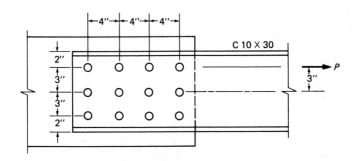

Solution. For C 10 × 30,

$$A = 8.82 \text{ in.}^2, \quad t_f = 0.436 \text{ in.}, \quad t_w = 0.673 \text{ in.}, \quad I_x = 103 \text{ in.}^4$$
$$\Sigma d^2 = \Sigma (h^2 + v^2) = 6(6 \text{ in.})^2 + 6(2 \text{ in.})^2 + 8(3 \text{ in.})^2 = 312 \text{ in.}^2$$
$$M = 3P$$

Maximum load occurs at a corner rivet.

$$V_M = \frac{Mh}{\Sigma d^2} = \frac{3P \times 6}{312} = 0.0577 P$$

$$H_M = \frac{Mv}{\Sigma d^2} = \frac{3P \times 3}{312} = 0.0288 P$$

$$V_S = 0$$

$$H_S = P/N = P/12 = 0.0833 P$$

$$R = \sqrt{V^2 + H^2}$$
$$= \sqrt{(0.0577 P)^2 + (0.0288 P + 0.0833 P)^2}$$
$$= 0.1261 P$$

Capacity of $\frac{7}{8}$-in. A502-1 rivets in shear

$$R_v = 10.5 \text{ k} \quad \text{(AISC, Table I-D)}$$

Capacity of one $\frac{7}{8}$-in. A502-1 rivet in bearing on 0.673-in. flange

From Table I-E, for 1-in. thickness (F_u = 58 ksi), allowable load = 60.9 k. Therefore, for 0.673-in. thickness, allowable load = 60.9 × 0.673 = 41 k.

Shear governs

$$P = \frac{10.5 \text{ k}}{0.1261} = 83.27 \text{ k}$$

Capacity of C 10 × 30

$$A_{net} = 8.82 \text{ in.}^2 - 3 \times 0.673 \text{ in.} \times \left(\frac{7}{8} \text{ in.} + \frac{1}{8} \text{ in.}\right) = 6.801 \text{ in.}^2$$

$$I_{net} = 103 \text{ in.}^4 - 0.673 \text{ in.} \times \left(\frac{7}{8} \text{ in.} + \frac{1}{8} \text{ in.}\right) \times (3 \text{ in.})^2 \times 2 = 90.9 \text{ in.}^4$$

$$S_{net} = \frac{90.9 \text{ in.}^4}{5 \text{ in.}} = 18.2 \text{ in.}^3$$

$P = 22 \times 8.82 = 194$ k

$P = 29 \times 0.85 \times 6.801 = 167.6$ k

$\therefore F_t = 29$ ksi

$$M = 3P, \quad F_b = 22.0 \text{ ksi}$$

$$\frac{P}{0.50 \times 58} + \frac{\frac{3P}{18.2}}{22.0} \leq 1.0 \quad \text{(AISCS H2)}$$

$0.006 P + 0.0075 P = 0.0135 P \leq 1.0$

$P \leq 74.1$ k

Channel capacity governs. Maximum load is 74.1 kips.

Example 5.11. Determine the size of A325-X high-strength bolts for this group of connectors.

BOLTS AND RIVETS 207

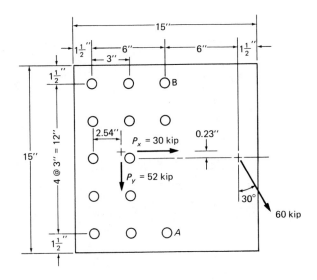

Solution. Locate the center of gravity with respect to the x and y axes.

$$\bar{x} = \frac{3 \times 6 \text{ in.} + 5 \times 3 \text{ in.}}{13} = 2.54 \text{ in.}$$

$$\bar{y} = \frac{3 \times 6 \text{ in.} + 3 \times 3 \text{ in.} - 2 \times 3 \text{ in.} - 3 \times 6 \text{ in.}}{13} = 0.23 \text{ in.}$$

$\Sigma v^2 = (5.77 \text{ in.})^2 \times 3 + (2.77 \text{ in.})^2 \times 3 + (0.23 \text{ in.})^2 \times 2$
$\qquad + (3.23 \text{ in.})^2 \times 2 + (6.23 \text{ in.})^2 \times 3 = 260.3 \text{ in.}^2$

$\Sigma h^2 = (2.54 \text{ in.})^2 \times 5 + (0.46 \text{ in.})^2 \times 5 + (3.46 \text{ in.})^2 \times 3 = 69.2 \text{ in.}^2$

$\Sigma d^2 = \Sigma v^2 + \Sigma h^2 = 260.3 \text{ in.}^2 + 69.2 \text{ in.}^2 = 329.5 \text{ in.}^2$

$P_x = 0.50 \times 60 \text{ k} = 30.0 \text{ k}$

$P_y = 0.866 \times 60 \text{ k} = 52.0 \text{ k}$

$M = 9.46 \text{ in.} \times 52.0 \text{ k} - 0.23 \text{ in.} \times 30.0 \text{ k} = 485 \text{ in.-k}$

$$V_s = \frac{52.0 \text{ k}}{13} = 4.0 \text{ k}$$

$$H_s = \frac{30.0 \text{ k}}{13} = 2.3 \text{ k}$$

Try Bolt A:

$$V_m = \frac{485 \text{ in.-k}}{329.5 \text{ in.}^2} \times 3.46 \text{ in.} = 5.09 \text{ k}$$

$$H_m = \frac{485 \text{ in.-k}}{329.5 \text{ in.}^2} \times 6.23 \text{ in.} = 9.17 \text{ k}$$

$V_s = 4.0$ kip
$V_m = 5.09$ kip
$H_s = 2.30$ kip $H_m = 9.17$ kip
○ A

$$R = \sqrt{(2.30 - 9.17)^2 + (5.09 + 4.0)^2} = 11.4 \text{ k}$$

Try Bolt B:

$$V_m = \frac{485 \text{ in.-k}}{329.5 \text{ in.}^2} \times 3.46 \text{ in.} = 5.09 \text{ k}$$

$$H_m = \frac{485 \text{ in.-k}}{329.5 \text{ in.}^2} \times 5.77 \text{ in.} = 8.49 \text{ k}$$

$V_s = 4.0$ kip
$V_m = 5.09$ kip
$H_s = 2.3$ kip $H_m = 8.49$ kip
○ B

$$R = \sqrt{(8.49 + 2.3)^2 + (5.09 + 4.0)^2} = 14.1 \text{ k (governs)}$$

A bolt of $\frac{7}{8}$ in. diameter is needed ($R_v = 18.0$ k).

Rivets and bearing bolts subjected to both shear and tension shall be designed such that the computed shear stresses f_v do not exceed the allowable values given in Table 5.1 (AISCS Table J3.2) and the computed tensile stresses f_t acting on the nominal body area A_b do not exceed the values computed in Table 5.2 (AISCS Table J3.3).

When allowable stresses are increased for wind or seismic loads, the constants in the equations in Table 5.2 shall be increased by $\frac{1}{3}$, but the coefficient applied to f_v shall not be increased.

For the case when A325 and A490 bolts are used in SC connections, the

Table 5.2. Allowable Tension Stress F_t for Fasteners in Bearing-type Connections (AISCS Table J3.3; reprinted with permission)

Description of Fasteners	Threads Included in Shear Plane	Threads Excluded from Shear Plane
A307 bolts	$26 - 1.8 f_v \leq 20$	
A325 bolts	$\sqrt{(44)^2 - 4.39 f_v^2}$	$\sqrt{(44)^2 - 2.15 f_v^2}$
A490 bolts	$\sqrt{(54)^2 - 3.75 f_v^2}$	$\sqrt{(54)^2 - 1.82 f_v^2}$
Threaded parts, A449 bolts over 1½-in. dia.	$0.43 F_u - 1.8 f_v \leq 0.33 F_u$	$0.43 F_u - 1.4 f_v \leq 0.33 F_u$
A502 Gr. 1 rivets	$30 - 1.3 f_v \leq 23$	
A502 Gr. 2 rivets	$38 - 1.3 f_v \leq 29$	

maximum allowable shear stress shall be multiplied by the reduction factor

$$\left(1 - \frac{f_t A_b}{T_b}\right)$$

where f_t is the average tensile stress due to the direct load applied to all of the bolts, T_b is the pretension load of the bolt given in Table 5.3 (AISCS Table J3.7), and A_b is the area of one bolt taken on the nominal shank. When allowable stresses are increased for wind or seismic loads, the reduced allowable shear stress shall be increased by one-third.

Table 5.3. Minimum Pretension for Fully-tightened Bolts, kips[a] (AISCS Table J3.7; reprinted with permission)

Bolts Size, in.	A325 Bolts	A490 Bolts
½	12	15
⅝	19	24
¾	28	35
⅞	39	49
1	51	64
1⅛	56	80
1¼	71	102
1⅜	85	121
1½	103	148

[a] Equal to 0.70 of minimum tensile strength of bolts, rounded off to nearest kip, as specified in ASTM specifications for A325 and A490 bolts with UNC threads.

Example 5.12. For the connections shown, determine P_{max} when

a) 10-$\frac{7}{8}$-in. A502 Grade 1 rivets are used (existing connection).
b) 10-$\frac{7}{8}$-in. A490 bolts are used with threads not excluded from the shear plane.

Note: Disregard the effect of prying. For prying design, see Sect. 7.5.

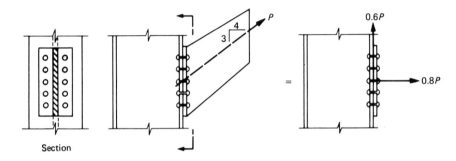

Solution.

$$P_h = \frac{4}{5} P = 0.8\, P$$

$$P_v = \frac{3}{5} P = 0.6\, P$$

$$f_v = P_v/A = \frac{0.6\, P}{10 \times 0.601} = 0.100\, P$$

$$f_t = P_h/A = \frac{0.8\, P}{10 \times 0.601} = 0.133\, P$$

a) $\quad F_t = 30 - 1.3\, f_v \leq 23$ ksi   (AISCS Table J3.3)

$0.133\, P = 30 - 1.3(0.100\, P) \leq 23$

$P = 114.1$ k

$f_v = 0.100\, P = 11.4$ ksi < 17.5 ksi   (AISCS Table J3.2)   ok

$f_t = 0.133\, P = 15.2$ ksi < 23 ksi   ok

b) $\quad F_t = \sqrt{(54)^2 - 3.75 f_v^2}$

$0.133\, P \leq \sqrt{(54)^2 - 3.75(0.1P)^2}$

BOLTS AND RIVETS   211

$$P = 230 \text{ k}$$
$$f_v = 0.1P = 23 \text{ ksi} < 28 \text{ ksi} \quad \text{ok}$$
$$f_t = 0.133P = 30.6 \text{ ksi} = F_t \quad \text{ok}$$

Example 5.13. Determine $P_{\max}$ for the bolt configuration given in Example 5.12 there are 10-$\frac{7}{8}$ in. diameter A490-SC bolts.

Solution.

$$A_b = 0.601 \text{ in.}^2$$

$$f_t = \frac{0.8\,P}{10 \times 0.601} = 0.133\,P$$

$$T_b = 49 \text{ k} \quad \text{(AISCS Table J3.7)}$$

$$\text{Reduction factor} = 1 - \frac{0.133\,P \times 0.601}{49} = 1 - 0.00163\,P$$

$$F_v = 21 \text{ ksi} \quad \text{(AISCS Table J3.2)}$$

$$(1 - 0.00163\,P) \times F_v \times A_b \times N = P$$

$$(1 - 0.00163\,P) \times 21 \times 0.601 \times 10 = P$$

$$P_{\max} = 104.5 \text{ k}$$

Example 5.14. Determine whether the connection shown is satisfactory using $\frac{7}{8}$-in. A307 bolts, each having an area of 0.6 in.^{2}

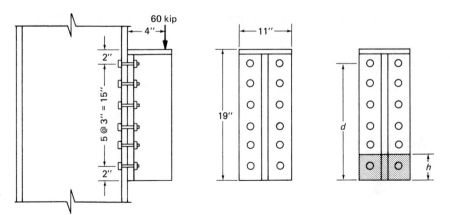

Solution. The centroid could be located by setting up a quadratic equation in terms of h, but is more conveniently obtained by trial and error. First assume the center of gravity of compression and tension areas at a distance h from the bottom, usually $h = (\frac{1}{6} \text{ to } \frac{1}{7})d$.

$$\text{Assume } h = 3 \text{ in.}$$

Check to see if center of gravity is sufficiently near assumed location.

Moment of compression area = 3 in. × 11 in. × 1.5 in. = 49.5 in.^{3}

$$\begin{aligned}
\text{Moment of tension area} &= (2 \times 0.6 \times 14) + (2 \times 0.6 \times 11) \\
&\quad + (2 \times 0.6 \times 8) + (2 \times 0.6 \times 5) \\
&\quad + (2 \times 0.6 \times 2) \\
&= 48 \text{ in.}^3
\end{aligned}$$

$$48.0 \text{ in.}^3 \approx 49.5 \text{ in.}^3$$

Say assumed center of gravity is ok.

Calculate moment of inertia of tension and compression areas.

$$I = \Sigma(I_0 + Ad^2)$$

For bolt areas, I_0 can be neglected.

$$I = \frac{11\,(3)^3}{12} + (33 \times 1.5^2) \times (2 \times 0.6 \times 14^2) + (2 \times 0.6 \times 11^2)$$
$$+ (2 \times 0.6 \times 8^2) + (2 \times 0.6 \times 5^2) + (2 \times 0.6 \times 2^2) = 591 \text{ in.}^4$$

Shear force on each bolt

$$\frac{F}{N_b} = \frac{60 \text{ k}}{12} = 5.0 \text{ k/bolt}$$

Shear stress on each bolt

$$\frac{P}{A} = \frac{5.0 \text{ k}}{0.6 \text{ in.}^2} = 8.33 \text{ ksi}$$

Tension stress on extreme top bolt

$$\frac{Mc}{I} = \frac{(60 \text{ k} \times 4 \text{ in.}) \times 14 \text{ in.}}{591 \text{ in.}^4} = 5.69 \text{ ksi}$$

For A307 bolts,

$$F_t = 26.0 - 1.8 f_v \leq 20 \text{ ksi}$$
$$= 26.0 - 1.8 (8.33) = 11.0 \text{ ksi} < 20.0 \text{ ksi}$$

Connection is satisfactory, because $f_v = 8.33$ ksi $< F_v = 10.0$ ksi and $f_t = 5.69$ ksi $< F_t = 11.0$ ksi.

Example 5.15. Check to see if this connection is adequate with $\frac{3}{4}$-in.-diameter connectors and

a) using A502 Grade 1 rivets (existing connection).
b) using A325-N bolts (threads not excluded from the shear plane).
c) using A325-SC bolts.

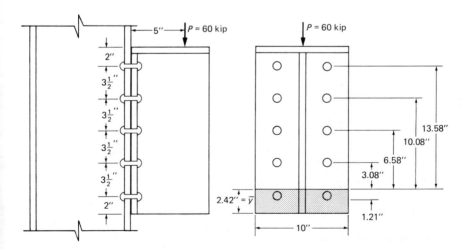

Solution. Assume neutral axis of connection ($\bar{y}$) as 2.42 in. from bottom of bracket.

214 STEEL DESIGN FOR ENGINEERS AND ARCHITECTS

Check assumption

$$\text{Moment of compression area} = 10 \text{ in.} \times 2.42 \text{ in.} \times 1.21 \text{ in.} = 29.28 \text{ in.}^3$$
$$\text{Moment of tension area} = 2 \times 0.442 \text{ in.}^2 \times (3.08 \text{ in.} + 6.58 \text{ in.}$$
$$+ 10.08 \text{ in.} + 13.58 \text{ in.})$$
$$= 29.45 \text{ in.}^3$$

Say assumption of neutral axis is ok.

$$I = \frac{10 \, (2.42)^3}{3} + 2 \times 0.442 \times (3.08^2 + 6.58^2 + 10.08^2 + 13.58^2)$$

$$= 346.74 \text{ in.}^4$$

$$f_v = \frac{60}{10 \times 0.442} = 13.57 \text{ ksi}$$

$$f_{t\text{top}} = \frac{(60 \times 5) \times 13.58}{346.74} = 11.75 \text{ ksi}$$

a) For A502-1 rivets,

$$F_v = 17.5 \text{ ksi} > 13.57 \text{ ksi} \quad \text{ok}$$
$$F_t = 30.0 - 1.3 \, (13.57) = 12.4 \text{ ksi} > 11.75 \text{ ksi} \quad \text{ok}$$

A502-1 rivets are adequate for loading shown.

b) For A325-N bolts,

$$F_v = 21 \text{ ksi} > 13.57 \text{ ksi} \quad \text{ok}$$
$$F_t = \sqrt{(44)^2 - 4.39 \, (13.57)^2} = 33.6 \text{ ksi} > 11.75 \text{ ksi} \quad \text{ok}$$

A325-N bolts are adequate for loading shown.

c) For A325-SC bolts.

Based on the discussion given in Commentary C-J3.6, the applied shear force is taken by the compressive area of the connection bearing against the flange of the column; thus, none of the bolts in tension carries any shear. The tensile stress in a bolt due to the pretension force is $T_b \times A_b = 28 \text{ k} \times 0.4418 \text{ in.}^2 =$

12.37 ksi. On the other hand, the maximum applied tensile stress was determined to be 11.75 ksi (see above). Consequently, because the maximum applied tensile stress is smaller than the pretension stress in the bolt, no change in bolt stress can occur. The tensile stress in the bolt can increase only if the applied stress becomes greater than the pretension stress; when this happens, separation will occur between the two plates that were originally in contact in the tensile zone.

Therefore, for this loading, the A325-SC bolts are adequate.

5.5 FASTENERS FOR HORIZONTAL SHEAR IN BEAMS

In a built-up member or a plate girder subjected to transverse loading, there is a tendency for the connected elements to slide horizontally if they are not fastened together; see Fig. 2.7. If the elements are connected continuously, horizontal shearing stress will develop in the connection. This shear, measured per unit length for the entire width, is called *shear flow* and is expressed as

$$q = \frac{VQ}{I} \qquad (5.1)$$

where V is the vertical shear force, I is the total moment of inertia of the member, and Q is the moment of the areas on one side of the sliding interface with respect to the centroid of the section. The spacing of fasteners is determined by dividing the shearing capacity of the fasteners to be used by the shear flow q. The maximum spacing between fasteners must also satisfy AISCS E4 and D2, as specified in AISCS B10. Where the component of a built-up compression member consists of an outside plate, the maximum spacing between bolts and rivets connecting the plate to the other member shall not exceed the thickness of the outside plate times $127/\sqrt{F_y}$ nor 12 in. When the fasteners are staggered, the maximum spacing on each gage line shall not exceed the thickness of the outside plate times $190/\sqrt{F_y}$ nor 18 in. (see Fig. 3.2). The spacing of fasteners connecting two rolled shapes in contact with each other shall not exceed 24 in. (AISCS E4).

The longitudinal spacing of bolts and rivets connecting a plate to a rolled shape or to another plate in a built-up tension member shall not exceed 24 times the thickness of the thinner plate nor 12 in. for painted members or unpainted members not subject to corrosion; the maximum spacing is 14 times the thinner of the two plates or 7 in. (whichever is smaller) for unpainted members of weathering steel subject to atmospheric corrosion (see Fig. 3.2). When two or more shapes in contact with one another form a built-up tension member, the longitudinal spacing of fasteners shall not exceed 24 in. (AISCS D2).

Example 5.16. Calculate the theoretical spacing of $\frac{3}{4}$-in.-diameter A325-N bolts for the beam shown. $I_{tot} = 461$ in.^{4}

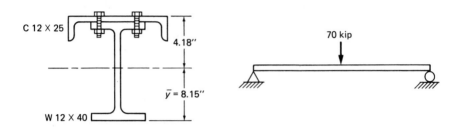

Solution.

$$A_{channel} = 7.35 \text{ in.}^2$$

$$Q = Ad = 7.35 \text{ in.}^2 \times (4.18 \text{ in.} - 0.674 \text{ in.}) = 25.77 \text{ in.}^3$$

$$V_{max} = 35 \text{ k}$$

$$q = \frac{VQ}{I} = \frac{35 \text{ k} \times 25.77 \text{ in.}^3}{461 \text{ in.}^4} = 1.96 \text{ k/in.}$$

Shear capacity of $\frac{3}{4}$-in. A325-N bolts

$$R_v = 9.3 \text{ k}$$

Spacing of bolts

$$\frac{N_b \times R_v}{q} = \frac{2 \times 9.3 \text{ k}}{1.96 \text{ k/in.}} = 9.49 \text{ in.}$$

Use two rows of bolts spaced at 9 in. $< s_{max} = 24$ in. ok
(Note: No hole deduction required in this case).

Example 5.17. For the beam shown, what is the required spacing of $\frac{3}{4}$-in. A490-N bolts at a section where the external shear is 100 kips?

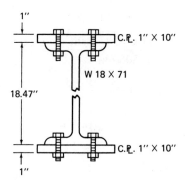

Solution. For W 18 × 71,

$$A = 20.8 \text{ in.}^2, \quad d = 18.47 \text{ in.}, \quad b_f = 7.64 \text{ in.},$$
$$t_f = 0.81 \text{ in.}, \quad I = 1170 \text{ in.}^4$$

Check requirements of AISCS B10:

a) Hole deduction:

$$A_{fg} = (7.64 \times 0.81) + (10 \times 1) = 16.19 \text{ in.}^2$$
$$A_{fn} = 16.19 - 2 \times (\tfrac{3}{4} + \tfrac{1}{8})(0.81 + 1.0) = 13.02 \text{ in.}^2$$
$$0.5 \times 58 \times 13.02 = 378 \text{ k} > 0.6 \times 36 \times 16.19 = 350 \text{ k}$$

Therefore, no deduction for bolt holes is required.

b) Maximum cover plate area:

$$\text{Total flange area} = 2 \times 16.19 = 32.38 \text{ in.}^2$$
$$\text{Total cover plate area} = 2 \times 10 \times 1 = 20 \text{ in.}^2$$
$$0.7 \times 32.38 = 22.67 \text{ in.}^2 > 20 \text{ in.}^2 \quad \text{ok}$$

$$I_{\text{tot}} = 1170 \text{ in.}^4 + \left(\frac{10 \text{ in. } (1 \text{ in.})^3}{12}\right) \times 2 + 2(1 \text{ in.} \times 10 \text{ in.} \times (9.74 \text{ in.})^2)$$
$$= 3069 \text{ in.}^4$$

$$Q = Ad = 1 \text{ in.} \times 10 \text{ in.} \times 9.74 \text{ in.} = 97.4 \text{ in.}^3$$

$$q = \frac{VQ}{I} = \frac{100 \text{ k} \times 97.4 \text{ in.}^3}{3069 \text{ in.}^4} = 3.17 \text{ k/in.}$$

Shear capacity of $\frac{3}{4}$-in. A490-N bolts

$$R_v = 12.4 \text{ k}$$

Spacing of bolts

$$\frac{N_b \times R_v}{q} = \frac{2 \times 12.4 \text{ k}}{3.17 \text{ k/in.}} = 7.82 \text{ in.}$$

Use two rows of bolts spaced at $7\frac{1}{2}$ in. $< s_{\max} = 12$ in. ok

Example 5.18. For the section shown, what is the required spacing of $\frac{3}{4}$-in. A325-N bolts if external shear = 150 kips?

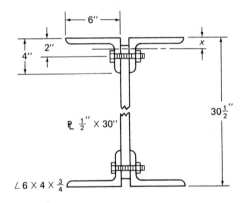

Solution. For 6 × 4 × $\frac{3}{4}$ in. angle,

$A = 6.94$ in.^{2}

$I = 8.68$ in.^{4}

$x = 1.08$ in.

$I = \Sigma(I_0 + Ad^2)$

$$= \frac{0.5 \text{ in.} \times (30)^3}{12} + 4(8.68 \text{ in.}^4) + 4(6.94 \text{ in.}^2)(15.25 \text{ in.} - 1.08 \text{ in.})^2$$

$= 6734$ in.^{4}

$Q = Ad = 2 \times 6.94 \text{ in.}^2 \times (15.25 \text{ in.} - 1.08 \text{ in.})$

$= 196.7 \text{ in.}^3$

$q = \dfrac{VQ}{I} = \dfrac{150 \text{ k} \times 196.7 \text{ in.}^3}{6734 \text{ in.}^4}$

$= 4.38 \text{ k/in.}$

Capacity of $\frac{3}{4}$-in. A325-N bolts in double shear

$$R_v = 18.6 \text{ k}$$

Spacing of bolts

$$\dfrac{R_v}{q} = \dfrac{18.6 \text{ k}}{4.38 \text{ k/in.}} = 4.25 \text{ in.}$$

Space bolts at $4\frac{1}{4}$ in. center to center.

5.6 BLOCK SHEAR FAILURE

Tests have shown that when a beam with a top coped flange has a high-strength bolted end connection, as shown in Fig. 5.5, failure may occur by tearing out of the shaded portion along a line through the holes. This type of failure, commonly referred to as *block shear failure*, can be attributed to a combination of shear and tensile stresses acting on the net shear and tensile areas, respectively.

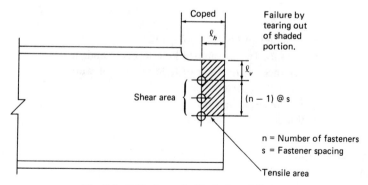

Fig. 5.5. Critical area for block shear failure.

According to AISCS J4, the allowable shear stress F_v acting on the net shear area A_v is given in (J4-1):

$$F_v = 0.30 F_u \tag{5.2}$$

and the allowable tensile stress F_t acting on the net tensile area A_t is given in (J4-2):

$$F_t = 0.50 F_u \tag{5.3}$$

The sum of the resistances due to shear and tension constitutes the total resistance to block shear R_{BS}:

$$R_{BS} = A_v F_v + A_t F_t \tag{5.4}$$

In general,

$$A_v = \left[l_v + (n-1)s - (n-1)d_h - \frac{d_h}{2} \right] t_w \tag{5.5a}$$

or

$$A_v = \left[l_v + (n-1)(s - d_h) - \frac{d_h}{2} \right] t_w \tag{5.5b}$$

and

$$A_t = \left(l_h - \frac{d_h}{2} \right) t_w \tag{5.6}$$

where d_h = diameter of the bolt hole, in.
l_h = distance from center of hole to beam end, in.
l_v = distance from center of hole to edge of web, in.
n = number of fasteners
s = fastener spacing, in.
t_w = thickness of beam web, in.

Using eqns. (5.2) to (5.6), we can write the total resistance to block shear as follows:

R_{BS} = total resistance to block shear, kips

$$= 0.3\, A_v F_u + 0.5\, A_t F_u$$

$$= 0.3 \left[l_v + (n-1)(s - d_h) - \frac{d_h}{2} \right] t_w F_u$$

$$+ 0.5 \left[l_h - \frac{d_h}{2} \right] t_w F_u \qquad (5.7)$$

Table I-G on p. 4-8 of the AISCM tabulates the coefficients in the above equation based on standard holes and a 3-in. fastener spacing. Note that in the case when the flange of the beam is not coped, experiments have shown that block shear does not occur.

Block shear failure is also evident in similar types of connections (see Figs. C-J4.2, C-J4.3, and C-J4.4, which are reproduced here as Fig. 5.6). As above, the sum of the resistance to shear and to tension gives the total resistance to this type of failure.

5.7 AISCM DESIGN TABLES

The tables in AISCM, pp. 4-3 through 4-8, can be used to determine areas, allowable tensile, shear, and bearing loads for rivets and bolts most commonly used in steel construction.

Tension Tables I-A and I-B list the allowable tensile stresses for various bolts, rivets, and grades of steels used to produce threaded fasteners and the allowable tensile loads for $\frac{5}{8}$- to $1\frac{1}{2}$-in-diameter fasteners of those materials.

Table I-C lists the ASTM Specifications for bolts, studs, and threaded round stock.

Shear Table I-D lists the allowable shear stresses and allowable loads for $\frac{5}{8}$- to $1\frac{1}{2}$-in-diameter rivets and bolts made from the materials mentioned above. To use these tables, the designer must locate the allowable load under the column for the fastener diameter in question and referenced to the row corresponding to the material of the fastener, type of connection, and the hole type. Note that the allowable loads for double shear (D) are exactly twice those for single shear (S).

Bearing Table I-E for F_u equal to 58, 65, 70, and 100 ksi of the connected part's material give the allowable bearing loads, in kips, of $\frac{3}{4}$, $\frac{7}{8}$, and 1-in.-diameter fasteners on material thicknesses from $\frac{1}{8}$ to 1 in., based on the fastener spacing given in the notes at the bottom of the table. As the bearing values in the tables are limited by the double shear capacity of A490-X bolts, some values in the tables are not listed. To determine these values, the designer must mul-

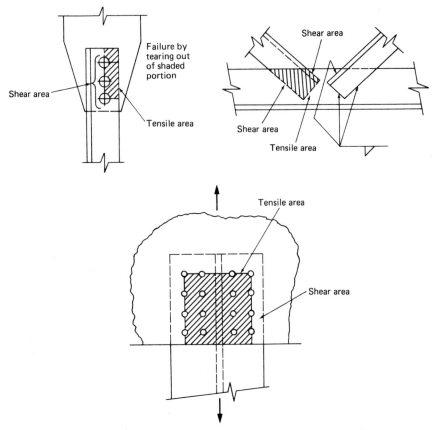

Fig. 5.6. Critical area for block shear failure in similar connections (AISCS Figs. C-J4.2, C-J4.3, C-J4.4; reprinted with permission).

tiply the value corresponding to 1-in.-thick material by the actual thickness of the connected material.

Table I-F gives the allowable bearing loads as a function of the bolt edge distance.

The block shear capacity for members with their top flange coped is given in Table I-G (see Sect. 5.6). For cases which are not specifically covered in the table, the equation provided at the bottom of the table (which is the same as eqn. (5.6)) can be utilized. Given are values of C_1 and C_2; the resistance to block shear in kips is equal to $(C_1 + C_2)F_u t_w$, which should be greater than or equal to the beam reaction.

"Framed Beam Connections," Tables II-A, II-B, and II-C, in AISCM, pp.

4-9 through 4-22, are provided to help determine the number of bolts required at ends of simply supported beams framing into other members when the reactions and the fastener type are known. In the general groupings of fastener types, under the column of the selected fastener diameter, the designer must locate the allowable load, in kips, that is equal to or slightly greater than the required connection capacity. Proceeding to the left from that load, the designer can determine the required number of bolts (n) and the length of the framing angles (L and L'). Along the top of the tables, the suggested thickness of these angles is tabulated according to the bolt type and diameter. If the length L (or L') exceeds the depth of the beam minus the flange thicknesses and fillets, the diameter of the bolts should be increased or the type changed to a stronger one to reduce the number of bolts required.

If and when the tabulated allowable load in Tables II-A and II-B is followed by a superscript, the allowable net shear on the framing angles must be checked. From Table II-C, the designer can determine the allowable load on two angles for a given thickness (t) and length (L), as determined from Tables II-A and II-B. If the tabulated load is greater than the beam reaction, then the angles are acceptable. Otherwise, the angle thickness should be increased. The designer should note that this table is for A36 double angles only. It is also important to note that a thickness limitation of $\frac{5}{8}$ in. for the framing angles is given in the tables to assure the flexibility of the connection.

For eccentric loads on fastener groups, the tables in AISCM, pp. 4-57 to 4-69, are extremely useful for determining the maximum load on the critical fastener in the group for a given vertical load. Using l, the value of a coefficient C can be obtained from the appropriate table. If the capacity of one fastener (R_v) is known, the allowable load P can be determined by multiplying R_v times the coefficient C ($P = C \times R_v$). Conversely, the required shear capacity of the fastener can be determined from $R_v = P/C$ when P is known and C is obtained from the tables. The designer should note that these "Eccentric Loads on Fastener Groups" tables are based on the ultimate method of analysis and that the longhand approach, using the elastic method described in AISCM, p. 4-59, will give more conservative results; however, it can easily be applied to situations where the load is not vertical and the bolt configuration is not regular. Examples using these tables were done previously in Sect. 5.4.

Example 5.19. For a W 36 × 260 beam with an end reaction of 390 kips, select the number of $\frac{7}{8}$-in.-diameter A490-X bolts required for framing the beam. Assume A572 Gr 50 steel, with $F_u = 65$ ksi for the beam and A36 steel connection angles.

Solution. For W 36 × 260,

$$t_w = 0.840 \text{ in.}$$

Using Table II-A under bolt type A490-X and bolt diameter $\frac{7}{8}$ in., select $n = 9$ corresponding to a shear force of 433 k.

From the table, suggested angle thickness is $\frac{5}{8}$ in. and length is $26\frac{1}{2}$ in.

Check bearing of nine bolts on 0.840-in. web for $F_u = 65$ ksi.

From Table I-E, for bolt diameter of $\frac{7}{8}$ in., bolt spacing of 3 in., and $F_u = 65$ ksi, allowable bearing load on 1-in.-thick material = 68.3 k.

Allowable bearing load = 9 × 0.84 in. × 68.3 k = 516.3 k > 390 k ok

Check bearing of 9 bolts on two $\frac{5}{8}$-in. angles with $F_u = 58$ ksi.

Assume the typical value of $1\frac{1}{4}$-in. vertical edge distance for the top bolt and a 3-in. fastener spacing (see the figure at the top of Table II). Since $1\frac{1}{4}$ in. is less than $1\frac{1}{2}d$ ($1\frac{5}{16}$ in.), use Table I-E to determine a value of 36.3 k/in. bearing value for the top bolt. The value for the remaining eight bolts is 60.9 k/in. Therefore, the capacity of the connection angles is

$$(\tfrac{5}{8}) \times [36.3 + (8 \times 60.9)] = 327.2 \text{ k} > 390/2 = 195 \text{ k} \quad \text{ok}$$

Because the beam is not coped, web tear out need not be checked.

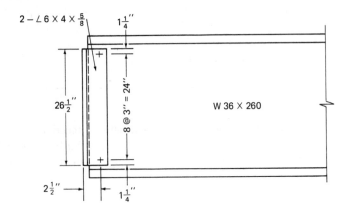

Example 5.20. Select the number of $\frac{3}{4}$-in.-diameter high-strength A325-N bolts needed for framing of a W 18 × 55 beam carrying a total load, symmetrically placed, of 100 kips. Assume A36 steel and that the beam will be coped.

Solution.

$$\text{Reaction of W 18} \times 55 = \frac{100 \text{ k}}{2} = 50 \text{ k}$$

For W 18 × 55,

$$t_w = 0.390 \text{ in.}$$

From Table II-A, for A325-N bolt, $\frac{3}{4}$-in. diameter, and an allowable shear of 55.7 k (>50 k required), select $n = 3$, angle length $l = 8\frac{1}{2}$ in., and angle thickness = $\frac{5}{16}$ in.

Check shear capacity of two 6 × 4 × $\frac{5}{16}$ in. angles, $8\frac{1}{2}$ in. long.

From Table II-C, allowable load on angle if $\frac{3}{4}$ in. bolts are used is 65.9 k > 50 k ok.

Note: since the 55.7 k value in Table II-A has no superscript, it is not required to check the shear capacity of the angle for $F_y = 36$ ksi.

Check bearing of bolts on beam web:
  Again, assume $1\frac{1}{4}$-in. vertical edge distance and 3-in. fastener spacing. Since $1\frac{1}{4}$ in. is greater than $1\frac{1}{2}d$ ($1\frac{1}{8}$ in.), read 52.2 k/in. from Table I-F for all three bolts. Therefore, the allowable load is

$$3 \times 52.2 \times (0.390) = 61.1 \text{ k} > 50 \text{ k} \quad \text{ok}$$

Check bearing of bolts on connection angles:

$$3 \times 52.2 \times (0.3125) = 48.9 \text{ k} > \frac{50}{2} = 25 \text{ k} \quad \text{ok}$$

Check web tearout (block shear failure):
  From above, $l_v = 1\frac{1}{4}$ in. Using 6 in. × 4 in. framing angles, determine $l_h = 4 - (4 - 2\frac{1}{2}) - \frac{1}{2} = 2$ in. Note that $2\frac{1}{2}$ in. is the usual gage for a 4-in. angle leg and that $\frac{1}{2}$ in. is the usual clearance setback.

From Table I-G, select $C_1 = 1.38$ for $l_v = 1\frac{1}{4}$ in. and $l_h = 2$ in. Also, select $C_2 = 0.99$ for $n = 3$ bolts with a diameter of $\frac{3}{4}$ in.

Web tearout resistance of beam = $(C_1 + C_2)F_u t_w$

$(1.38 + 0.99)(58)(0.390) = 53.6 \text{ k} > 50 \text{ k}$ ok

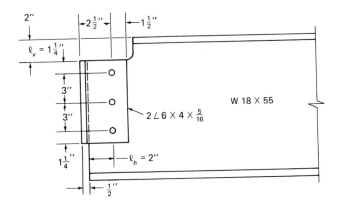

PROBLEMS TO BE SOLVED

5.1 through 5.4. Calculate the capacity of the tension members and connections shown.

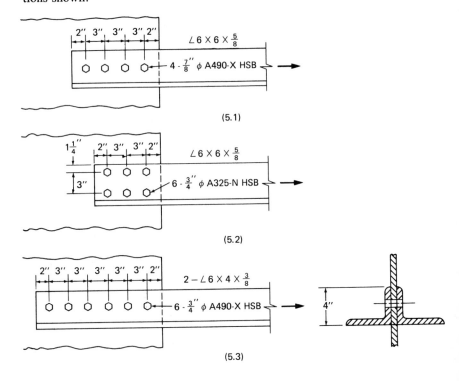

(5.1)

(5.2)

(5.3)

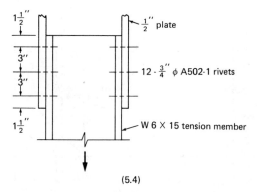

(5.4)

5.5. Determine the stagger s such that the loss due to the holes is equal to the loss of two holes only. Also determine the number of $\frac{7}{8}$-in. ϕ A325-SC bolts required to carry the load.

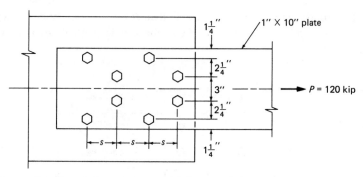

5.6. Determine P and the minimum thickness of the plate for the connection shown. a) Use the elastic method. b) Use AISCM design tables.

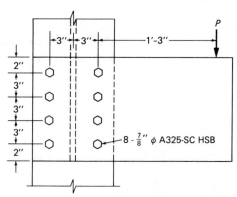

5.7. In the existing construction shown, A502-1 rivets of $\frac{7}{8}$-in. diameter were used. Inspection revealed that the rivets denoted by A are not usable. Determine the allowable load for the connection with only five good rivets.

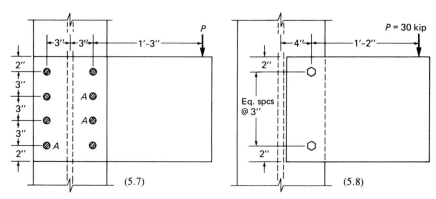

5.8. Determine the number of $\frac{3}{4}$-in. ϕ A325-X high-strength bolts required to carry the 30-k load shown. a) Use the elastic method. b) Use AISCM design tables.

5.9. Determine the maximum allowable load P for the connection shown. Assune six $\frac{3}{4}$-in. ϕ A490-SC bolts. a) Use the elastic method. b) Use AISCM design tables.

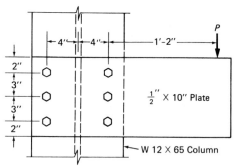

5.10. For the connection shown, determine the maximum allowable load P if a) the load passes through the centroid of the bolt group and b) the load passes through the second line of bolts from the top. Assume eight 1-in.-diameter A325-X bolts. Neglect the prying effect.

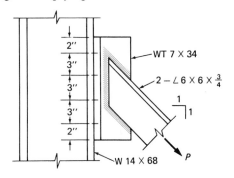

5.11. For the connection shown, determine the maximum allowable load P, assuming eight 1-in.-diameter A325-X bolts. Neglect the prying effect.

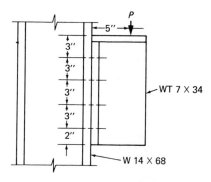

5.12 and 5.13. Calculate the maximum allowable spacing of $\frac{7}{8}$-in. ϕ A490-X bolts for the beams shown. Maximum shear force in the beams is 160 kips. F_y = 50 ksi.

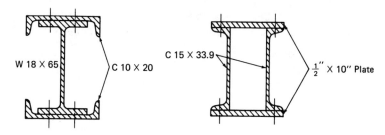

5.14. A W 27 × 84 is reinforced on the bottom flange with a $\frac{1}{2}$ × 9 in. plate to carry the loading shown. Determine a) the required spacing of $\frac{3}{4}$-in.-diameter A325-N bolts to connect the plate adequately to the wide flange and b) the required number of $\frac{3}{4}$-in.-diameter A325-SC bolts in the termination region. Use A572 Gr 50 steel.

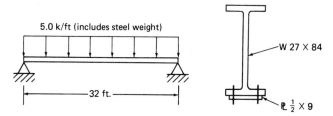

5.15. A W 18 × 65 beam carrying a reaction of 75 kips frames into a W 14 × 68 column flange. Assuming $\frac{3}{4}$-in. ϕ A490-X bolts and F_y = 36 ksi, completely design and sketch the framed beam connection, using AISC design tables.

5.16. Rework Problem 5.15, assuming that the top flange of the W 18 × 65 beam is coped to frame into the web of a W 24 × 62 girder.

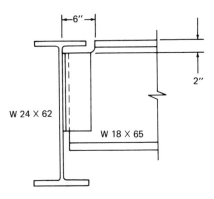

5.17. A W 24 × 76 beam supports a concentrated load of 300 kips at one of its third points. Assuming 50-ksi steel for the beam, 36-ksi steel for the framing angles, and $\frac{7}{8}$-in. ϕ A325-X bolts, design the end connection for both ends of the beam. The beam need not be coped. Use same diameter bolts for both ends.

6
Welded Connections

6.1 WELDED CONNECTIONS

Welding is the joining of two pieces of metal by the application of heat or through fusion. Most welds are of the fusion type, consisting of the shielding arc welding process. The surfaces of the members to be connected are heated, allowing the parts to fuse together, usually with the addition of another molten metal. When the molten metal solidifies, the bond between the members is completed.

Arc welding is performed by either an electric arc or a gas process. Shielding is added to control the molten metal and improve the weld properties by preventing oxidation and is usually provided by slag around the electrode used for joining the metals.

Weld symbols used on structural drawings have been developed by the American Welding Society and are reprinted in many manuals, including the AISCM (p. 4-155). Any combination of symbols can be used to identify any set of conditions regarding the weld joint. Included in the information is weld type, size, length, and special processes.

6.2 WELD TYPES

The different types of welds are fillet, penetration (groove), plug, and slot welds, the most common being fillet and penetration welds.

Fillet welds are triangular in shape, with each leg connected along the surface of one member. When the hypotenuse is concave, the outer surface is in tension after shrinkage occurs and often cracks. The outer surfaces of convex-shaped welds are in compression after shrinkage and usually do not crack. For this reason, convex fillet welds are preferred (see Fig. 6.2).

The fillet weld size is defined as the side length (leg) of the largest triangle that can be inscribed within the weld cross section. In most cases, the legs are

232 STEEL DESIGN FOR ENGINEERS AND ARCHITECTS

Fig. 6.1. Crown Hall, Illinois Institute of Technology campus, Chicago.

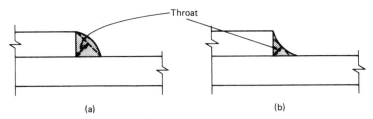

Fig. 6.2. Types of fillet welds: (a) convex fillet weld; (b) concave fillet weld.

of equal length. Welds with legs of different lengths are less efficient than those with equal legs (see Fig. 6.3).

Groove welds are used to join the ends of members that align in the same plane. They are commonly used for column splices, butting of beams to columns, and joining of flanges of plate girders (see Fig. 6.4).

If a groove weld is of greater thickness than the members joined, the weld is said to have *reinforcement*. This reinforcement adds little strength to the weld

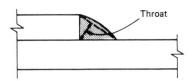

Fig. 6.3. Unequal leg fillet weld.

WELDED CONNECTIONS   233

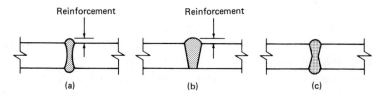

Fig. 6.4. Types of groove welds: (a) square groove weld; (b) single-V groove weld; (c) double-V partial penetration groove weld.

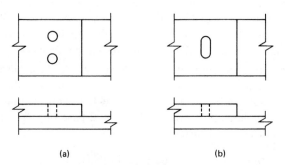

Fig. 6.5. Plug and slot welds: (a) plug weld; (b) slot weld.

but has the advantage of ease of application. This reinforcement, however, is not advantageous for connections subjected to fatigue. Tests have shown that stress concentrations develop in the reinforcement and contribute to early failure. For this condition, reinforcement is added and then ground flush with the joined members.

Plug and slot welds are formed in openings in one of two members and are joined to the top of the other member. Occasionally they may be used to "stitch" together cover plates to another rolled section. Another common practice is to add plug welds when the fillet weld is not of adequate length (see Fig. 6.5).

6.3 INSPECTION OF WELDS

Failure of welds usually originates at an existing crack or imperfection caused by shrinkage, poor workmanship, or slag inclusion. For this reason, it is advantageous to grind smooth the surface of welds and inspect for imperfections. Inspection of welds for cracks can be done in one of the following ways.

Use of dye penetrants. The surface of the weld, after being cleaned of slag, is painted with a dye dissolved in a penetrant. Due to capillary action, the dye enters the cracks. The surface is then cleaned of excess penetrant, but the dye in the cracks remains, and show up as fine lines.

Magnetic inspection. This method is based on the knowledge that discontinuities disturb the uniformity of magnetic fields. In this method a strong magnet is moved along the weld, and the field generated is studied. Any sudden variation indicates the existence of a flaw.

Magnetic particle testing. This method is also based on the property of cracks in welds to disturb magnetic fields. In this method, iron powder is scattered over the weldment, and a strong magnetic field is generated in the member. Cracks in the weld cause the particles to form a pattern that can be seen.

Radiographic inspection. An x-ray generating source placed on one side of the welded surface exposed a film placed on the other side with an image of the weld and any discontinuities in it. After the film is developed, it is studied to determine the quality of the weld and locate any flaws.

Ultrasonic testing. An ultrasonic transducer, which injects ultrasonic pulses into the weldment, is carried along the weld. Reflections of the pulses off the front surface and the back surface of the weld are projected on the screen of a cathode ray tube as peaks from a horizontal base line. The presence of a third peak indicates the existence of a flaw.

Although all five methods can be used in the field and in the shop for the inspection of welds, magnetic and radiographic inspection are seldom used in the field due to the greater amount of equipment and power required.

6.4 ALLOWABLE STRESSES ON WELDS

AISCS Table J2.5 (reproduced here as Table 6.1) lists allowable weld stress and required strength levels for various types of welds. For fillet welds, the type most commonly used, the allowable weld shear stress is limited to 0.30 times the nominal tensile strength of the weld metal. In addition, the electrode must be at a strength level equal to or less than the "matching" weld metal. According to the American Welding Society (AWS) Structural Welding Code AWS D1.1, for A36 steel, E60 or E70 electrodes are permitted; for A572 Gr 50 and A588 Gr 50, E70 electrodes must be used. The "E" indicates electrode, and the number following E indicates the ultimate tensile capacity of the weld in ksi.

In the sections which follow, full and partial penetration welds will not be discussed. The reader is referred to the relevant areas in AISCS J2 and pp. 4-152ff for information on prequalified welds of these types.

6.5 EFFECTIVE AND REQUIRED DIMENSIONS OF WELDS

According to AISCS J2.2a, the area of fillet welds is considered as the effective length of the weld times the effective throat thickness. The effective throat

Table 6.1. Allowable Stress on Welds[f] (AISCS Table J2.5; reprinted with permission).

Type of Weld and Stress[a]	Allowable Stress	Required Weld Strength Level[b,c]
Complete-penetration Groove Welds		
Tension normal to effective area	Same as base metal	"Matching" weld metal shall be used.
Compression normal to effective area	Same as base metal	
Tension or compression parallel to axis of weld	Same as base metal	Weld metal with a strength level equal to or less than "matching" weld metal is permitted.
Shear on effective area	$0.30 \times$ nominal tensile strength of weld metal (ksi)	
Partial-penetration Groove Welds[d]		
Compression normal to effective area	Same as base metal	
Tension or compression parallel to axis of weld[e]	Same as base metal	Weld metal with a strength level equal to or less than "matching" weld metal is permitted.
Shear parallel to axis of weld	$0.30 \times$ nominal tensile strength of weld metal (ksi)	
Tension normal to effective area	$0.30 \times$ nominal tensile strength of weld metal (ksi), except tensile stress on base metal shall not exceed $0.60 \times$ yield stress of base metal	
Fillet Welds		
Shear on effective area	$0.30 \times$ nominal tensile strength of weld metal (ksi)	Weld metal with a strength level equal to or less than "matching" weld metal is permitted.
Tension or compression Parallel to axis of weld[e]	Same as base metal	
Plug and Slot Welds		
Shear parallel to faying surfaces (on effective area)	$0.30 \times$ nominal tensile strength of weld metal (ksi)	Weld metal with a strength level equal to or less than "matching" weld metal is permitted.

[a]For definition of effective area, see Sect. J2.
[b]For "matching" weld metal, see Table 4.1.1, AWS D1.1.
[c]Weld metal one strength level stronger than "matching" weld metal will be permitted.
[d]See Sect. J2.1b for a limitation on use of partial-penetration groove welded joints.
[e]Fillet welds and partial-penetration groove welds joining the component elements of built-up members, such as flange-to-web connections, may be designed without regard to the tensile or compressive stress in these elements parallel to the axis of the welds.
[f]The design of connected material is governed by Chapters D through G. Also see Commentary Sect. J2.4.

Table 6.2. Minimum Size Fillet Weld (AISC Table J2.4; reprinted with permission).

Material Thickness of Thicker Part Joined (Inches)	Minimum[a] Size of Fillet Weld (Inches)
To $\frac{1}{4}$ inclusive	$\frac{1}{8}$
Over $\frac{1}{4}$ to $\frac{1}{2}$	$\frac{3}{16}$
Over $\frac{1}{2}$ to $\frac{3}{4}$	$\frac{1}{4}$
Over $\frac{3}{4}$	$\frac{5}{16}$

[a]Leg dimension of fillet welds. Single-pass welds must be used.

thickness for fillet welds is defined as the shortest distance from the root to the face of the weld. For welds with equal legs, the distance is equal to $\sqrt{2}/2$ or 0.707 times the length of the legs. Reinforcement, if used, is not to be added to the throat thickness. The effective length of the weld is the width of the part joined.

The size of fillet welds shall be within the range of minimum and maximum (AISCS J2.2b) weld sizes. The minimum weld size is determined by the thicker of the parts joined, as given in Table 6.2. Nevertheless, it need not exceed the thickness of the thinner part unless a larger size is required by calculated stress.

The maximum size of a fillet weld is determined by the edge thickness of the member along which the welding is done. Along edges of material less than $\frac{1}{4}$ in. thick, the maximum size may be equal to the thickness of the material. Along the edges of material $\frac{1}{4}$ in. thick or more, the maximum weld size is $\frac{1}{16}$ in. less than the thickness of the material, unless the weld is built up to obtain full throat thickness.

Wherever possible, returns shall be included for a distance not less than twice the nominal size of the weld. If returns are used, the effective length of the returns is added to the effective weld length.

The minimum length of fillet welds shall be not less than four times the nominal size, or else the size of the weld shall be considered not to exceed one-fourth of its effective length. For intermittent welds, the length shall in no case be less than $1\frac{1}{2}$ in. All of these requirements, as well as those for lap joints, are shown in Fig. 6.6.

For plug and slot welds, the effective shearing area is the nominal cross-sectional area of the hole or slot in the plane of the faying (connected surface) member (AISCS J2.3a). Fillet welds in slots or holes are not to be considered as plug or slot welds and must be calculated as fillet welds. Size requirements for plug and slot welds are given in AISCS J2.3b and shown in Fig. 6.7.

In material $\frac{5}{8}$ in. or less in thickness, the weld thickness for plug or slot welds shall be the same as the material. In material over $\frac{5}{8}$ in. thick, the weld must be at least one-half the thickness of the material, but not less than $\frac{5}{8}$ in.

WELDED CONNECTIONS 237

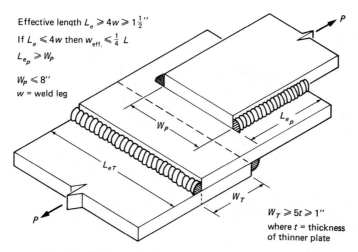

Fig. 6.6. Minimum effective length and overlap of lap joints.

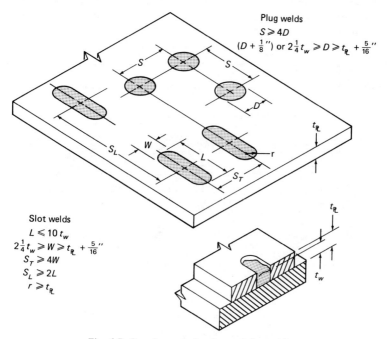

Fig. 6.7. Requirements for plug and slot welds.

238 STEEL DESIGN FOR ENGINEERS AND ARCHITECTS

Welds may be used in combination with bolts, provided the bolts are of the friction type (AISCS J.10) and are installed prior to welding.

6.6 DESIGN OF SIMPLE WELDS

Fillet welds are usually designed such that the shear on the effective weld area is less than that allowed (see Table 6.1). Once the type and size of weld to be used have been determined, the required weld length is determined by

$$l_w = \frac{P}{F_w} \quad (6.1)$$

where P is the design load and F_w is the weld capacity per inch of weld. The weld length is taken to the next greater $\frac{1}{4}$ in. The allowable shearing stresses are 21 ksi for E70 and 18 ksi for E60 electrodes. Since the stress is calculated through the effective throat for the length of the weld, the weld strength per inch of length is determined at the allowable stress times the effective throat.

The weld strength is usually calculated for 1-in. length of $\frac{1}{16}$-in. fillet weld. Thus the stress value of various welds is determined by multiplying the weld stress per sixteenth by the number of sixteenths. Returns should be added to fillet welds when feasible. In some cases, it is easier to weld continuously from one corner to the other. In any case, the effective weld length is the sum of all the weld lengths, including returns (AISCS J2.2b).

Example 6.1. Determine the strength of a $\frac{5}{16}$-in. fillet weld.

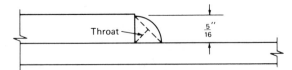

Solution. The allowable stress of E70 electrodes is 0.30 the nominal tensile strength of the weld metal (AISC, Table J2.5).

Allowable weld stress = 0.30 (70 ksi) = 21.0 ksi

The weld strength per inch (F_w) is the allowable stress times the effective throat. For a $\frac{5}{16}$-in. weld, the throat is

$$(\sin 45°) \times (\tfrac{5}{16} \text{ in.}) = 0.707 \times \tfrac{5}{16} \text{ in.} = 0.221 \text{ in.}$$

F_w = Allowable weld stress $\times$ t_{throat} = 21.0 ksi $\times$ 0.221 in. = 4.64 k/in.

WELDED CONNECTIONS 239

The effective throat thickness for equal leg welds per sixteenth inch of weld is

$$(\tfrac{1}{16} \text{ in.}) \times (0.707) = 0.0442 \text{ in.}$$

Multiplying the throat by the allowable stress gives the weld strength per inch per sixteenth. For E70 electrodes, the allowable stress is

$$0.0442 \text{ in.} \times 21.0 \text{ ksi} = 0.928 \text{ k/in./sixteenth inch}$$

From this, the weld strength can be determined by multiplying the number of sixteenths in the weld leg. As above

$$5 \text{ sixteenths} \times 0.928 \text{ k/in./sixteenth} = 4.64 \text{ k/in.}$$

For E60 electrodes, the weld strength is

$$0.928 \, (\tfrac{60}{70}) = 0.795 \text{ k/in./sixteenth inch}$$

Example 6.2. A $\tfrac{5}{16}$-in. fillet weld is used to connect two plates as shown. Determine the maximum load that can be applied to the connection.

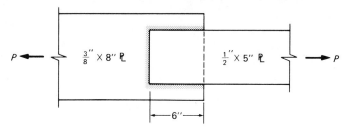

Solution. For welding

$$P_{\max} = l_{\text{weld}} \times F_w$$

For connected members

$$P_{\max} = A_g \times 0.6 \, F_y \quad \text{or}$$

$$P_{\max} = A_n \times 0.5 \, F_u \quad \text{(AISCS D1)}$$

Since $A_n = A_g$ and $0.5 \, F_u > 0.6 \, F_y$, use

$$P_{\max} = A \times 0.6 \, F_y$$

240 STEEL DESIGN FOR ENGINEERS AND ARCHITECTS

Using the general relationship shown in Example 6.1

F_w = 5 sixteenths × 0.928 k/in./sixteenth = 4.64 k/in.

$P_{max} = l_w \times F_w$ = (6 in. + 5 in. + 6 in.) × 4.64 k/in. = 78.9 k

For $\frac{1}{2}$ in. × 5-in. bar

$P_{max} = A \times 0.6 F_y$ = ($\frac{1}{2}$ in. × 5 in.) × (22.0 ksi) = 55.0 k

For $\frac{3}{8}$-in. × 8-in. bar

$P_{max} = A \times 0.6 F_y$ = ($\frac{3}{8}$ in. × 8 in.) × (22.0 ksi) = 66.0 k

The connections capacity is 55.0 kips resulting from the $\frac{1}{2}$ in. × 5 in. member allowable stress.

Example 6.3. For the members shown below, determine:

a) the minimum fillet weld length and corresponding weld size.
b) the minimum weld size and corresponding fillet weld length.

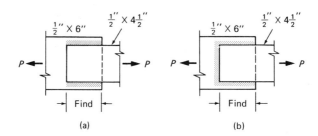

(a)   (b)

Solution.

- Case a):

  1. Use U = 0.75
     Minimum weld length = plate width = 4.5 in. (AISCS B3)
     Plate capacity:

     P_{max} = lesser of $\begin{cases} 22 \times 2.25 = 49.5 \text{ k} \\ 29 \times 0.75 \times 2.25 = 48.9 \text{ k (governs)} \end{cases}$

From eqn. (6.1):

$$F_w = \frac{48.9}{2 \times 4.5} = 5.43 \text{ k/in.}$$

Define D = number of sixteenths of an inch in fillet weld size. Then,

$$5.43 \text{ k/in.} = D \times 0.928 \text{ k/in./sixteenth}$$
$$D = 5.9 \text{ sixteenths, Use } D = 6 \rightarrow \text{weld size} = \tfrac{3}{8} \text{ in.}$$

2. Minimum weld size = $\tfrac{3}{16}$ in. (AISCS Table J2.4)

$$D = 3$$
$$F_w = 3 \times 0.928 = 2.78 \text{ k/in.}$$

Try $U = 0.87$:

$$P_{max} = \text{lesser of} \begin{cases} 22 \times 2.25 = 49.5 \text{ k (governs)} \\ 29 \times 0.87 \times 2.25 = 56.8 \text{ k} \end{cases}$$

$$\text{Weld length} = \frac{49.5}{2 \times 2.78} = 8.9 \text{ in.} \quad \text{Use 9 in.}$$

Check initial assumption:

$$2 \times 4.5 = 9 > 8.9 > 1.5 \times 4.5 = 6.75 \quad \text{(AISCS B3)}$$

Therefore, $U = 0.87$, as assumed.

- Case b):

$$U = 1$$

Plate capacity:

$$P_{max} = \text{lesser of} \begin{cases} 22 \times 2.25 = 49.5 \text{ k (governs)} \\ 29 \times 2.25 = 65.3 \text{ k} \end{cases}$$

1. Use maximum allowed weld size to minimize weld length:
From AISCS J2.2b, use $w = \tfrac{7}{16}$ in. $\rightarrow D = 7$.

$$F_w = 7 \times 0.928 = 6.5 \text{ k/in.}$$

$$\text{Total weld length} = \frac{49.5}{6.5} = 7.6 \text{ in.}$$

$$\text{Weld length on each side} = \frac{7.6 - 4.5}{2} = 1.55 \text{ in.}$$

$$1.55 \text{ in.} > 2 \times \frac{7}{16} = 0.875 \text{ in., which is the minimum return length (AISCS J2.2b)}$$

Therefore, use $1\frac{5}{8}$-in. weld length on each side.

2. Minimum $D = 3$

$$F_w = 3 \times 0.928 = 2.78 \text{ k/in.}$$

$$\text{Total weld length} = \frac{49.5}{2.78} = 17.8 \text{ in.}$$

$$\text{Weld length each side} = \frac{17.8 - 4.5}{2} = 6.65 \text{ in.}$$

Use $6\frac{3}{4}$-in. weld length on each side.

Example 6.4. For the member below, determine the required length of weld when the minimum-size fillet weld is used.

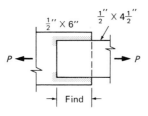

Solution. Assume that $U = 0.87$
From Example 6.3:

$$P_{max} = 49.5 \text{ k}, \quad \text{minimum } D = 3, \quad \text{total weld length} = 17.8 \text{ in.}$$

$$\text{Minimum return length} = 2 \times (\tfrac{3}{16}) = 0.375 \text{ in.}$$

$$\text{Weld length each side} = \frac{17.8 - (2 \times 0.375)}{2} = 8.53 \text{ in.}$$

Check initial assumption for U:

$$2 \times 4.5 = 9 > 8.53 > 1.5 \times 4.5 = 6.75 \quad \text{ok}$$

Use $8\frac{5}{8}$-in. weld length on each side.

Example 6.5. Determine the allowable tensile capacity of the connection shown. Use a $\frac{3}{8}$-in. fillet weld with E60 electrodes.

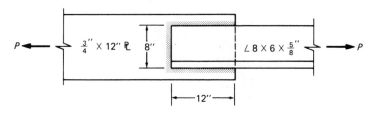

Solution.

Angle capacity:

$$P_{max} = \begin{cases} 22 \times 8.36 = 183.9 \text{ k (governs)} \\ 29 \times (0.85 \times 8.36) = 206.1 \text{ k} \end{cases}$$

Plate capacity:

$$P_{max} = (0.75 \times 12) \times 22 = 198 \text{ k}$$

Weld capacity:

For E60 welds,

$$F_w = 6 \times 0.795 \text{ k/in./sixteenth} = 4.77 \text{ k/in.}$$

$$l_w = 12 + 8 + 12 = 32 \text{ in.}$$

$$P_{max} = 4.77 \times 32 = 152.6 \text{ k}$$

Since the weld capacity is less than the steel capacity, the allowable tensile capacity of the connection is 152.6 k.

Note: Although the weld lengths on each side are equal, the effects of the eccentricity can be omitted (see AISCS J1.9).

244 STEEL DESIGN FOR ENGINEERS AND ARCHITECTS

At times, combinations of welds may be necessary to develop sufficient strength. For weld combinations, the effective capacity of each type shall be separately computed, with reference to the axis of the group, in order to determine the allowable capacity of the combination (AISCS J2.5).

Example 6.6. A 12 × 30 channel (A572 Gr 50 steel) is to be connected to another heavy member (A36 steel) using $\frac{7}{16}$-in. E70 fillet welds. Due to clearance limitations, the two members can overlap only 5 in., as shown. Determine the distance l so that the connection capacity is 170 kips.

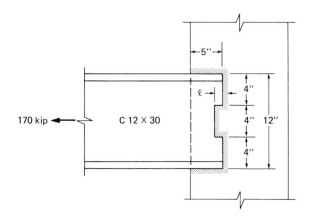

Solution.

$$l_w = P_{max}/F_w$$

$$F_w = 7 \times 0.928 = 6.5 \text{ k/in.}$$

$$l_w = \frac{170 \text{ k}}{6.5 \text{ k/in.}} = 26.15 \text{ in.}$$

The maximum length outside the slot is

$$(5 \text{ in.} + 5 \text{ in.} + 4 \text{ in.} + 4 \text{ in.}) = 18.0 \text{ in.}$$

The fillet weld length required in the slot is

$$26.15 \text{ in.} - 18.0 \text{ in.} = 8.15 \text{ in.}$$

$$\text{slot} = 4 \text{ in.} + 2l = 8.15$$

$$l = 2.08 \text{ in. use } 2\tfrac{1}{8} \text{ in.}$$

Example 6.7. Redesign the channel connection in Example 6.6 as a combination fillet and slot weld to support the same applied load. Use 4 in. overlap, as shown.

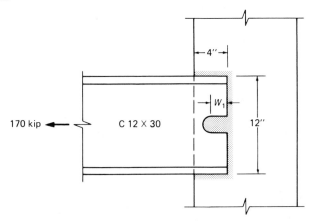

Solution. Because the web thickness is $\frac{1}{2}$ in., the maximum fillet weld size is $\frac{7}{16}$ in. (AISCS J2.2b).

$$F_w = 7 \times 0.928 \text{ k/in.}/16\text{th} = 6.5 \text{ k/in.}$$

Allowable slot weld stress $= 0.30 \times 70.0$ ksi $= 21.0$ ksi (AISCS, Table J2.5).

Assume that the width of the slot weld is 1 in.

Capacity of fillet weld

$$P_{max} = 6.5 \text{ k/in.} \times (4 \text{ in.} + 12 \text{ in.} - 1 \text{ in.} + 4 \text{ in.}) = 123.5 \text{ k}$$

Balance to be resisted by slot weld

$$P_{bal} = 170.0 \text{ k} - 123.5 \text{ k} = 46.5 \text{ k}$$

Design of slot weld

The thickness of slots welds in material $\frac{5}{8}$ in. or less in thickness shall be equal to the thickness of the material (AISCS J2.3).

$$\therefore \text{ Weld thickness} = \tfrac{1}{2} \text{ in. (thickness of material)}$$

Minimum width of slot weld $= \frac{1}{2}$ in. $+ \frac{5}{16}$ in. $= \frac{13}{16}$ in.

Maximum width of slot weld $= 2\frac{1}{4} \times \frac{1}{2}$ in. $= 1\frac{1}{8}$ in.

246 STEEL DESIGN FOR ENGINEERS AND ARCHITECTS

Maximum length of slot weld = $10 \times \frac{1}{2}$ in. = 5 in.

Effective shearing area = width × length

Try 1-in.-wide slot weld

$$l_w = \frac{46.5 \text{ k}}{1.0 \text{ in.}} \bigg/ 21.0 \text{ ksi} = 2.21 \text{ in.} \quad \text{say } 2\frac{1}{4} \text{ in.}$$

Try $\frac{13}{16}$ in. wide slot weld

$$l_w = \frac{46.5 \text{ k}}{\frac{13}{16} \text{ in.}} \bigg/ 21.0 \text{ ksi} = 2.73 \text{ in.} \quad \text{say } 2\frac{3}{4} \text{ in.}$$

Try $1\frac{1}{8}$-in.-wide slot weld

$$l_w = \frac{46.5 \text{ k}}{\frac{18}{16} \text{ in.}} \bigg/ 21.0 \text{ ksi} = 1.97 \text{ in.} \quad \text{say 2 in.}$$

Use any slot weld above in combination with $\frac{7}{16}$-in. fillet weld.

When calculated stress is less than the minimum size weld, intermittent fillet welds may be used. The effective length of any segment of fillet weld shall be not less than four times the weld size nor $1\frac{1}{2}$ in. (AISCS J2.2b).

Example 6.8. A built-up beam consisting of $1\text{-}\frac{1}{2}$ in. × 14 in. flanges and $\frac{1}{2}$ in. × 42 in. web is subject to a maximum shear of 160 kips. Determine the fillet weld size for a continuous weld and an intermittent weld.

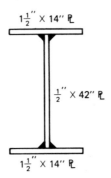

Solution. The stress in the weld is found by the shear flow at the intersection of the web and flanges. The required weld size is the shear flow divided by the allowable weld stress per inch per sixteenth inch weld leg.

$$q_v = \frac{VQ}{I}$$

$V = 160$ kips

$Q = A\bar{y} = (1.5 \text{ in.} \times 14 \text{ in.}) \times 21.75 \text{ in.} = 456.8 \text{ in.}^3$

$$I = \frac{0.5 \text{ in.} \times (42 \text{ in.})^3}{12} + 2 \times [(1.5 \text{ in.} \times 14 \text{ in.}) \times (21.75 \text{ in.})^2]$$

$= 22960 \text{ in.}^4$

$$q_v = \frac{(160 \text{ k} \times 456.8 \text{ in.}^3)}{22960 \text{ in.}^4} = 3.18 \text{ k/in.}$$

$F_w = 0.928$ k/in./sixteenth weld leg

$$D = \frac{3.18 \text{ k/in.}}{2 \times 0.928 \text{ k/in.}} = 1.72 \text{ sixteenths} \quad \text{use } \frac{1}{8} \text{ in. weld}$$

The minimum weld size for a thickness of $\frac{3}{4}$ in. or more (thicker part joined) is $\frac{5}{16}$ in.

Use $\frac{5}{16}$-in. E70 continuous weld.

If intermittent fillet weld is to be used, assuming $\frac{3}{8}$-in. weld:

$F_w = 6 \times 0.928$ k/in./sixteenth $= 5.57$ k/in.

$q_v = 3.18$ k/in. $\times 12$ in./ft $= 38.15$ k/ft

$$l_w = \frac{38.15 \text{ k/ft}}{2 \times 5.57 \text{ k/in.}} = 3.42 \text{ in./ft} \quad \text{say } 3\frac{1}{2} \text{ in.}$$

Use $\frac{3}{8}$-in. weld $3\frac{1}{2}$ in. long, 12 in. on center.

Note: The use of intermittent fillet weld is acceptable for quasi-static loading. In case of fatigue loading—which is beyond the scope of our study here—intermittent fillet welds become for all practical purposes unusable. See Appendix A.

Example 6.9. A W 24×104 has 1 in. $\times$ 10 in. cover plates attached to the flanges by $\frac{5}{16}$ in. E70 fillet welds as shown. Determine the length of connecting welds for a location where a 110-kip shear occurs.

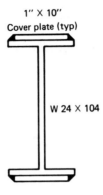

1" × 10" Cover plate (typ)

W 24 × 104

Solution. The shear flow at the connection of cover plates is

$$q_v = \frac{VQ}{I}$$

where $V = 110$ kips

$Q = A\bar{y} = (1 \text{ in.} \times 10 \text{ in.}) \times 12.53 \text{ in.} = 125.3 \text{ in.}^3$

$I = 3100 \text{ in.}^4 + 2 \times (1 \text{ in.} \times 10 \text{ in.}) \times (12.53 \text{ in.})^2 = 6240 \text{ in.}^4$

$q_v = \dfrac{110 \text{ k} \times 125.3 \text{ in.}^3}{6240 \text{ in.}^4} = 2.21 \text{ k/in.}$

$F_w = 2 \times (5 \times 0.928 \text{ k/in./sixteenth}) = 9.28 \text{ k/in.}$

$l_w = \dfrac{2.21 \text{ k/in.} \times 12 \text{ in./ft}}{9.28 \text{ k/in.}} = 2.86 \text{ in./ft}$ say 3 in. at 12-in. spacing

Use $\tfrac{5}{16}$-in. E70 welds 3 in. long, 12 in. on center. See note, Example 6.8.

Example 6.10. Design welded partial length cover plates for the loading shown using a W 24 × 76 section. Assume full lateral support.

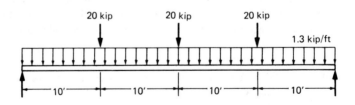

Solution. Wide-flange beam properties

$$W\ 24 \times 76$$
$$A = 22.4 \text{ in.}^2$$
$$d = 23.92 \text{ in.}$$
$$b_f = 8.99 \text{ in.}$$
$$I = 2100 \text{ in.}^4$$
$$S = 176 \text{ in.}^3$$

Shear and moment diagrams

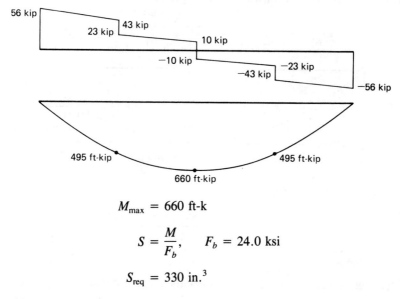

$$M_{max} = 660 \text{ ft-k}$$
$$S = \frac{M}{F_b}, \quad F_b = 24.0 \text{ ksi}$$
$$S_{req} = 330 \text{ in.}^3$$

Available capacity of W 24 × 76

$$S_{WF} = 176 \text{ in.}^3$$

Additional capacity required

$$S_{pl} = 330 \text{ in.}^3 - 176 \text{ in.}^3 = 154 \text{ in.}^3$$

Determine the moment of inertia of the entire section to determine required plate size.

250 STEEL DESIGN FOR ENGINEERS AND ARCHITECTS

$$I = I_{WF} + 2A_{pl}\, d^2 = 2100 \text{ in.}^4 + 2(bt)\left(\frac{23.92 \text{ in.} + t}{2}\right)^2$$

$$= 2100 \text{ in.}^4 + \frac{1}{2}(23.92 \text{ in.} + t)^2 (bt)$$

$$S = \frac{I}{c} = \frac{4200 \text{ in.}^4 + (23.92 \text{ in.} + t)^2 (bt)}{23.92 \text{ in.} + 2t} = 330 \text{ in.}^3$$

Selecting $t = 1$ in., $b = 7$ in., S becomes 329.7 in.^{3}

To determine the theoretical cutoff points, superimpose the moment capacity and moment diagrams.

Moment capacity of beam without cover plates

$$M_{WF} = \frac{S_{WF} \times F_b}{12 \text{ in.}/\text{ft}} = \frac{176 \text{ in.}^3 \times 24 \text{ ksi}}{12 \text{ in.}/\text{ft}} = 352 \text{ ft-k}$$

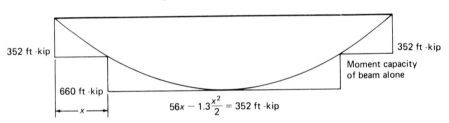

$$0.65\, x^2 - 56\, x + 352 = 0;$$

$$x = \frac{28 \pm \sqrt{28^2 - (.65 \times 352)}}{.65} = 6.83 \text{ ft, } 49.17 \text{ ft.}$$

Determine welds

Maximum shear flow at cover plates occurs at cutoff points.

$$V = 56 \text{ k} - (6.83 \text{ ft} \times 1.3 \text{ k/ft}) = 47.12 \text{ k}$$

$$I = I_0 + Ad^2 = 2100 \text{ in.}^4 + 2(1 \text{ in.} \times 7 \text{ in.})\left(\frac{23.93 \text{ in.}}{2} + \frac{1 \text{ in.}}{2}\right)^2$$

$$= 4274 \text{ in.}^4$$

$$Q = A_{pl}\, d = (1 \text{ in.} \times 7 \text{ in.})\left(\frac{23.92 \text{ in.}}{2} + \frac{1 \text{ in.}}{2}\right) = 87.22 \text{ in.}^3$$

$$q = \frac{VQ}{I} = \frac{47.12 \text{ k} \times 87.22 \text{ in.}^3}{4274 \text{ in.}^4} = 0.96 \text{ k/in.}$$

WELDED CONNECTIONS 251

Use minimum intermittent fillet weld. For 1-in. cover plates, the minimum size fillet weld = $\frac{5}{16}$ in. See note, Example 6.8.

$$F_w = 0.928 \text{ k/in./sixteenth} \times 5 \text{ sixteenths} \times 2 \text{ sides} = 9.28 \text{ k/in.}$$

Force to be carried per foot = 0.96 k/in. × 12 in./ft = 11.54 k/ft

$$\frac{11.54 \text{ k/ft}}{9.28 \text{ k/in.}} = 1.24 \text{ in./ft}$$

Minimum weld length = 4 × weld size but not less than $1\frac{1}{2}$ in.

$$4 \times \frac{5}{16} \text{ in.} = 1\frac{1}{4} \text{ in.} \quad \text{use } 1\frac{1}{2} \text{ in. intermittent welds} \quad \text{(AISCS J2.2b)}$$

Partial-length cover plates must be extended past the theoretical cutoff point for a distance capable of developing the cover plate's portion of flexural stress in the beam at the theoretical cutoff point (AISCS B10). In addition, welds connecting cover plates must be continuous for a distance dependent on plate width and weld type and must be capable of developing the flexural stresses in that length.

The AISC Commentary B10 shows that the force to be developed by fasteners in the cover plate extension is

$$H = \frac{MQ}{I} \quad \text{(C-B10-1)}$$

$$H = \frac{MQ}{I} = \frac{352 \text{ ft-k} \times 12 \text{ in./ft} \times 87.22 \text{ in.}^3}{4274 \text{ in.}^4} = 86.20 \text{ k}$$

Design welds in plate extension to be continuous along both edges of cover plate, but no weld across end of plate.

$$a' = 2 \times b = 14 \text{ in.} \quad \text{(AISCS B10)}$$

Weld capacity = 2 × 14 in. × (5 × 0.928 k/in.) = 129.9 k > 86.2 k ok

Plate extension is adequate. Use partial cover plates as shown in sketch.

252 STEEL DESIGN FOR ENGINEERS AND ARCHITECTS

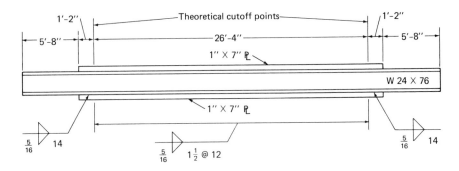

Example 6.11. For the continuous beam shown, choose a wide-flange beam to resist positive bending, and design a welded partial cover plate where required for negative bending. Assume full lateral support. In the analysis, neglect the additional stiffness created by the cover plate.

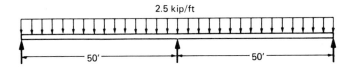

Solution. Shear and moment diagrams

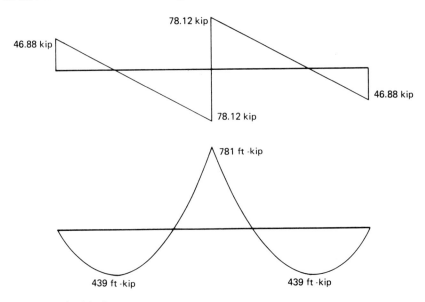

Selection of wide flange

Maximum positive moment = 439 ft-k

WELDED CONNECTIONS 253

Here the effect of redistribution of moments has *not* been used. See Chapter 3 for discussion of redistribution (AISCS F1).

Select W 24 × 94 (M_{max} = 444-ft-k)

$$A = 27.7 \text{ in.}^2$$
$$d = 24.31 \text{ in.}$$
$$b_f = 9.065 \text{ in.}$$
$$I = 2700 \text{ in.}^4$$
$$S = 222 \text{ in.}^3$$

Total section modulus required

$$\text{Maximum moment} = 781 \text{ ft-k}$$

$$S = \frac{M}{F_b} = 391 \text{ in.}^3 = \frac{I}{c}$$

$$I = 2700 \text{ in.}^4 + 2(bt)\left(\frac{24.31 \text{ in.} + t}{2}\right)^2$$

$$S = \frac{2700 \text{ in.}^4 + \frac{1}{2}bt(24.31 \text{ in.} + t)^2}{\frac{24.31 \text{ in.}}{2} + t} = 391 \text{ in.}^3$$

Try $t = 1\frac{3}{8}$ in.

$$b = \frac{(12.155 \text{ in.} + t) \times 391 \text{ in.}^2 - 2700 \text{ in.}^4}{\frac{1}{2}t(24.31 \text{ in.} + t)^2} = 5.71 \text{ in.} \quad \text{say 6 in.}$$

$$I = 2700 \text{ in.}^4 + 2(6 \text{ in.} \times 1.375 \text{ in.}) \times \left(\frac{24.31 \text{ in.}}{2} + \frac{1.375 \text{ in.}}{2}\right)^2$$

$$= 5421 \text{ in.}^4$$
$$S = 400 \text{ in.}^3 \quad \text{ok}$$

To determine theoretical cutoff point, assume that beam has constant moment of inertia. Actual beam behavior is such that a slight difference in moments will occur with varying I.

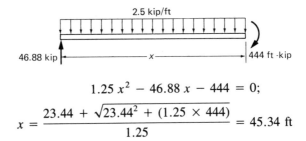

$$1.25 x^2 - 46.88 x - 444 = 0;$$

$$x = \frac{23.44 + \sqrt{23.44^2 + (1.25 \times 444)}}{1.25} = 45.34 \text{ ft}$$

Maximum shear flow at cover plates occurs at support.

$V = 78.1$ k

$$Q = A_{pl} \, d = (1.375 \text{ in.} \times 6 \text{ in.}) \left(\frac{24.31 \text{ in.}}{2} + \frac{1.375 \text{ in.}}{2} \right) = 106.0 \text{ in.}^3$$

$$q = \frac{VQ}{I} = \frac{78.1 \text{ k} \times 106.0 \text{ in.}^3}{5421 \text{ in.}^4} = 1.53 \text{ k/in.}$$

Minimum fillet weld size = $\frac{5}{16}$ in.

$$F_w = 0.928 \text{ k/in./sixteenth} \times 5 \text{ sixteenths} \times 2 \text{ sides} = 9.28 \text{ k/in.}$$

Force to be carried per foot = 1.53 k/in. × 12 in./ft = 18.36 k/ft

$$\frac{18.36 \text{ k/ft}}{9.28 \text{ k/in.}} = 1.98 \text{ in.} \quad \text{say 2 in. length of intermittent weld}$$

Minimum intermittent weld length = $1\frac{1}{2}$ in. < 2 in. ok

Plate extension

Use continuous welds along both edges of cover plate and across end of plate.

$$a' = 1.5 \times b = 9 \text{ in.} \quad (\tfrac{5}{16} \text{ in.} < \tfrac{3}{4} \times 1.375 \text{ in.} = 1.03 \text{ in.})$$

Weld capacity = $[(2 \times 9 \text{ in.}) + 6 \text{ in.}] \times (5 \times 0.928 \text{ k/in.}) = 111.4$ k

Required force

$$\frac{MQ}{I} = \frac{444 \text{ ft-k} \times 12 \text{ in./ft} \times 106 \text{ in.}^3}{5421 \text{ in.}^4} = 104 \text{ k} < 111.4 \text{ k} \quad \text{ok}$$

6.7 REPEATED STRESSES (FATIGUE)

When a weld is subjected to repeated stresses, the centroids of the member and weld shall coincide (AISCS J1.9). For nonsymmetrical members, unless provision is made for eccentricity, welds of different lengths are necessary. See Appendix A for further details.

Example 6.12. One leg of a 6 × 6 × $\frac{5}{8}$-in. angle is to be connected with side welds and a weld at the end of the angle to a plate behind. Design the connection to develop full capacity of the angle. Balance the fillet welds around the center of gravity of the angle, using the maximum size weld and E60 electrodes.

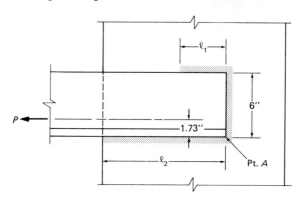

Solution. To have the shortest connection, select the maximum size weld.

$$\text{Maximum weld size} = \frac{5}{8} \text{ in.} - \frac{1}{16} \text{ in.} = \frac{9}{16} \text{ in.}$$

$$P_{max} = A_g \times 0.60 \, F_y = 7.11 \text{ in.}^2 \times 22.0 \text{ ksi} = 156.4 \text{ k}$$

$$F_w = 9 \times 0.795 \text{ k/in./sixteenth} = 7.16 \text{ k/in.}$$

$$l_w = \frac{156.4 \text{ k}}{7.16 \text{ k/in.}} = 21.84 \text{ in.}$$

Summing moments of weld fillets about point A

1.73 in. × 21.84 in. = 6 in. (l_1) + 3 in. (6 in.)

$l_1 = (37.78 - 18.0)/6 = 3.3$ in.   say $3\frac{1}{2}$ in.

$l_2 = 21.84$ in. $- 6$ in. $- 3.5$ in. $= 12.34$ in.   say $12\frac{1}{2}$ in.

256 STEEL DESIGN FOR ENGINEERS AND ARCHITECTS

Note: This case is typical of a connection with repeated variations of stresses. Though desirable, it is not required to design for eccentricity between the gravity axes in statically loaded members, but this is required in members subject to fatigue loading (AISCS J1.9).

Example 6.13. A $7 \times 4 \times \frac{3}{8}$-in. angle is to be connected to develop full capacity as shown to plate using $\frac{5}{16}$-in. welds. Determine l_1 and l_2 so the centers of gravity for the welds and angle coincide.

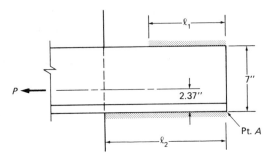

Solution.

$$P_{max} = A_g \times F_t = 3.98 \text{ in.}^2 \times 22.0 \text{ ksi} = 87.56 \text{ k}$$

$$F_w = 5 \times 0.928 \text{ k/in.} = 4.64 \text{ k/in.}$$

$$l_w = 87.56 \text{ k}/4.64 \text{ k/in.} = 18.87 \text{ in.}$$

Summing moments of weld fillets about point A

$$2.37 \text{ in.} \times 18.87 \text{ in.} = 7.0 \text{ in.} \times l_1$$

$$l_1 = 44.72 \text{ in.}^2/7.0 \text{ in.} = 6.39 \text{ in.} \quad \text{say } 6\tfrac{1}{2} \text{ in.}$$

$$l_2 = 18.87 \text{ in.} - 6\tfrac{1}{2} \text{ in.} = 12.37 \text{ in.} \quad \text{say } 12\tfrac{1}{2} \text{ in.}$$

Lengths of welds as determined may be used. However, to satisfy AISC end return requirements, see the following example.

Example 6.14. Same as Example 6.13, but include end returns.

Solution. Side or end fillet welds terminating at ends or sides, respectively, of parts or members shall, wherever practicable, be returned continuously around the corners for a distance not less than two times the nominal size of the weld (AISCS J2.2b).

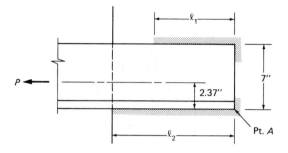

Solution.

$$P = 87.56 \text{ k}$$

$$l_w = 18.87 \text{ in.}$$

$$\text{Minimum end return} = 2 \times \tfrac{5}{16} \text{ in.} = \tfrac{5}{8} \text{ in.}$$

$$l_1 + l_2 + (2 \times \tfrac{5}{8} \text{ in.}) = 18.87 \text{ in.}$$

$$l_1 + l_2 = 17.62 \text{ in.}$$

Summing moments about point A

$$2.37 \text{ in.} \times 18.87 \text{ in.} = \tfrac{5}{16} \text{ in.} (\tfrac{5}{8} \text{ in.}) + (7 \text{ in.} - \tfrac{5}{16} \text{ in.})(\tfrac{5}{8} \text{ in.}) + 7(l_1)$$

$$l_1 = (44.72 - 0.20 - 4.18)/7 = 5.76 \text{ in.} \quad \text{say } 5\tfrac{3}{4} \text{ in.}$$

$$l_2 = 17.62 \text{ in.} - 5\tfrac{3}{4} \text{ in.} = 11.87 \quad \text{say } 12 \text{ in.}$$

6.8. ECCENTRIC LOADING

When eccentric loads are carried by welds, either shear and torsion or shear and bending may result in the weld (see Figs. 6.8 and 6.9, respectively). For the case of shear and torsion, the force per unit length is determined by

$$f = \frac{Td}{J} \tag{6.2}$$

where T is the torque, d is the distance from the center of gravity of the weld group to the point in consideration and J is its polar moment of inertia. The polar moment of inertia is given as the sum of the moment of inertia in the x and y directions, i.e., $J = I_x + I_y$. The direction of the force is perpendicular

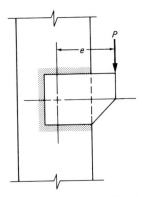

Fig. 6.8. Shear and torsion condition in welds.

to the line connecting the point and the center of gravity of the weld group. For ease of computation, the distance d is usually reduced to its horizontal (h) and vertical (v) components.

If the horizontal and vertical components of the force applied to the centroid are, respectively, P_H and P_V, the moment M and the total length of weld l in inches, the stresses due to the direct force are

$$f_{h,P} = \frac{P_H}{l} \qquad f_{v,P} = \frac{P_V}{l} \tag{6.3}$$

Stresses due to the moment are

$$f_{h,M} = \frac{Mv}{J} \qquad f_{v,M} = \frac{Mh}{J} \tag{6.4}$$

The resultant shear force is then computed by taking the square root of the squares of the components,

$$f_r = \sqrt{(f_{h,P} + f_{h,M})^2 + (f_{v,P} + f_{v,M})^2} \tag{6.5}$$

When computing the weld size, a 1-inch weld is usually used so that the weld area is easily computed. The weld size required is then determined by dividing the resultant shear force by the allowable shear stress of the weld.

Example 6.15. Determine the maximum eccentric load P in kips acting as shown. Weld size = $\frac{1}{4}''$.

WELDED CONNECTIONS 259

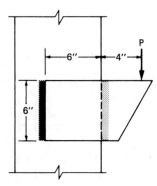

Solution.

Determine J of the weld group:

$$I_x = 2 \times \frac{1 \times (6)^3}{12} = 36 \text{ in.}^4$$

$$I_y = 2 \times 6 \times 1 \times (3)^2 = 108 \text{ in.}^4$$

$$J = I_x + I_y = 144 \text{ in.}^4$$

$T = 7 \text{ in.} \times P$

Direct shear stress:

$$f_v = \frac{P}{2 \times 6} = 0.0833P$$

$$f_h = 0$$

Torsional shear stress:

$$f_v = \frac{7P \times 3}{144} = 0.1458P$$

$$f_h = \frac{7P \times 3}{144} = 0.1458P$$

Resultant shear stress:

$$f_r = \sqrt{(0.1458P)^2 + (0.0833P + 0.1458P)^2} = 0.2716P$$

$$f_r = F_w = 4 \times 0.928 = 3.71 \text{ k/in.}$$

$$P = \frac{3.71}{0.2716} = 13.7 \text{ k}$$

260 STEEL DESIGN FOR ENGINEERS AND ARCHITECTS

Example 6.16. Determine the minimum weld size to resist the load shown.

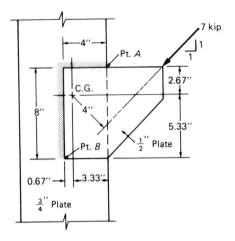

Solution. It is not obvious which corner is critically stressed. Therefore, the stresses at both *A* and *B* must be computed.

$$f = \frac{Td}{J}, \quad T = 7 \text{ k} \times 4.0 \text{ in.} = 28 \text{ in.-k}$$

$$J = I_x + I_y$$

$$J = \frac{(2.67 \text{ in.})^3}{3} + \frac{(5.33 \text{ in.})^3}{3} + (4 \text{ in.}^2 \times (2.67 \text{ in.})^2) + \frac{(0.67 \text{ in.})^3}{3}$$

$$+ \frac{(3.33 \text{ in.})^3}{3} + (8 \text{ in.}^2 \times (0.67 \text{ in.})^2) = 101.33 \text{ in.}^4$$

The shearing stress of the applied force is broken down into horizontal and vertical components and combined with the torsional components.

- Direct shear stress

$$f_v = \frac{0.707 \times 7 \text{ k}}{1 \text{ in.} \times (4 \text{ in.} + 8 \text{ in.})} = 0.412 \text{ k/in.}^2$$

$$f_h = \frac{0.707 \times 7 \text{ k}}{1 \text{ in.} \times (4 \text{ in.} + 8 \text{ in.})} = 0.412 \text{ k/in.}^2$$

WELDED CONNECTIONS

- Torsional components

    Point A

    $$f_v = \frac{28 \text{ in.-k} \times 3.33 \text{ in.}}{101.33 \text{ in.}^4} = 0.92 \text{ k/in.}^2$$

    $$f_h = \frac{28 \text{ in.-k} \times 2.67 \text{ in.}}{101.33 \text{ in.}^4} = 0.74 \text{ k/in.}^2$$

See sketch for force orientation.

$$f_r = \sqrt{(0.412 + 0.92)^2 + (0.74 - 0.412)^2} = 1.37 \text{ k/in.}^2$$

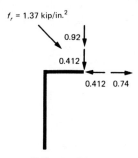

Weld stress—Point A

Point B

$$f_v = \frac{28 \text{ in.-k} \times 0.67 \text{ in.}}{101.33 \text{ in.}^4} = 0.19 \text{ k/in.}^2$$

$$f_h = \frac{28 \text{ in.-k} \times 5.33 \text{ in.}}{101.33 \text{ in.}^4} = 1.47 \text{ k/in.}^2$$

See sketch for force orientation

$$f_r = \sqrt{(0.19 - 0.412)^2 + (0.412 + 1.47)^2} = 1.90 \text{ k/in.}^2$$

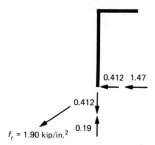

Weld stress—Point B

262 STEEL DESIGN FOR ENGINEERS AND ARCHITECTS

The most stressed point of the weld is Point B.

For a 1-in. weld

$$F_w = 16 \times 0.928 \text{ k/in.} = 14.85 \text{ k/in.}$$

$$w_{size} = \frac{1.90 \text{ k/in.}^2}{14.85 \text{ k/in.}} = 0.13 \text{ in.} \quad \text{use } \frac{1}{4} \text{ in. minimum weld size}$$

Example 6.17. What load P can be supported by the connection shown? The weld is $\frac{5}{16}$ in. E60 electrode.

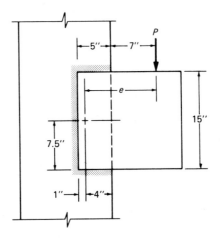

Solution.

$$f_v = \frac{Th}{J}, \quad T = Pe; \quad f_h = \frac{Td}{J}$$

Locating the C.G. of the weld group

$$\frac{5 \text{ in.} \times 2.5 \text{ in.} \times 2}{(5 \text{ in.} \times 2) + 15 \text{ in.}} = 1 \text{ in.}$$

Assuming a 1-in. weld

$$J = (I_x + I_y)_{weld} = \left(\frac{1(15)^3}{12}\right) + (1 \times 5)\left(\frac{15}{2}\right)^2 \times 2 + (1 \times 15)(1)^2$$

$$+ 2 \times \left[\frac{(1)^3}{3} + \frac{(4)^3}{3}\right] = 902.08 \text{ in.}^4$$

$$e = 7 \text{ in.} + 4 \text{ in.} = 11 \text{ in.}$$

Solving for components

shear
$$f_v = \frac{P}{(5 \text{ in.} + 15 \text{ in.} + 5 \text{ in.}) \times 1 \text{ in.}} = 0.040\, P$$

torque
$$f_v = \frac{11 \text{ in.} \times P \times 4 \text{ in.}}{902.08 \text{ in.}^4} = 0.0488\, P$$

$$f_h = \frac{11 \text{ in.} \times P \times 7.5 \text{ in.}}{902.08 \text{ in.}^4} = 0.0915\, P$$

resultant
$$f_r = P\sqrt{(0.0488 + 0.040)^2 + (0.0915)^2} = 0.1275\, P$$

Since f_r is determined assuming a 1-in. weld, conversion for $\tfrac{5}{16}$-in. weld is necessary.

$$F_w = 5 \times 0.795 \text{ k/in.} = 3.98 \text{ k/in.}$$

$0.1275\, P = 3.98 \text{ k/in.}$

$$P = \frac{3.98 \text{ k/in.}}{0.1275\,(1/\text{in.})} = 31.2 \text{ k}$$

For welds subjected to shear and bending, the shear is assumed to be uniform throughout the member. The weld is designed to carry the shear and maximum bending stress. The resulting shear force is the square root of the sum of the squares of the shear force and the bending force. Referring to Fig. 6.9, the shear stress due to the applied load P is

$$f_v = \frac{P}{l} \quad \text{(vertical)} \tag{6.5}$$

where l is the total weld length. In addition, the bending stress is given by

$$f_b = \frac{M}{S} \quad \text{(horizontal)} \tag{6.6}$$

264 STEEL DESIGN FOR ENGINEERS AND ARCHITECTS

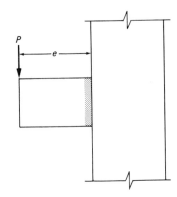

Fig. 6.9. Shear and bending condition in welds.

where $M = Pe$

$$S = \frac{I}{c} = \text{section modulus of the weld group.}$$

The resultant stress on the weld system is then

$$f_r = \sqrt{(f_v)^2 + (f_b)^2} \tag{6.7}$$

As before, calculations are usually performed for a 1-in. weld for convenience.

Example 6.18. Determine the required weld size for the connection shown below. Assume a $\frac{5}{16}$-in. weld for return length purposes only. Also, the bracket plate and column strengths will not govern.

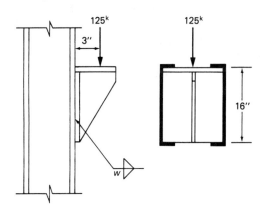

WELDED CONNECTIONS 265

Return length = $2 \times \frac{5}{16} = 0.625$ in.

For a 1-in. weld:

$$I = 2 \times \left(\frac{1 \times 16^3}{12}\right) + 4 \times [(1 \times 0.625) \times 8^2] = 843 \text{ in.}^4$$

$$S = \frac{843}{8} = 105.4 \text{ in.}^3$$

$$M = 125 \text{ k} \times 3 \text{ in.} = 375 \text{ in.-k}$$

$$f_v = \frac{125}{(2 \times 16) + (4 \times 0.625)} = 3.62 \text{ k/in.} \tag{6.5}$$

$$f_b = \frac{375}{105.4} = 3.56 \text{ k/in.} \tag{6.6}$$

$$f_r = \sqrt{(3.62)^2 + (3.56)^2} = 5.08 \text{ k/in.} \tag{6.7}$$

Required weld size:

$$D = \frac{5.08}{0.928} = 5.47 \text{ sixteenths}$$

Use $\frac{3}{8}$-in. weld.

The above example uses the customary method of taking the shear stress uniformly distributed over the entire length of the welds when the load is applied at the centroid of the weld group (see Fig. 6.10b). However, as shown in Fig. 6.10c, the actual shear stress would have a maximum value at midheight of $VQ/I = 5.54$ k/in.; at the ends, the stresses would be zero.

It is important to note that the actual maximum shear stress and maximum bending stress do not occur at the same location on the weld group (see Fig. 6.10c and 6.10d). Therefore, at the ends, the total stress is 3.56 k/in., while at midheight, the total stress is 5.54 k/in. The required weld size would then be

$$D = \frac{5.54}{0.928} = 5.97 \text{ sixteenths;} \quad \text{Use } \tfrac{3}{8}\text{-in. weld.}$$

In general, taking the shear stress as uniform simplifies the computations.

266 STEEL DESIGN FOR ENGINEERS AND ARCHITECTS

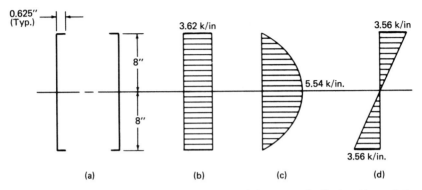

Fig. 6.10. (a) Weld group of Example 6.18; (b) assumed shear stress distribution; (c) actual shear stress distribution; (d) bending stress distribution.

6.9 AISCM DESIGN TABLES

The AISCM, in its "Connections" chapter (Part 4), provides tables for computing required weld sizes for connections subjected to eccentric loads. From these tables, weld dimensions, properties, and maximum allowable load can be determined (e.g., see Fig. 6.11).

Eccentric loads on weld groups tables in previous AISC manuals were based on elastic theory and the method of summing force vectors, as introduced in Sect. 6.8 of this text. The AISCM now offers more liberal values for weld design based on ultimate strength design. The saving in welding is due to plastic

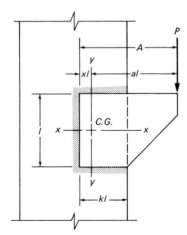

Fig. 6.11. One of several standard weld patterns that can be found in the AISCM and can be designed using tables.

WELDED CONNECTIONS 267

behavior, where yield stress is reached in the entire weld before failure occurs. Nevertheless, the elastic method is still recognized (see Alternate Method 1— Elastic, p. 4-72). It is important to note that the elastic method produces conservative results and can readily be applied to any weld configuration with loads in any directions.

When a weld pattern is standard, i.e., a configuration is tabulated in the AISCM, Part 4, the weld can be designed using the AISCM eccentric loads on weld groups tables. For welds unsymmetrical about the y-y axis, when the values of l, kl, and a are known, the center of gravity of the weld pattern can be determined from xl, where x is given in the tables based on k. Once a is known, the coefficient C is determined from the tables. Because the tables are calculated for E70 electrodes (the most commonly used), another coefficient C_1 is obtained from p. 4-72 and is used in conjunction with the tables for electrodes other than E70. As was discussed previously for the bolted connections, Alternate Method 2 (p. 4-73) is also provided for the welded connections to extend the AISCM tables for eccentric vertical loads to eccentric loads inclined at any angle from the vertical.

Example 6.19. Using the AISCM for eccentric loads on weld groups, determine the maximum allowable value of P for the given weld. Use a $\tfrac{5}{16}$-in. a weld of E70 electrodes.

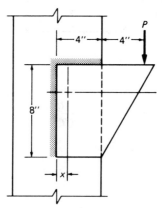

Solution.

$$P = CC_1 Dl$$

The weld formation is found in Table XXV. For the given conditions, $kl = 4$ in., $l = 8$ in., $xl + al = 8$ in., $D = 5$, $C_1 = 1.0$.

$$k = 4 \text{ in.}/l = 4 \text{ in.}/8 \text{ in.} = 0.5$$

268 STEEL DESIGN FOR ENGINEERS AND ARCHITECTS

The value for x under $k = 0.5$ is $x = 0.083$

$$al = 8 - xl; \quad a = 1 - 0.083 = 0.917$$

Using the chart for $k = 0.5$ and $a = 0.917$,

a	C	
0.90	.409	
0.917		$C = .403$ (through interpolation)
1.00	.371	

$$P = .403 \, (1.0) \, 5 \, (8) = 16.1 \text{ k}$$

Example 6.20. For the welded connection below, determine the minimum size weld required to carry the applied load. Use the AISC eccentric load tables and E60 electrodes.

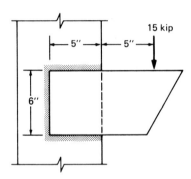

Solution.

$$D = \frac{P}{CC_1 l}$$

The weld Table XXIII is to be used.

Given values:

$$kl = 5 \text{ in.}$$
$$l = 6 \text{ in.}$$
$$xl + al = 10 \text{ in.}$$
$$P = 15 \text{ k}$$
$$C_1 = 0.857 \text{ (from table on p. 4-7 for E60 electrode)}$$
$$k = \tfrac{5}{6} = 0.83$$

Interpolating for $k = 0.83$, the value of x is determined to be 0.259.

$$al = 10 - xl; \quad a = \tfrac{10}{6} - 0.259; \quad a = 1.4$$

For $k = 0.83$ and $a = 1.4$, $C = 0.689$ (through interpolation)

$$D = \frac{P}{CC_1 l} = \frac{15\,k}{0.689\,(0.857)\,6} = 4.2 \quad \text{Use 5}$$

Minimum weld size is $\tfrac{5}{16}$ in.

Example 6.21. Rework Example 6.17, solving for P using the AISC eccentric loads table. Welds are $\tfrac{5}{16}$ in. E60 electrodes.

Solution.

$$P = CC_1 Dl$$

The weld Table XXIII is to be used.

Given values:

$kl = 5$ in.

$l = 15$ in.

$xl + al = 12$ in.

$D = 5$

$C_1 = 0.857$

$k = \tfrac{5}{15} = 0.33$

Interpolating for $k = 0.33$, the value of x is determined to be 0.066.

$$al = 12 - xl; \quad a = \tfrac{12}{15} - 0.066; \quad a = 0.734$$

For $k = 0.33$ and $a = 0.734$, $C = 0.630$ (double interpolation)

$$P = 0.630(0.857)\,5\,(15) = 40.5\,k$$

The value obtained shows an increase of 30% when compared with the elastic analysis method. Thus, a major saving can be realized for common weld groups

270 STEEL DESIGN FOR ENGINEERS AND ARCHITECTS

found in the AISCM. The elastic method is convenient to use on unusual weld groups and in all cases is conservative.

PROBLEMS TO BE SOLVED

Note: Unless specific mention is made, E70 weld is to be used.

6.1. Determine the strengths of $\frac{5}{16}$-in. and $\frac{1}{2}$-in. fillet welds 12 in. long for E60 and E70 electrodes.

6.2. A $\frac{3}{8}$-in. fillet weld is used to connect two plates as shown. Determine the maximum load that can be applied to the connection.

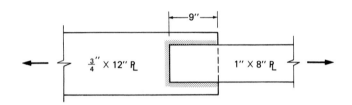

6.3. For the members shown, determine the lengths of $\frac{5}{16}$-in. fillet welds required to develop full tensile strength of the plates.

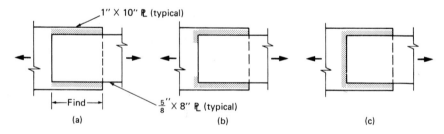

6.4. Determine the allowable tensile capacity of the connection shown. Use $\frac{5}{16}$-in. fillet weld with E60 electrodes.

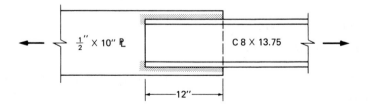

6.5. A 1-in. × 8-in. plate is to be connected to another member using $\frac{1}{2}$-in., E60 fillet welds. Due to clearance limitations, the two members can overlap only 5 in., as shown. Determine the maximum load P.

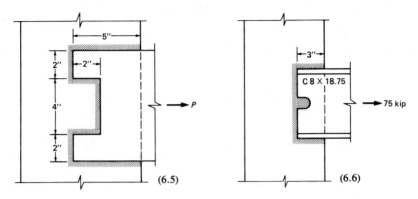

6.6. A channel is to be connected to a plate as shown. For clearance reasons, the channel cannot overlap the plate by more than 3 in. Design a combination fillet and slot weld to carry the applied load.

6.7. A welded beam consisting of two 2-in. × 16-in. flanges and $\frac{1}{2}$-in. × 40-in. web is subject to a maximum shear of 270 kips. Determine the fillet weld sizes for continuous welds and for intermittent welds.

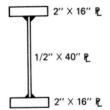

6.8. A W 21 × 101 has two 1-in. × 8-in. cover plates attached to the flanges by $\frac{3}{8}$-in. fillet welds, as shown. Determine the length of intermittent fillet welds required at a location where the shear is 140 kips.

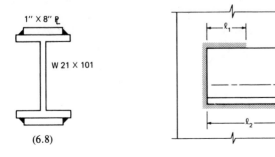

6.9. One leg of an $8 \times 8 \times \frac{1}{2}$-in. angle is to be connected with side welds and a weld at the end of the angle to a plate behind. Design the connection to develop full tensile capacity of the angle. Balance the fillet welds around the center of gravity of the angle, using $\frac{7}{16}$-in. weld and E60 electrodes.

6.10. A $6 \times 4 \times \frac{1}{2}$-in. angle is to be connected to develop full tensile capacity, as shown, to a plate using $\frac{3}{8}$-in. fillet welds. Determine l_1 and l_2 so that the centers of gravity for the welds and the angle coincide.

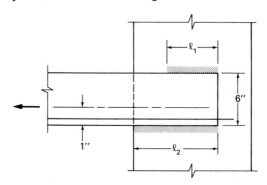

6.11. Rework Problem 6.10, but use end returns.

6.12. Design partial length cover plates for a W 21×101 rolled section for the loading shown. Assume full lateral support.

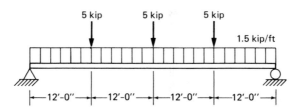

6.13. For the continuous beam shown, choose a wide-flange beam to resist the maximum positive moment, and design a welded partial cover plate where required for the negative moments. Assume full lateral support.

Note: To simplify the analysis of the continuous beam, neglect the variations of stiffness due to the cover plates. Also, omit moment redistribution.

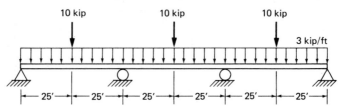

6.14. The weld shown is to carry an eccentric load of 20 kips acting 12 in. from the vertical fillet. Determine the minimum weld size required to carry the load. Do not use any tables.

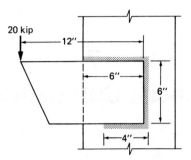

6.15. Determine the minimum weld size required to resist the load shown.

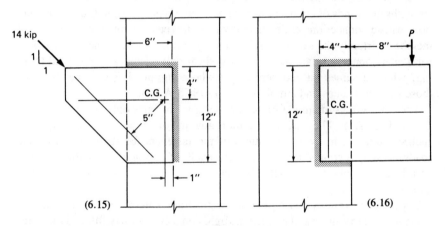

6.16. What load P can be supported by the connection shown if $\frac{3}{8}$-in. weld and E60 electrodes are used? Do not use any tables.

6.17. Rework Problem 6.16 using the appropriate AISCM tables.

6.18. Rework Example 6.18, without returns, using the appropriate AISCM tables.

7
Special Connections

7.1 BEAM-COLUMN CONNECTIONS

Three types of constructions are allowed by the AISC. For each type, the strengths and kinds of connections are specified. Type 1 is rigid frame construction, whose connections are sufficiently rigid to keep the rotation of connected members virtually the same at each joint. Type 2 is simple frame construction, for which connections are made to resist shear only. Type 3 is semirigid framing, which assumes an end connection whose rigidity is somewhat between those of rigid frames and simple frames (AISCS A2.2).

The connections of simple frames are designed to carry only shear, i.e., Type 2 (see Fig. 7.1). They allow a certain amount of elastic rotation and some inelastic rotation to avoid overstress of the fasteners for such cases as wind loading. Simple connections are usually made of two angles framing the web of a beam or only a seat angle. For the design of such connections, refer to Chapters 5 and 6.

Type 1 rigid frames and type 3 semirigid frames have connections that are designed to carry shear and moments. Rigid connections carry full end moments and semirigid ones carry an end moment that reduces the simply supported midspan moment. However, semirigid connections are not rigid enough to completely prevent a rotation or change in angle between the connected members (see Fig. 7.3).

The AISC, in Section A2.2, states that semirigid construction will be permitted only upon evidence that the connections to be used are capable of furnishing, as a minimum, a predictable portion of full end restraints. The exact moment to be carried by the end connection is difficult to determine. Charts would be necessary relating the moment-end rotation relationship for each type of connection. A conservative design approach is to assume a fully rigid end connection for a moment less than the fully fixed one and to design the connection for that moment. The beam is then designed for the positive moment resulting from the developed negative moment at the connection (see Fig. 7.5).

SPECIAL CONNECTIONS 275

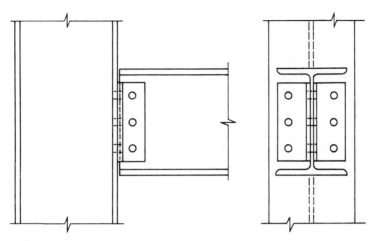

Fig. 7.1. Simple frame bolted connection. Bolts may be replaced by welds, most commonly to the beam web.

With the use of rigid connections, the AISC allows redistribution of moments. Compact beams and plate girders, which are rigidly framed to columns (or are continuous over supports), may be proportioned for $\frac{9}{10}$ of the negative moments produced by gravity loading, provided that the maximum positive moment is increased by $\frac{1}{10}$ of the average negative moments. The negative moment reduction may be used for proportioning the column for the combined

Fig. 7.2. A beam-to-column moment connection with welded butt plate.

276 STEEL DESIGN FOR ENGINEERS AND ARCHITECTS

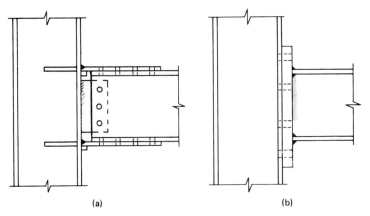

Fig. 7.3. Moment connections offering rigid and semirigid action: (a) shop-welded flange plates connected to the column flange with a field-bolted beam; (b) end plate shop-welded to the beam and then field-bolted to the column flange.

Fig. 7.4. A beam-to-column moment connection with flange angles.

SPECIAL CONNECTIONS 277

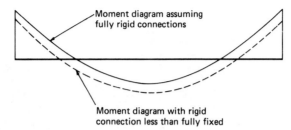

Fig. 7.5. Redistribution of moments for semirigid connections.

axial and bending loads, provided that the axial stress f_a does not exceed 0.15 F_a (AISCS F1.1) (also see "Continuous Beams" in Chapter 2).

7.2 SINGLE-PLATE SHEAR CONNECTIONS

Recent research has demonstrated that the single-plate connection, as shown in Fig. 7.6, can be considered a Type 2 connection, i.e., it will transmit only shear to the supporting member. However, due to the inherent rigidity of this connection, the bolts and welds must be designed for both shear and moment. The following example is based on the procedure given on p. 4-52 of the AISCM. For a similar analysis and tabulation, see p. 3-51 and Appendix C of

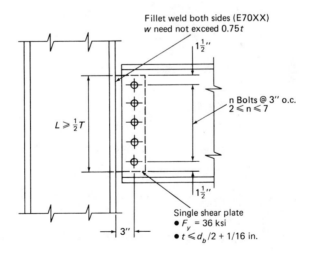

d_b = diam. of bolt, in.
t = thickness of shear plate, in.
T = distance between web toes of fillet at top and bottom of web, in.

Fig. 7.6. Typical single-plate shear connection.

278 STEEL DESIGN FOR ENGINEERS AND ARCHITECTS

"Engineering for Steel Construction," American Institute of Steel Construction (Chicago, 1984).

Example 7.1. A W 27 × 84 spans 24 ft, and its compression flange has full lateral support. Design the single-plate connection for the maximum allowable uniform load. Use $\frac{7}{8}$-in.-diameter A325-N bolts and E70 electrodes.

Solution. For a W 27 × 84:

$$S_x = 213 \text{ in.}^3$$

$$M_{all} = \frac{24 \text{ ksi} \times 213 \text{ in.}^3}{12} = 426 \text{ ft-k}$$

$$w_{all} = \frac{8 \times 426 \text{ ft-k}}{(24 \text{ ft})^2} = 5.92 \text{ k/ft}$$

$$R_{max} = 5.92 \text{ k/ft} \times 24 \text{ ft}/2 = 71 \text{ k}$$

1) Determine required number of bolts, n:

$$R_{max} = C \times R_v \quad \text{(Table XI, p. 4-62 of AISCM)}$$

where R_v = single shear bolt capacity, kips

$$= 12.6 \text{ k for } \tfrac{7}{8}\text{-in. A325-N} \quad \text{(Table I-D, p. 4-5 of AISCM)}$$

$$C = \frac{71 \text{ k}}{12.6 \text{ k}} = 5.63$$

From Table XI for $l = 3$ in. and $b = 3$ in., required number of bolts $n = 6.59$. Use 7 bolts.

$$L = (2 \times 1.5 \text{ in.}) + (6 \times 3 \text{ in.}) = 21 \text{ in.} > \tfrac{1}{2} \times 24 \text{ in.} = 12 \text{ in.} \quad \text{ok}$$

2) Determine required plate thickness t based on the net shear fracture capacity of the plate R_{ns}:

$$R_{ns} = 0.3 F_u [L - n(d_b + \tfrac{1}{16})]t = R_{max}$$

$$21 \text{ in.} - 7(0.875 \text{ in.} + 0.0625 \text{ in.}) = 14.4 \text{ in.}$$

$$71 \text{ k} = 0.3 \times 58 \text{ ksi} \times 14.4 \text{ in.} \times t \rightarrow t = 0.2834 \text{ in.}$$

Use $t = \frac{5}{16}$ in.

Note: $t = 0.3125$ in. $< \left(\dfrac{0.875 \text{ in.}}{2}\right) + 0.0625$ in. $= 0.5$ in. ok

3) Check plate capacity in yielding R_0[1]:

$$R_0 = 0.4 \, F_y \times L \times t$$
$$= 0.4 \times 36 \text{ ksi} \times 21 \text{ in.} \times 0.3125 \text{ in.} = 95 \text{ k} > 71 \text{ k} \quad \text{ok}$$

4) Check plate bearing capacity P_b (considering moment on bolts):
From Table I-E (p. 4-6 of AISCM):

$$P_b = 5.63 \times 60.9 \text{ k/bolt/in.} \times 0.3125 \text{ in.} = 107 \text{ k} > 71 \text{ k} \quad \text{ok}$$

5) Determine required fillet weld size, D in sixteenths, to develop R_0:

$$D = \frac{R_0}{L \times C}$$

where C is taken from Table XIX (p. 4-75 of AISCM) using $al = 3$ in. or $n \times 1$ in., whichever is larger, and $k = 0$.

Governing $al = 7 \times 1$ in. $= 7$ in. $\rightarrow a = \frac{7}{21} = 0.333$

From Table XIX, $C = 1.07$

$$D = \frac{95 \text{ k}}{21 \text{ in.} \times 1.07} = 4.2 \quad \text{Use } \tfrac{1}{4}\text{-in. weld.}[2]$$

Note that Table X-F on p. 4-56 of the AISCM could also be used in this case. The above calculations demonstrate how the allowable loads in these tables are obtained. Also, the bearing capacity and block shear (if coped) of the beam web should be checked.

7.3 TEE FRAMING SHEAR CONNECTIONS

Although not presented in the AISCM, a very useful type of shear connection is one that utilizes a single tee. A schematic representation of this connection

---

[1] Note that if the capacity is considered with respect to the elastic behavior, the maximum shear stress would be $1.5 R_0 / Lt$.
[2] Fillet weld size w need not exceed $0.75 \times t = 0.75 \times 0.3125$ in. $= 0.23$ in. Also, the thickness of the supporting member should be considered.

280 STEEL DESIGN FOR ENGINEERS AND ARCHITECTS

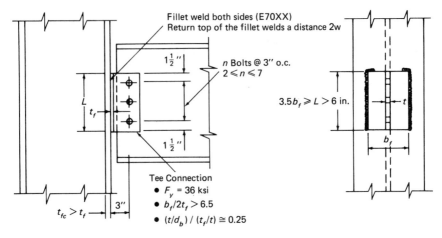

d_b = diam. of bolt, in.
b_f = flange width of tee, in.
t_f = flange thickness of tee, in.
t_{fc} = flange thickness of column, in.
t = web thickness of tee, in.

Fig. 7.7. Typical tee framing shear connection.

is given in Fig. 7.7. As with the single plate, it is also assumed in this case that no moment is transferred to the supporting member; the bolts and welds must be designed for the moment. It is important to note that the single-tee connection may offer somewhat more lateral moment resistance than a single-plate connection, which offers very little.

The design procedure presented in the following example is similar to the one for single-plate shear connections. Tabulated values for maximum loads can be found in "Allowable Stress Design of Simple Shear Connections," American Institute of Steel Construction (Chicago, 1990).

Example 7.2. Rework Example 7.1 using a tee framing shear connection.

Solution.

$$\text{From Ex. 7.1, } R_{\max} = 71 \text{ k}$$

1) Determine required number of bolts n:

$$R_{\max} = C \times R_v$$

From Ex. 7.1, $R_v = 12.6$ k, $C = 5.63$ (Table XI)

SPECIAL CONNECTIONS 281

Use seven bolts.

$$L = 21 \text{ in.} > 6.0 \text{ in.}$$

2) Determine required tee stem thickness t based on the net shear fracture capacity of the tee stem R_{ns}:

$$R_{ns} = 0.3 \, F_u [L - n(d_b + \tfrac{1}{16})] t = R_{max}$$

From Ex. 7.1, $t_{req} = 0.2834$ in. Try $t = \tfrac{5}{16}$ in.

3) Check tee stem capacity in yielding R_0:

$$R_0 = 0.4 \, F_y \times L \times t = 95 \text{ k} > 71 \text{ k} \quad \text{ok}$$

4) Check tee stem bearing capacity P_b:

From Ex. 7.1, $P_b = 107 \text{ k} > 71 \text{ k} \quad \text{ok}$

5) Determine required fillet weld size, D in sixteenths, to develop R_0:

$$D = \frac{R_0}{L \times C}$$

where C is taken from Table XIX (p. 4-75 of AISCM) using $al = 3$ in. and $k = 0$.

$$a = \frac{3 \text{ in.}}{21 \text{ in.}} = 0.143$$

From Table XIX, $C = 1.52$

$$D = \frac{95 \text{ k}}{21 \text{ in.} \times 1.52} = 2.98 \quad \text{Use } \tfrac{3}{16}\text{-in. weld.}$$

6) Selection of tee:
   The selection of the tee will be guided by the design assumptions given in Fig. 7.7.
   a) $(t/d_b)/(t_f/t) \cong 0.25$
      Given $t = 0.3125$ in., $d_b = 0.875$ in., solve for t_f:

$$t_f = \frac{(0.3125 \text{ in.}/0.875 \text{ in.})}{0.25} \times 0.3125 \text{ in.} = 0.45 \text{ in.}$$

b) $L \leq 3.5b_f$

$$b_f = \frac{21 \text{ in.}}{3.5} = 6.0 \text{ in. (min.)}$$

Try WT 8 × 18:

$$t = 0.295 \text{ in.} > 0.2834 \text{ in.} \quad \text{ok}$$
$$b_f = 6.985 \text{ in.} > 6.0 \text{ in.} \quad \text{ok}$$
$$t_f = 0.43 \text{ in.} \cong 0.45 \text{ in.} \quad \text{ok}$$
$$\frac{b_f}{2t_f} = \frac{6.985 \text{ in.}}{2 \times 0.43 \text{ in.}} = 8.1 > 6.5 \quad \text{ok}$$

Use WT 8 × 18 × 1 ft, 9 in.

Note: The thickness of the column flange t_{fc} should be greater than $t_f = 0.43$ in. Otherwise, the size of the tee should be modified. Also, the bearing capacity and block shear of the beam web (if coped) should be checked.

7.4 DESIGN OF MOMENT CONNECTIONS

The requirements for rigid and semirigid connections are given in AISCS J1. Usual connection details consist of plates attached to the beam flanges and to the column flange or web and angles connecting the beam web to the column flange or web. An alternative approach, when joining the beam to the column flange, is to weld a plate to the end of the beam and then to bolt the plate to the column (see Fig. 7.3b). Stiffeners may be required to prevent the column section from either buckling or yielding due to the force applied to the column by the bending moment of the joint.

Both rigid and semirigid connections assume that the web connections carry the entire shear force and that the top and bottom flange connections carry the end moment. Fasteners are designed for the combined effects of shear, tensile forces resulting from the moment induced by the rigidity of the connection, or a combination of both. The number of web fasteners required is determined from

$$N_b \geq \frac{R}{R_v} \tag{7.1}$$

SPECIAL CONNECTIONS 283

Fig. 7.8. Corner detail of beam-to-column connections. Note how the large T section provides moment connection for the beam on the right while providing tension connection for the four-angle wind bracing member.

where R is the end reaction and R_v is the shear capacity of one fastener. If web angles are chosen to transfer shear, they may be determined as presented in Chapter 5 and may be selected from the AISCM framed beam connections tables. If a single plate is used, only shear has to be considered, as the moment is assumed to be carried by the moment plates (see Fig. 7.3a).

Beam end moments are assumed to be resisted totally by couples acting in the top and bottom flanges of the beam. Bolted and riveted connections are put into single shear by the tensile force at the top and compressive force at the bottom connections. Flange forces developed are determined by dividing the moment by the moment arm d or

$$T = \frac{12\,M}{d} \qquad (7.2)$$

where M is the support moment in ft-kips and T is the flange force in kips. Weld length and size must similarly be designed to resist the force. For flange plate connections, the plate is designed as a tension member.

Alternatively, an end plate connection can be provided in lieu of web and flange plate connections where the end plate is shop-welded to the beam and field-bolted to the column flange. High-strength bolts are arranged near the top flange to transfer the flexural tensile force in the flange and additionally elsewhere to help resist the beam end shear. The welds connecting the end plate to the beam are usually fillet welds, though they may be full penetration type. The tension flange weld is sized to develop the full force due to bending, and the web weld is designed to carry the beam reaction. In addition, the end plate must be designed to resist the bending stress developed between connecting bolts.

The need for column web stiffeners must be investigated for the moment-induced force factored according to the type of loading case. Web stiffeners are required on both the tension and compression flanges whenever the value of A_{st} is positive (AISCS K1.8) where

$$A_{st} = \frac{P_{bf} - F_{yc}t_{wc}(t_b + 5k)}{F_{yst}} \quad (7.3a)$$

or

$$A_{st} = \frac{P_{bf} - P_{wi}t_b - P_{wo}}{F_{yst}} \quad (7.3b)$$

where $P_{wi} = F_{yc}t_{wc}$, kips/in.
$P_{wo} = 5F_{yc}t_{wc}k$, kips
t_{wc} = thickness of column web, in.
t_b = thickness of beam flange or connection plate delivering the concentrated force to the column, in.

The factored force $P_{bf} = 5T/3$ when T is due to live and dead loads only, and $P_{bf} = 4T/3$ when T is due to live and dead loads plus wind or earthquake forces, unless local building codes dictate otherwise. The yield stress subscripts c and st refer to the column and stiffener sections, respectively. Also, column stiffeners must be provided opposite the compression flange connection when (AISCS K1.6)

$$d_c > \frac{4100t_{wc}^3\sqrt{F_{yc}}}{P_{bf}} \quad (7.4a)$$

or

$$P_{bf} > P_{wb} \quad (7.4b)$$

SPECIAL CONNECTIONS 285

where $P_{wb} = 4100 t_{wc}^3 \sqrt{F_{yc}/d_c}$ = maximum column web resisting force at beam compression flange, kips
d_c = depth of column web clear of fillets, in.

Note that in this case, the stiffeners shall be designed as columns in accordance with AISCS K1.8 and E2. Web stiffeners must be provided opposite the connection of the tension flange when (AISCS K1.2):

$$t_f < 0.4 \sqrt{\frac{P_{bf}}{F_{yc}}} \quad (7.5a)$$

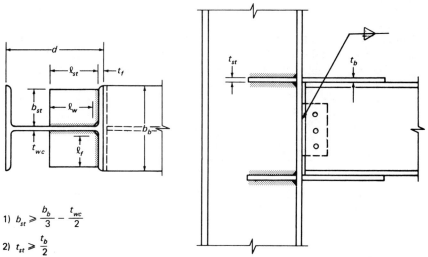

1) $b_{st} \geq \dfrac{b_b}{3} - \dfrac{t_{wc}}{2}$

2) $t_{st} \geq \dfrac{t_b}{2}$

3) $\ell_{st} \geq \dfrac{d}{2} - t_f$ for one-sided loading (if required by AISCS K1.2 or (K1-9))

   $\ell_{st} = d - 2 t_f$ for both sides loaded

4) $\dfrac{b_{st}}{t_{st}} \leq \dfrac{95}{\sqrt{F_y}}$ (AISCS B5)

5) Welds:
   (a) Welds joining stiffeners to the column web shall be sized to carry the force in the stiffener caused by the unbalanced moments on opposite sides of the column (AISCS K1.8).
   (b) Stiffeners shall be welded to the loaded flange if the load normal to the flange is tensile. When the load normal to the flange is compressive, the stiffener shall either bear on or be welded to the loaded flange (AISCS K1.8).

Fig. 7.9. Column web stiffener requirements.

or

$$P_{bf} > P_{fb} \tag{7.5b}$$

where $P_{fb} = F_{yc}t_f^2/0.16$ = maximum column web resisting force at beam tension flange, kips

t_f = thickness of column flange, in.

Regardless of the force or formula, stiffeners, when required, must meet the minimum area and dimensional requirements given in AISCS K1.8. These criteria are depicted in Fig. 7.9. Recommended design procedures for various types of moment connections are given on pp. 4-100 through 4-125 of the AISCM. The following examples illustrate these design procedures.

Example 7.3. Design bolted flange plate moment connections (single shear plate shop-welded, field-bolted[3]) for the situation shown. Use $\frac{3}{4}$-in. A325-SC bolts and an allowable beam bending stress F_b = 22 ksi.

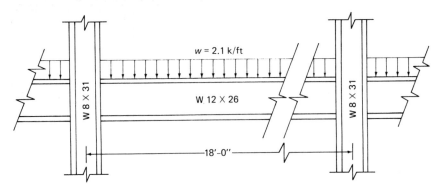

Solution. For W 8 × 31

d = 8.00 in., t_w = 0.285 in., b_f = 7.995 in., t_f = 0.435 in., $k = \frac{15}{16}$ in.

For W 12 × 26

d = 12.22 in., t_w = 0.230 in., b_f = 6.490 in.,

t_f = 0.380 in., $k = \frac{7}{8}$ in.

---

[3]Welded to column, bolted to beam. See AISCM p. 4-109.

Frame analysis has yielded $M^- = 48.60$ ft-k and $M^+ = 36.45$ ft-k

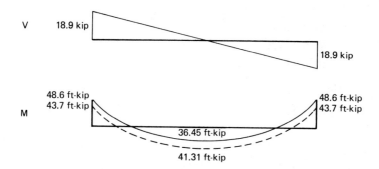

Redistributing moments for rigidly framed members (AISCS F1.1):

$$M^- = 0.9 \times 48.6 \text{ ft-k} = 43.7 \text{ ft-k}$$
$$M^+ = (0.1 \times 48.6 \text{ ft-k}) + 36.45 \text{ ft-k} = 41.31 \text{ ft-k}$$

Check beam flange area reduction for two rows of $\frac{3}{4}$-in. connecting bolts (AISCS B10):

$$A_{fg} = 6.49 \text{ in.} \times 0.38 \text{ in.} = 2.47 \text{ in.}^2$$
$$A_{fn} = 2.47 \text{ in.}^2 - [2(0.75 \text{ in.} + 0.125 \text{ in.}) \times 0.38 \text{ in.}] = 1.81 \text{ in.}^2$$
$$0.5 F_u A_{fn} = 0.5 \times 58 \text{ ksi} \times 1.81 \text{ in.}^2 = 52.5 \text{ k}$$
$$0.6 F_y A_{fg} = 0.6 \times 36 \text{ ksi} \times 2.47 \text{ in.}^2 = 53.4 \text{ k}$$

Since 53.4 k > 52.5 k, the effective tension flange area is

$$A_{fe} = \frac{5}{6}\left(\frac{58}{36}\right)(1.81) = 2.43 \text{ in.}^2$$

$$I_n = 204 \text{ in.}^4 - \left[(2.47 \text{ in.}^2 - 2.43 \text{ in.}^2)\left(\frac{12.22 \text{ in.} - 0.38 \text{ in.}}{2}\right)^2\right]$$
$$= 202.6 \text{ in.}^4$$

$$S_n = \frac{202.6 \text{ in.}^4}{6.11 \text{ in.}} = 33.2 \text{ in.}^3$$

$$S_{\text{req}} = \frac{43.7 \text{ ft-k} \times 12 \text{ in./ft}}{22 \text{ ksi}} = 23.8 \text{ in.}^3 < 33.2 \text{ in.}^3 \quad \text{ok}$$

288 STEEL DESIGN FOR ENGINEERS AND ARCHITECTS

Web connection (single shear plate):

Shear capacity of $\tfrac{3}{4}$-in. A325-SC in single shear = 7.51 k (Table I-D)

$$\text{No. of bolts required} = \frac{18.9 \text{ k}}{7.51 \text{ k}} = 2.52 \quad \text{Use 3 bolts.}$$

Design of shear plate:

Try 9-in. shear plate 3 in. center-to-center of bolts, with $1\tfrac{1}{2}$ in. at ends.

$$l_n = 9 \text{ in.} - 3(.75 \text{ in.} + .125 \text{ in.}) = 6.375 \text{ in.}$$

$$F_v = 0.30\, F_u = 17.4 \text{ ksi} \quad \text{(AISCS J4)}$$

$$t_{pl} \geq \frac{R}{l_n F_v} = \frac{18.9 \text{ k}}{6.375 \text{ in.} \times 17.4 \text{ ksi}} = 0.170 \text{ in.}$$

Try $\tfrac{1}{4}$-in. plate.

Check bearing on beam web:

For $F_u = 58$ ksi, $1\tfrac{1}{2}$-in. edge distance, 3-in. bolt spacing, and $t = 0.23$ in.:
1.5(0.75 in.) = 1.125 in. < 1.5 in. edge distance

From Table I-E:

Allowable load = 3 × 52.2 k/in. × 0.23 in. = 36 k > 18.9 k   ok

Check bearing on plate:

From Table I-E:

Allowable load = 3 × 13.1 k = 39.3 k > 18.9 k   ok

Plate fillet weld to column flange (E70XX weld)

$$D \geq \frac{18.9 \text{ k}}{2(0.928 \text{ k/in.}) \times 9 \text{ in.}} = 1.13$$

Since column flange thickness is 0.435 in., use minimum weld size of $\tfrac{3}{16}$ in. (Table J2.4).

Design of flange plates

Flange force due to moment

$$T = \frac{12 M}{d} = \frac{12 \text{ in./ft.} \times 48.6 \text{ ft-k}}{12.22 \text{ in.}} = 47.7 \text{ k}$$

$$A_n \geq \frac{T}{0.5 F_u} = \frac{47.7 \text{ k}}{29 \text{ ksi}} = 1.64 \text{ in.}^2$$

$$A_g \geq \frac{T}{0.6 F_y} = \frac{47.7 \text{ k}}{22 \text{ ksi}} = 2.17 \text{ in.}^2$$

Try $\frac{3}{8} \times 6\frac{1}{2}$-in. plate

$A_g = 0.375$ in. $\times$ 6.5 in. $= 2.44$ in.^{2} > 2.17 in.^{2} ok

$A_n = 2.44$ in.^{2} $- 2(.75 + .125) \times .375$ in. $= 1.78$ in.^{2} > 1.64 in.^{2} ok

Number of bolts required (use oversize holes to assist in field assembly):

$$N_b = \frac{T}{R_v} = \frac{47.7 \text{ k}}{6.63 \text{ k}} = 7.2$$

Use minimum four bolts in each of two rows $\left(3\frac{1}{2} \text{ in. gage}\right)$

Check for column web stiffeners

$$A_{st} \geq \frac{P_{bf} - F_{yc} t_{wc}(t_b + 5 k)}{F_{yst}} \quad \text{(AISCS K1.8)}$$

$$P_{bf} = \frac{5}{3} T = 79.5 \text{ k}^4$$

$$t_{wc} = 0.285 \text{ in.}, \quad k = \frac{15}{16} \text{ in.}$$

$$t_b = t_{pl} = 0.375 \text{ in.}$$

---

[4]Could have also used the force at the stiffeners:

$$P_{bf} = \frac{5}{3} \times \frac{M}{(d + t_b)} = \frac{5 \times 48.6 \times 12}{3 \times (12.22 + 0.375)} = 77.2 \text{ k}$$

290 STEEL DESIGN FOR ENGINEERS AND ARCHITECTS

$$A_{st} \geq \frac{79.5 - 36 \times 0.285\left(.375 + 5 \times \frac{15}{16}\right)}{36} = 0.77 \text{ in.}^2$$

Column web stiffeners are required

Design of stiffeners (AISCS K1.8)

1) $b_{st} + \dfrac{t_{wc}}{2} \geq \dfrac{b_b}{3}$

$$b_{st} \geq \frac{b_b}{3} - \frac{t_{wc}}{2} = \frac{6.5 \text{ in.}}{3} - \frac{0.285 \text{ in.}}{2} = 2.02 \text{ in.}$$

2) $t_{st} \geq \dfrac{t_b}{2} = \dfrac{0.375 \text{ in.}}{2} = 0.19 \text{ in.}$

3) $l_{st} = \dfrac{8.0 \text{ in.}}{2} - 0.435 \text{ in.} = 3.57 \text{ in.}$

Try $3\tfrac{1}{2}$ in. × $\tfrac{1}{4}$ in. × 0 ft, 4 in. Clip corner $\tfrac{3}{4}$ in. × $\tfrac{3}{4}$ in.

$$A_{st} = 2(.25 \times 3.5) = 1.75 \text{ in.}^2 > 0.77 \text{ in.}^2 \quad \text{ok}$$

4) $\dfrac{3.5 \text{ in.}}{0.25 \text{ in.}} \leq 15.83 \quad \text{ok} \quad \text{(AISCS B5)}$

5) Weld length

From Table J2.4:

Min. weld size to column web: $\tfrac{3}{16}$ in.

Min. weld size to column flange: $\tfrac{3}{16}$ in.

Determine the forces at the column web to be resisted by stiffener welds:

$$P_{bf} - P_{wi}t_b - P_{wo} = 79.5 \text{ k} - (10 \text{ k/in.} \times 0.375 \text{ in.}) - 48 \text{ k} = 27.8 \text{ k}$$

where P_{wi} and P_{wo} were obtained from the bottom of the Column Load Tables on p. 3-31.

Determine the forces at the tension flange to be resisted by stiffener welds:

$$P_{bf} - P_{fb} = 79.5 \text{ k} - 43 \text{ k} = 36.5 \text{ k}$$

where P_{fb} is also obtained from p. 3-31.

SPECIAL CONNECTIONS

Also, $P_{wb} = 93 \text{ k} > 79.5 \text{ k}$

Thus, tension flange force governs.

Load in stiffener = 36.5 k/2 = 18.3 k each side.

$$l_w = \frac{\text{Load in stiffener}}{2 \times D \times 0.928 \times \text{load factor}} = \frac{18.3}{2 \times 3 \times 0.928 \times 1.67}$$
$$= 2.0 \text{ in.}$$

Use weld length of 4 in. − 0.75 in. = $3\frac{1}{4}$ in.

$$l_f = l_w = 2.0 \text{ in.}$$

Use weld length of 3.5 in. − 0.75 in. = $2\frac{3}{4}$ in.

As an alternative, it is permissible to finish the two compression flange stiffeners to bear instead of welding to the flange since the end moments are not reversible:

$$F_p = 0.9 \times 36 \text{ ksi} \times 1.67 = 54.1 \text{ ksi}$$

$$f_p = \frac{18.3 \text{ k}}{(3.5 \text{ in.} - 0.75 \text{ in.}) \times 0.25 \text{ in.}} = 26.6 \text{ ksi} < 54.1 \text{ ksi} \quad \text{ok}$$

Use end moment connection as shown.[5]

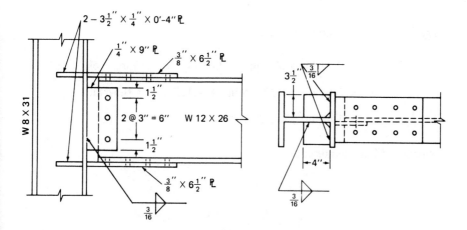

---

[5]Since beams frame into both sides of the column, the stiffener length should actually be $d - 2t_f$ = 7.13 in. (see Fig. 7.9). This example illustrates the required stiffener and weld lengths if one beam frames into one side of the column only.

Example 7.4. For the previous example, design a semirigid moment connection, for a total distributed load of 1.8 k/ft, such that the negative and positive moments are equal. Use $\frac{3}{4}$-in. A325-N bolts (instead of SC) to allow slippage of connection, obtaining semirigid action.

Solution.

$$R = \frac{1}{2} \times 1.8 \text{ k/ft} \times 18 \text{ ft} = 16.2 \text{ k}$$

$$M^- = M^+ = \frac{wL^2}{16} = 36.45 \text{ ft-k}$$

The web shear connection will not change. Design flange plates for the reduced moment value.

$$T = \frac{12 M}{d} = \frac{12 \text{ in./ft} \times 36.45 \text{ ft-k}}{12.22 \text{ in.}} = 35.8 \text{ k}$$

$$A_n \geq \frac{T}{0.5 F_y} = \frac{35.8 \text{ k}}{29 \text{ ksi}} = 1.23 \text{ in.}^2$$

$$A_g \geq \frac{T}{0.6 F_y} = \frac{35.8 \text{ k}}{22 \text{ ksi}} = 1.63 \text{ in.}^2$$

Try $\frac{5}{16}$-in. × $6\frac{1}{2}$-in. moment plate:

$$A_g = 0.31 \text{ in.} \times 6.5 \text{ in.} = 2.02 \text{ in.}^2 > 1.63 \text{ in.}^2 \quad \text{ok}$$

$$A_n = 2.02 \text{ in.}^2 - 2(0.75 \text{ in.} + 0.125 \text{ in.}) \times 0.31 \text{ in.}$$

$$= 1.48 \text{ in.}^2 > 1.23 \text{ in.}^2 \quad \text{ok}$$

Number of bolts required

$$R_v = 9.3 \text{ k}$$

$$N_b = \frac{R}{R_v} = \frac{35.8 \text{ k}}{9.3 \text{ k}} = 3.85$$

Use two bolts in each of two rows ($3\frac{1}{2}$-in. gage)

Check for column web stiffeners

$$A_{st} \geq \frac{P_{bf} - F_{yc}t_{wc}(t_b + 5k)}{F_{yst}}$$

$$P_{bf} = \frac{5}{3}T = 59.7 \text{ k}$$

$$A_{st} \geq \frac{59.7 - 36 \times 0.285\left(0.313 + 5 \times \frac{15}{16}\right)}{36} = 0.23 \text{ in.}^2$$

Column web stiffeners are required.

1) $b_{st} \geq \dfrac{b_b}{3} - \dfrac{t_{wc}}{2} = 2.02$ in.

2) $t_{st} \geq \dfrac{t_b}{2} = 0.16$ in.

3) $l_{st} \geq \dfrac{d}{2} - t_f = 3.57$ in.

Try 2 in. × $\frac{3}{16}$ in. × 0 ft, 4 in. Clip corner $\frac{3}{4}$ in. × $\frac{3}{4}$ in.

$$A_{st} = 2(0.188 \text{ in.} \times 2.0 \text{ in.}) = 0.75 \text{ in.}^2 > 0.23 \text{ in.}^2 \quad \text{ok}$$

4) $\dfrac{2.0}{0.188} \leq 15.83$ ok

5) Weld length

$$P_{bf} = 59.7 \text{ k} < P_{wb} = 93 \text{ k}$$

Force at column web $= P_{bf} - P_{wi}t_b - P_{wo}$

$$= 59.7 - (10 \times 0.31) - 48 = 8.6 \text{ k}$$

Force at tension flange $= P_{bf} - P_{fb} = 59.7 - 43 = 16.7$ k (governs)

$$l_w = \frac{16.7/2}{2 \times 3 \times 0.928 \times 1.67} = 0.9 \text{ in.}$$

Use weld length = 4 in. − 0.75 in. = $3\frac{1}{4}$ in.

$l_f = l_w = 0.9$ in.

Use weld length = 2 in. − 0.75 in. = $1\frac{1}{4}$ in.

Use end moment connection as shown.

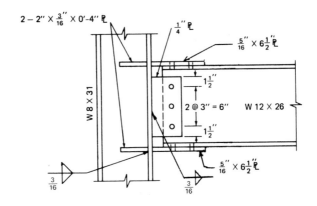

Example 7.5. A wide-flange beam spanning 35 ft supports a uniform load of 1.3 k/ft. Design the beam and semirigid connections assuming that the maximum positive moment and maximum negative moment will be equal. Use web angles and tee connections to a W 14 × 90 column in the weak axis with $\frac{3}{4}$-in. A325-X bolts. Assume full lateral support.

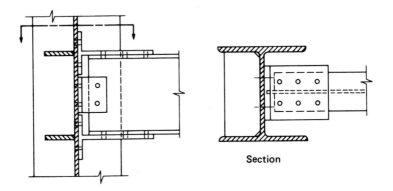

Section

Solution. For W 14 × 90 column

$$d = 14.02 \text{ in.}, \quad b_f = 14.52 \text{ in.}, \quad t_f = 0.71 \text{ in.},$$

SPECIAL CONNECTIONS

$t_w = 0.44$ in., $k = 1\frac{3}{8}$ in.

$$M_{max} = \frac{wL^2}{16} = \frac{1.3 \text{ k/ft} \times (35.0 \text{ ft})^2}{16} = 99.5 \text{ ft-k}$$

$$S_{req} = \frac{M}{F_b} = \frac{12 \text{ in./ft} \times 99.5 \text{ ft-k}}{24.0 \text{ ksi}} = 49.8 \text{ in.}^3$$

Use W 18 × 35 beam, $S_x = 57.6$ in.^{3}

$d = 17.70$ in., $b_f = 6.00$ in., $t_f = 0.425$ in., $t_w = 0.30$ in.

At the connection, design the web angles to carry full shear and the end moment by the top and bottom connections.

$$R = \frac{wL}{2} = \frac{1.3 \text{ k/ft} \times 35 \text{ ft}}{2} = 22.8 \text{ k}$$

Referring to Framed Beam Connections, a total number of two $\frac{3}{4}$-in. A325-X bolts are required per line:

$R_{max} = 53.0$ k   (AISC, Table II-A)

$R_{max} = 33.7$ k ($\frac{1}{4}$ in. angle)   (AISC, Table II-C)

22.8 k < 33.7 k   ok

Use two 4 in. × 4 in. × $\frac{1}{4}$ in. beam web framing angles $5\frac{1}{2}$ in. long.

Flange tee connections

Flange force due to moment

$$T = \frac{12 M}{d} = \frac{12 \text{ in./ft} \times 99.5 \text{ ft-k}}{17.70 \text{ in.}} = 67.5 \text{ k}$$

See Example 7.11 for the design of tee section, as it also involves the prying effect.

Use WT 10.5 × 36.5, $8\frac{1}{2}$ in. long

Check for column web stiffener

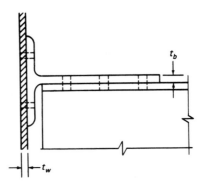

$$A_{st} \geq \frac{P_{bf} - F_{yc}t_{wc}(t_b + 5k)}{F_{yst}}$$

$$P_{bf} = \frac{5}{3}T = 112.5 \text{ k}$$

$$t_{wc} = 0.44 \text{ in.}$$

$$t_b = 0.455 \text{ in.}$$

$k = 0$ (connection is to web of column)

$$A_{st} \geq \frac{112.5 - 36 \times 0.44 \times (0.455 + 0)}{36} = 2.92 \text{ in.}^2$$

Column web stiffener is required

Design of stiffener

$$b_{st} \geq \frac{b_b}{3} - \frac{t_{wc}}{2} = \frac{8.5 \text{ in.}}{3} - \frac{0.44 \text{ in.}}{2} = 2.61 \text{ in.}$$

$$t_{st} \geq \frac{t_b}{2} = \frac{0.46 \text{ in.}}{2} = 0.23 \text{ in.}$$

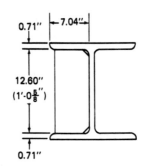

Try 4 in. × $\frac{5}{16}$ in. × $12\frac{5}{8}$ in.

$$A_{st} = 0.3125 \text{ in.} \times 12.625 \text{ in.} = 3.94 \text{ in.}^2 > 2.92 \text{ in.}^2 \quad \text{ok}$$

$$\frac{12.625 \text{ in.}}{0.3125 \text{ in.}} = 40.4 > 15.83 \quad \text{N.G.}$$

$$t_{min} = \frac{12.625 \text{ in.}}{15.83} = 0.80 \text{ in.}$$

Try 4 in. × $\frac{7}{8}$ in. × $12\frac{5}{8}$ in.

Weld length

Try $\frac{5}{16}$ in. weld (min. weld size)

$$P_{bf} - P_{wi}t_b - P_{wo} = 112.5 - (16 \times 0.455) - 0 = 105.2 \text{ k}$$

$$\text{Total weld length required} = \frac{105.2}{5 \times 0.928 \times 1.67} = 13.6 \text{ in.}$$

Along column flanges:

$$\frac{13.6 \text{ in.}}{4} = 3.40 \text{ in.} \quad \text{Use } 3\frac{1}{2} \text{ in.}$$

Along column web:

$$\frac{13.6 \text{ in.}}{2} = 6.80 \text{ in.} \quad \text{Use 7 in. minimum}$$

Assume 1-in. notch in corners of stiffener plate.

Use $4\frac{1}{2}$ in. × $\frac{7}{8}$ in. × 1 ft, $\frac{5}{8}$ in. stiffener with continuous $\frac{5}{16}$-in. weld and 1 in. × 1 in. notch at corners.

Example 7.6. Design an end plate moment connection to resist the maximum negative moment force and maximum end reaction without beam web reinforcement for a W 12 × 65 beam to a W 14 × 159 column. Use A490-N bolts, F_y = 36.0 ksi.

298 STEEL DESIGN FOR ENGINEERS AND ARCHITECTS

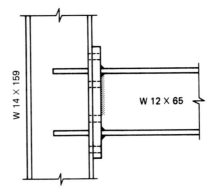

Solution.

$$F_b = 24.0 \text{ ksi}, \quad F_v = 14.4 \text{ ksi}$$

For W 14 × 159

$$d = 14.98 \text{ in.}, \quad b_f = 15.57 \text{ in.}, \quad t_f = 1.19 \text{ in.}, \quad t_w = 0.745 \text{ in.}$$

For W 12 × 65

$$d = 12.12 \text{ in.}, \quad b_f = 12.00 \text{ in.}, \quad t_f = 0.605 \text{ in.}, \quad t_w = 0.390 \text{ in.}$$

Maximum beam moment

$$M = F_b \times S = 24 \text{ ksi} \times 87.9 \text{ in.}^3 = 2110 \text{ in.-k} = 175.8 \text{ ft-k}$$

Maximum shear force

The effective area in resisting shear is the overall depth times the web thickness.

$$A_w = 0.390 \text{ in.} \times 12.12 \text{ in.} = 4.73 \text{ in.}^2$$
$$V = F_v \times A_w = 14.4 \text{ ksi} \times 4.73 \text{ in.}^2 = 68.1 \text{ k}$$

Flange force due to moment

$$T = \frac{12 M}{d - t_{fb}} = \frac{12 \text{ in./ft} \times 175.8 \text{ ft-k}}{12.12 \text{ in.} - 0.605 \text{ in.}} = 183.2 \text{ k}$$

SPECIAL CONNECTIONS

Tensile flange force is resisted by four bolts in tension at tension flange

$$\text{Force per bolt} = \frac{183.2 \text{ k}}{4} = 45.8 \text{ k/bolt}$$

From Table I-A, use $1\frac{1}{8}$ in diam. bolt (allow. load = 53.7 k).

Try eight bolts to resist reaction, four at the tension flange and four at the compression flange.

Number of bolts required to carry shear

$$R_v = 27.8 \text{ k} \quad (\text{AISC, Table I-D})$$

$$N_b = \frac{V}{R_v} = \frac{68.1 \text{ k}}{27.8 \text{ k}} = 2.45 < 8 \quad \text{ok}$$

Beam flange welds must also resist tension force.

For E70XX weld, capacity per sixteenth is

$$21 \text{ ksi} \times 0.707/16 = 0.928 \text{ k/in./sixteenth}$$

$$D \geq \frac{T}{0.928 \times [2(b_f + t_f) - t_w]} = \frac{183.2 \text{ k}}{0.928[2(12.0 + 0.605) - 0.39]} = 7.95$$

Use $\frac{1}{2}$-in. weld.

End plate design (the design follows the procedure on p. 4-119 of AISCM).

For proper welding, recommended plate width $b_p = b_f + 1 = 12 + 1 = 13$ in.

For the bolt arrangement shown,

$$P_f = \frac{5.5 \text{ in.} - 0.605 \text{ in.}}{2} = 2.45 \text{ in.} > d_b + \frac{1}{2} \text{ in.} = 1.63 \text{ in.} \quad \text{ok}$$

$$P_e = P_f - \left(\frac{d_b}{4}\right) - 0.707w$$

$$= 2.45 \text{ in.} - \left(\frac{1.125 \text{ in.}}{4}\right) - 0.707(0.50 \text{ in.}) = 1.82 \text{ in.}$$

300 STEEL DESIGN FOR ENGINEERS AND ARCHITECTS

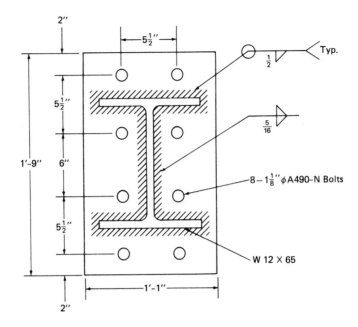

$M_e = \alpha_m T P_e / 4$

$\alpha_m = C_a C_b (A_f/A_w)^{1/3} (P_e/d_b)^{1/4}$

$C_a = 1.13 \quad (F_y = 36 \text{ ksi}; \text{ p. 4-119 of AISCM})$

$C_b = (b_f/b_p)^{1/2} = 0.96$

$A_f/A_w = 1.706$

$\alpha_m = 1.13 (0.96)(1.706)^{1/3}(1.82/1.125)^{1/4} = 1.46$

$M_e = 1.46 \times 183.2 \text{ k} \times 1.82 \text{ in.}/4 = 121.7 \text{ in.-k}$

Required plate thickness

$$t_p \geq \left(\frac{6 M_e}{b_p F_b}\right)^{1/2} = \left(\frac{6 \times 121.7 \text{ in.-k}}{13 \text{ in.} \times 27.0 \text{ ksi}}\right)^{1/2} = 1.44 \text{ in.}$$

Try $1\frac{1}{2}$-in. plate thickness

Beam web to end plate weld:

Minimum weld size = $\frac{5}{16}$ in.

SPECIAL CONNECTIONS 301

Required weld to develop maximum web tension stress ($0.60 F_y = 22$ ksi) in web near flanges:

$$D = \frac{22.0 \text{ ksi} \times t_{wb}}{2 \times 0.928} = \frac{22.0 \times 0.39}{2 \times 0.928} = 4.6$$

Use $\frac{5}{16}$-in. fillet welds continuously on both sides of the web.

Check if column stiffeners are required:

$$P_{bf} = \tfrac{5}{3} T = 305.3 \text{ k}$$

a) Check for column web yielding opposite the compression flange:

$$\begin{aligned}
P_{bf} \text{ (allow)} &= F_{yc} t_{wc}(t_{fb} + 6k + 2t_p + 2w) \\
&= 36 \times 0.745 [0.605 + (6 \times 1.875) + (2 \times 1.5) \\
&\quad + (2 \times 0.50)] \\
&= 425.3 \text{ k} > 305.3 \text{ k} \quad \text{Stiffeners opposite compression} \\
&\qquad\qquad\qquad\qquad\qquad\quad \text{flange are not required.}
\end{aligned}$$

b) Check for column web buckling:

$$P_{bf} \text{ (allow)} = P_{wb} = 904 \text{ k} \quad \text{(Column Load Table, p. 3-22)}$$

$P_{bf} < P_{wb}$ No stiffeners required.

c) Check for column flange bending opposite the tension flange: Stiffeners are required if $t_{fc} < t_p$ (required) based on the following assumptions:

$b_p = 2.5 \times (P_f + t_{fb} + P_f) = 2.5 \times (2.45 + 0.605 + 2.45) = 13.8$ in.

$P_{ec} = (g/2) - k_1 - (d_b/4)$ where g = usual gage for column flange

$\quad\ = (5.5/2) - 1.0 - (1.125/4) = 1.47$ in.

$C_b = 1.0, \quad A_f/A_w = 1.0$

$\alpha_m = 1.13 \times 1.0 \times (1.0)^{1/3} \times (1.47/1.125)^{1/4} = 1.21$

$M_e = 1.21 \times 183.2 \times (1.47/4) = 81.3$ in.-k

Required flange thickness:

$$t_p = \sqrt{\frac{6 \times 81.3}{27 \times 13.8}} = 1.14 \text{ in.} < 1.19 \text{ in.} \quad \text{Stiffeners are not required.}$$

Use $1\frac{1}{2}$-in. end plate moment connection.

7.5 MOMENT-RESISTING COLUMN BASE PLATES

Columns subject to both an axial compressive force and an applied moment produce a combination of axial compressive stress and bending stress on the base plate. With the moment transposed to an eccentric load equal to the axial load P_c, the stress diagram is determined by superposition.

Consider a uniaxial bending condition along the Y-Y plane producing a combined axial and bending stress (see Fig. 7.10). This stress multiplied by the

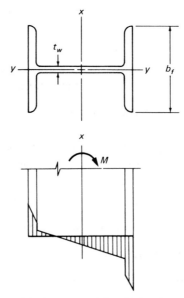

Fig. 7.10. Distribution of reaction (force per unit length) in column due to combined axial and flexural stress.

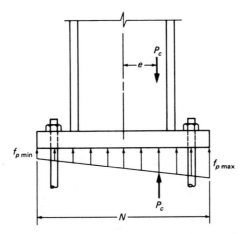

Fig. 7.11. Bearing stress of base plate when $e \leq N/6$.

width of the flange and the thickness of the web yields a force distribution that is greatest at the column flanges. To ensure that the flanges, which are of a larger cross-sectional area, carry the greater stress, the column is set with eccentricity lying within the plane of the web.

When the eccentricity is less than $\frac{1}{6} N$, where N is the base plate dimension parallel to the column web, there is no uplift of the plate at the support. For this condition, the stress in the base plate is

$$f_c = \frac{P_c}{A} \pm \frac{P_c e}{S} \qquad (7.6)$$

where A is the base plate area and S is the section modulus of the base plate, equal to $BN^2/6$ (see Fig. 7.11).

Uplift of the base plate will occur if the eccentricity is greater than $\frac{1}{6} N$. The uplift is resisted by the anchor hold-down bolts. At the extreme edge of the bearing plate, the bearing stress is maximum and decreases linearly across the plate for a distance Y (see Fig. 7.12). An approximate method of determining Y is to assume the center of gravity to be fixed at a point coinciding with the concentrated compressive force of the column flange (see Fig. 7.13).

When a more exact method of determining the uplift is desired, an approach similar to the design of a reinforced concrete section is employed. Assuming plate dimensions that can be estimated by the above method, the value for Y

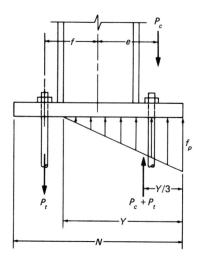

Fig. 7.12. Uplift of base plate when $e > N/6$.

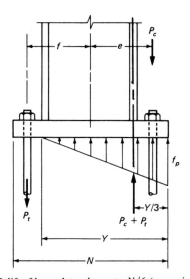

Fig. 7.13. Uplift of base plate when $e > N/6$ (approximate method).

can be determined by the cubic equation (see Fig. 7.12)

$$Y^3 + K_1 Y^2 + K_2 Y + K_3 = 0 \qquad (7.7)$$

where $K_1 = 3(e - N/2)$, $K_2 = (6nA_s/B)(f + e)$, $K_3 = -K_2(N/2 + f)$, and $n = E_s/E_c$.

Once Y is found, the tensile force on the hold-down bolts (P_t) is

$$P_t = -P_c \frac{N/2 - Y/3 - e}{N/2 - Y/3 + f} \tag{7.8}$$

and the bearing stress f_p is equal to $2(P_c + P_t)/YB$.[6]

Example 7.7. Design a base plate for a W 14 × 74 column to resist an axial force of 350 kips. Then check the base for the same axial load and a 90-ft-kip wind-induced bending moment applied in the major axis. Assume that the base plate is set on the full area of concrete support with a compressive strength f'_c = 3000 psi.

Solution.

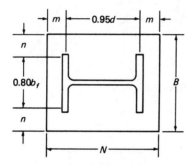

For a W 14 × 74

$$d = 14.17 \text{ in.}, \quad b_f = 10.07 \text{ in.}$$

$$F_p = 0.35 f'_c = 0.35 \times 3000 \text{ psi} = 1.05 \text{ ksi}$$

$$A_{\text{req}} = \frac{P}{F_p} = \frac{350.0 \text{ k}}{1.05 \text{ ksi}} = 333.33 \text{ in.}^2$$

---

[6]For derivation of eqns. 7.7 and 7.8, refer to O. W. Blodgett, *Design of Welded Structures*. The James F. Lincoln Arc Welding Foundation, Cleveland, Ohio (1966), pp. 3.3-6 through 3.3-19.

Assume $N = 21$ in.

$$B = \frac{A}{N} = \frac{333.33 \text{ in.}^2}{21 \text{ in.}} = 15.87 \text{ in.}$$

Use $B = 16$ in.

$$f_{p\,\text{actual}} = \frac{350 \text{ k}}{21 \text{ in.} \times 16 \text{ in.}} = 1.04 \text{ ksi} < 1.05 \text{ ksi} \quad \text{ok}$$

$$m = \frac{N - 0.95\,d}{2} = \frac{21 - 0.95(14.17)}{2} = 3.76 \text{ in.}$$

$$n = \frac{B - 0.80\,b_f}{2} = \frac{16 - 0.80(10.07)}{2} = 3.97 \text{ in.} \quad \text{(governs)}$$

$$n' = \tfrac{1}{4}\sqrt{db_f} = \tfrac{1}{4}\sqrt{14.17 \times 10.07} = 2.99 \text{ in.}$$

$$t = 3.97 \text{ in.} \sqrt{\frac{1.04 \text{ ksi}}{0.25 \times 36 \text{ ksi}}} = 1.35 \text{ in.}$$

Use base plate 16 in. $\times$ $1\tfrac{1}{2}$ in. $\times$ 1 ft, 9 in.

Check concrete bearing for applied moment.

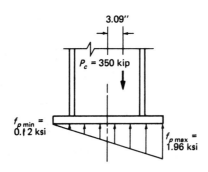

$$M = 90 \text{ ft-k}$$

$$e = \frac{M}{P_c} = \frac{90 \text{ ft-k} \times 12 \text{ in.}/\text{ft}}{350 \text{ k}} = 3.09 \text{ in.}$$

$$N/6 = 3.50 \text{ in.} > 3.09 \text{ in.}$$

No uplift occurs

$$A = 21 \text{ in.} \times 16 \text{ in.} = 336 \text{ in.}^2$$

$$S = \frac{BN^2}{6} = \frac{16 \text{ in.} \times (21 \text{ in.})^2}{6} = 1176 \text{ in.}^3$$

$$f_p = \frac{P_c}{A} \pm \frac{P_c e}{S} = \frac{350 \text{ k}}{336 \text{ in.}^2} \pm \frac{350 \text{ k} \times 3.09 \text{ in.}}{1176 \text{ in.}^3} = 1.04 \pm 0.92$$

$$f_p = 1.96 \text{ ksi}, \quad 0.12 \text{ ksi}$$

1.96 ksi > 1.05 ksi × 1.33 = 1.40 ksi (AISCS A5.2)

Base plate will not carry applied moment.

Example 7.8. Design a base plate to carry the axial force and bending moment in the preceding example without uplift.

Solution.

$$P_c = 350 \text{ k}$$

$$M = 90 \text{ ft-k} \times 12 \text{ in.}/\text{ft} = 1080 \text{ in.-k}$$

$$F_p = \frac{P_c}{A} \pm \frac{M}{S} < 1.05 \text{ ksi} \times 1.33 = 1.40 \text{ ksi}$$

Try $N = 24$ in., $B = 19$ in.

$$A = 24 \text{ in.} \times 19 \text{ in.} = 456 \text{ in.}^2$$

$$S = \frac{BN^2}{6} = \frac{19 \text{ in.} \times (24 \text{ in.})^2}{6} = 1824 \text{ in.}^3$$

$$f_p = \frac{350 \text{ k}}{456 \text{ in.}^2} \pm \frac{1080 \text{ in.-k}}{1824 \text{ in.}^3} = 0.77 \text{ ksi} \pm 0.59 \text{ ksi} = 1.36 \text{ ksi}, \quad 0.18 \text{ ksi}$$

$$f_{p\max} = 1.36 \text{ ksi} < 1.05 \text{ ksi} \times 1.33 = 1.40 \text{ ksi} \quad \text{ok}$$

$$m = \frac{N - 0.95 d}{2} = \frac{24 - 0.95(14.17)}{2} = 5.27 \text{ in.}$$

$$n = \frac{B - 0.80 b}{2} = \frac{19 - 0.80(10.07)}{2} = 5.47 \text{ in.}$$

$n' = 2.99$ in. (see Ex. 7.7)

$$t = 5.47 \text{ in.} \sqrt{\frac{1.36 \text{ ksi}}{0.25 \times 36 \text{ ksi} \times 1.33}} = 1.84 \text{ in.}$$

Use base plate 19 in. × $1\frac{7}{8}$ in. × 2 ft, 0 in.

Example 7.9. Using the approximate method, design a base plate for a W 14 × 74 column to resist a combined axial force of 350 kips and bending moment due to gravity loads in the major axis of 250 ft-kips. Assume the footing of 3000-psi concrete to be quite large, making $F_p = 0.70 f'_c$. Three hold-down bolts on each side of the column flange are 9 in. from the column center line. Use A36 steel and A307 bolts. Verify solution with a more precise approach.

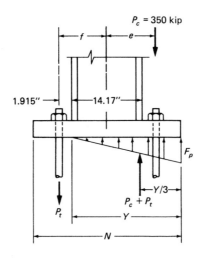

Solution. For a W 14 × 74

$$d = 14.17 \text{ in.}, \quad b_f = 10.07 \text{ in.}, \quad t_f = 0.785 \text{ in.}$$

$$F_p = 0.70 \times 3000 \text{ psi} = 2100 \text{ psi} = 2.1 \text{ ksi}$$

Assume compressive reaction from base is applied at the center of the flange, and its value is

$$P = P_c + P_t$$

$$e = \frac{M}{P_c} = \frac{250 \text{ ft-k} \times 12 \text{ in./ft}}{350 \text{ k}} = 8.57 \text{ in.}$$

Summing moments with respect to compression flange

$$P_t = \frac{350 \text{ k} \times \left(8.57 \text{ in.} - \dfrac{14.17 \text{ in.}}{2} + \dfrac{0.785 \text{ in.}}{2}\right)}{\left(\dfrac{14.17 \text{ in.}}{2} - \dfrac{0.785 \text{ in.}}{2} + 9.0 \text{ in.}\right)} = 41.88 \text{ k}$$

Total volume of compressive stress

$$\frac{1}{2} \times F_p \times Y \times B = P_c + P_t$$

$$Y = \frac{2 \times (41.88 \text{ k} + 350 \text{ k})}{2.1 \text{ ksi} \times B}$$

Try $B = 21$ in.

$$Y = \frac{2 \times 391.88 \text{ k}}{2.1 \text{ ksi} \times 21 \text{ in.}} = 17.77 \text{ in.}$$

Distance from center of flange to end of plate is $Y/3$.

$$\frac{Y}{3} = 5.923 \text{ in.}$$

Length of base plate

$$N = 2 \times 5.923 \text{ in.} + 14.17 \text{ in.} - 0.785 \text{ in.} = 25.23 \text{ in.} \quad \text{say } 26 \text{ in.}$$

Recalculating $Y/3$

$$\frac{Y}{3} = \frac{26 \text{ in.} - 14.17 \text{ in.} + 0.785 \text{ in.}}{2} = 6.31 \text{ in.}$$

$$m = \frac{N - 0.95 \, d}{2} = \frac{26 \text{ in.} - 0.95 \times 14.17 \text{ in.}}{2} = 6.27 \text{ in.}$$

$$n = \frac{B - 0.80 \, b_f}{2} = \frac{21 \text{ in.} - 0.80 \times 10.07 \text{ in.}}{2} = 6.47 \text{ in.}$$

$n' = 2.99$ in. (see Ex. 7.7)

$$f_p = \frac{2 \times (P_c + P_t)}{3 \times Y/3 \times B} = \frac{2 \times (350 \text{ k} + 41.88 \text{ k})}{3 \times 6.31 \text{ in.} \times 21 \text{ in.}} = 1.97 \text{ ksi} < 2.1 \text{ ksi} \quad \text{ok}$$

Assume bearing stress to be uniformly distributed for a conservative calculation of plate thickness.

$$t = 6.47 \text{ in.} \times \sqrt{\frac{1.97 \text{ ksi}}{0.25 \times 36 \text{ ksi}}} = 3.02 \text{ in.} \quad \text{use } 3\frac{1}{4} \text{ in.}$$

Checking stress in anchor bolts

three 1-in. ϕ A307 bolts have capacity of

$$T = 3 \times 0.7854 \text{ in.}^2 \times 20 \text{ ksi} = 47.1 \text{ k} > 41.88 \text{ k} \quad \text{ok}$$

Verify using "Reinforced Concrete Beam" approach. Y is the solution of the cubic equation.

$$Y^3 + K_1 Y^2 + K_2 Y + K_3 = 0, \text{ where}$$

$$K_1 = 3\left(e - \frac{N}{2}\right)$$

$$K_2 = \frac{6 n A_s}{B}(f + e)$$

$$K_3 = -K_2\left(\frac{N}{2} + f\right)$$

$$f = 1.915 \text{ in.} + \frac{14.17 \text{ in.}}{2} = 9.0 \text{ in.}$$

$e = 8.57$ in.

$B = 21$ in.

$N = 26$ in.

$n = 9$

$A_s = 3 \times 0.785 \text{ in.}^2 = 2.355 \text{ in.}^2$

$$f_p = \frac{2(P_c + P_t)}{Y \times B}$$

$$P_t = -P_c \frac{\frac{N}{2} - \frac{Y}{3} - e}{\frac{N}{2} - \frac{Y}{3} + f}$$

$$K_1 = 3 \times \left(8.57 \text{ in.} - \frac{26.0 \text{ in.}}{2}\right) = -13.29 \text{ in.}$$

$$K_2 = \frac{6 \times 9 \times 2.355 \text{ in.}^2}{21 \text{ in.}} \times (9.0 \text{ in.} + 8.57 \text{ in.}) = 106.40 \text{ in.}^2$$

$$K_3 = -106.4 \text{ in.}^2 \times \left(\frac{26.0 \text{ in.}}{2} + 9.0 \text{ in.}\right) = -2340.8 \text{ in.}^3$$

$$Y^3 - 13.29 \text{ in. } Y^2 + 106.40 \text{ in.}^2 \, Y - 2340.8 \text{ in.}^3 = 0$$

By trial and error

$$Y = 15.88 \text{ in.}$$

$$\frac{Y}{3} = 5.29 \text{ in.}$$

$$P_t = -350 \text{ k} \times \frac{\frac{26 \text{ in.}}{2} - \frac{15.88 \text{ in.}}{3} - 8.57 \text{ in.}}{\frac{26 \text{ in.}}{2} - \frac{15.88 \text{ in.}}{3} + 9.0 \text{ in.}} = 18.0 \text{ k}$$

$$f_p = \frac{2(18.0 + 350 \text{ k})}{15.88 \text{ in.} \times 21 \text{ in.}} = 2.21 \text{ ksi} \quad 5\% \text{ overstress} \quad \text{say ok}$$

We see that the compressive stresses are greater, but there is smaller tension in the bolts compared with the previous approximation.

7.6 FIELD SPLICES OF BEAMS AND PLATE GIRDERS

Field splices may become necessary to connect beams or plate girders when long spans are desired. If the shipment or erection of a member is difficult, or if the required length of a member is not readily available, splicing may be the solution.

A splice must develop the strength required by the stresses at the point of splice (AISCS J7). This is usually done with a web splice designed to carry

312 STEEL DESIGN FOR ENGINEERS AND ARCHITECTS

shear and a portion of the moment. The balance of the moment is carried by flange splices. An approximate alternative is to design web splices to carry the total shear and flange splices to carry the total moment.

For groove welded splices in plate girders and beams, AISCS J7 states that the weld must develop the full strength of the smaller spliced section.

Bolted splice plates are usually fastened to the outside surface of the flanges. Plates may be used on both the outside and the inside of the flanges to place the fasteners in double shear. The number of bolts is determined by the axial force resulting from the design moment.

Web splices are usually placed symmetrically over the height between fillets (T) on beams and between welds on plate girders. They transfer the total shear

Fig. 7.14. A column splice of two different rolled sections made with splice plates.

SPECIAL CONNECTIONS 313

and a part of the moment. The amount of moment M_w carried by the web splices is proportional to the total bending moment M:

$$M_w = M \frac{I_w}{I_{section}} \qquad (7.9)$$

where I_w is the gross moment of inertia of the web splice plates and $I_{section}$ is the gross moment of inertia of the web splice plates plus the flange splice plates. The remainder of the moment is carried by the flange splice plates. Once M_w and V are known, the number of bolts required on the web splices is determined in the same way as that for eccentric loading on fastener groups (see Chapter 5).

Example 7.10. Design a bolted beam splice for a W 30 × 108 at a point where the shear is 100 kips and the moment is 300 ft-kips. Use A325-X bolts.

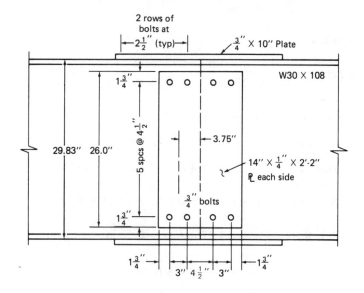

Solution. Preliminary design

Web splice

Design the web splice (plates on both sides of the web) to carry the entire shear and a portion of the moment.

314 STEEL DESIGN FOR ENGINEERS AND ARCHITECTS

Use plate height to toe of fillets

$$29\tfrac{7}{8} \text{ in.} - 2(1\tfrac{9}{16} \text{ in.}) = 26\tfrac{3}{4} \text{ in.} \quad \text{use plate} = 26 \text{ in.}$$

Try web splice with bolt spacing as shown.
For $\tfrac{3}{4}$-in. ϕ A325-X bolts

$$\text{Single shear} = 13.3 \text{ k} \quad \text{double shear} = 26.5 \text{ k}$$

Bolt shear capacity in web =

$$12 \text{ bolts} \times 26.5 \text{ k} = 318 \text{ k} > 100 \text{ k} \quad \text{ok}$$

Shear capacity of web
Try two plates 14 in $\times$ $\tfrac{1}{4}$ in. $\times$ 2 ft, 2 in.

$$f_v = \frac{100 \text{ k}}{13.0 \text{ in.}^2} = 7.7 \text{ ksi} < 14.4 \text{ ksi} \quad \text{ok}$$

Flange splice

$$\text{Force in flange} = \frac{M}{d - t_f} = \frac{300 \text{ ft-k} \times 12 \text{ in./ft}}{29.83 \text{ in.} - 0.76 \text{ in.}} = 123.8 \text{ k}$$

$$\text{Gross plate area required} = P/F_t = 5.63 \text{ in.}^2$$

$$\text{Effective net area required} = P/F_t = 4.27 \text{ in.}^2$$

$$\text{Max. } A_n = 0.85 \, A_g = 4.78 \text{ in.}^2 > 4.27 \text{ in.}^2 \quad \text{(AISCS B3)}$$

$$\text{Net plate width} = 10 \text{ in.} - 2(\tfrac{3}{4} \text{ in.} + \tfrac{1}{8} \text{ in.}) = 8.25 \text{ in.}$$

$$A = 8.25 \text{ in.} \times t = 4.27 \text{ in.}^2 \quad t = 0.52 \text{ in.} \quad \text{say } \tfrac{3}{4} \text{ in.}$$

Check for deduction of bolt holes (AISCS B10):

$$A_{fg} = (10.475 \times 0.76) + (0.75 \times 10) = 15.5 \text{ in.}^2$$

$$A_{fn} = 15.5 - [2 \times (0.75 + 0.125) \times (0.76 + 0.75)] = 12.9 \text{ in.}^2$$

$$0.5(58)(12.9) = 374 \text{ k} > 0.6(36)(15.5) = 335 \text{ k} \quad \text{(B10-1)}$$

No reduction in section properties is required.

SPECIAL CONNECTIONS

Final design

Considering that the bending moment at the splice carried by web and cover plates is proportioned to their moments of inertia

$$I_{\text{web plates}} = 0.25 \times \frac{(26)^3}{12} \times 2 = 732 \text{ in.}^4$$

$$I_{\text{flange plates}} = 10 \times 0.75 \times \left(\frac{29.83 + 0.75}{2}\right)^2 \times 2$$

$$= 3507 \text{ in.}^4$$

Moment carried by web = $300 \text{ ft-k} \times \left(\frac{732 \text{ in.}^4}{732 \text{ in.}^4 + 3507 \text{ in.}^4}\right) = 51.8 \text{ ft-k}$

Moment carried by flange = $300 - 51.8 = 248.2 \text{ ft-k}$

$$\text{Force in flange} = \frac{248.2 \text{ ft-k} \times 12 \text{ in./ft}}{29.83 \text{ in.} - 0.76 \text{ in.}} = 102.5 \text{ k}$$

Check web bolts:

Moment due to shear eccentricity = $100 \text{ k} \times (2.25 \text{ in.} + 1.50 \text{ in.})$

$$= 375 \text{ in-k}$$

Total moment on web bolts = $375 \text{ in-k} + (51.8 \text{ ft-k} \times 12 \text{ in./ft})$

$$= 996.6 \text{ in-k}$$

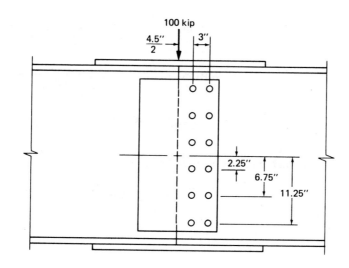

316 STEEL DESIGN FOR ENGINEERS AND ARCHITECTS

Force in extreme bolt

$$\Sigma d^2 = (1.5)^2 \times 12 + [(2.25)^2 + (6.75)^2 + (11.25)^2] \times 4$$
$$= 27.0 \text{ in.}^2 + 709 \text{ in.}^2 = 736 \text{ in.}^2$$

$$F = \sqrt{\left(\frac{100}{12} + \frac{996.6}{736} \times 1.5\right)^2 + \left(\frac{996.6}{736} \times 11.25\right)^2}$$

$$= 18.4 \text{ k} < 26.5 \text{ k} \quad \text{ok}$$

Bolts in flange plates

$$\frac{102.5 \text{ k}}{13.3 \text{ k}} = 7.7 \text{ bolts} \quad \text{use eight bolts}$$

Bolts must be staggered as shown so that loss is equal to two holes.

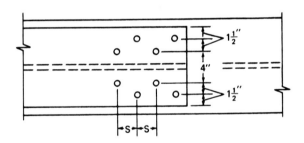

$$10 \text{ in.} - \left(2 \times \frac{7}{8} \text{ in.}\right) = 10 \text{ in.} - \left(4 \times \frac{7}{8} \text{ in.}\right) + \left(2 \times \frac{s^2}{4g}\right)$$

$$= 10 \text{ in.} - \left(4 \times \frac{7}{8} \text{ in.}\right) + \left(\frac{2 \times s^2}{4 \times 1.5}\right)$$

$$s = 2.29 \text{ in.} \quad \text{say } 2\frac{3}{8} \text{ in. minimum}$$

Check 3 in. minimum distance between bolts.

$$s = \sqrt{3^2 - 1.5^2} = 2.60 \text{ in.}$$

Use $s = 2\frac{5}{8}$ in.

Bending stress at splice location:

$$f_b = \frac{300 \text{ ft-k} \times 12 \text{ in./ft} \times (14.92 \text{ in.} + 0.75 \text{ in.})}{732 \text{ in.}^4 + 3507 \text{ in.}^4} = 13.3 \text{ ksi} \quad \text{ok}$$

7.7 HANGER-TYPE CONNECTIONS

When designing hanger-type connections (e.g., structural tee or double-angle hangers), prying action must be considered. The tensile force 2T which pulls on the web of the tee section in Fig. 7.15 causes the flange to deflect away from the supporting member in the vicinity of the web. The portions of the flange farthest from the web, however, bear against the supporting member. The bearing forces Q at the edges of the tee flange are commonly referred to as the *prying forces*.

In general, bending moments develop in the leg of the hanger, and additional tensile stresses occur in the bolts as a result of the prying action. By increasing the thickness of the flange and/or decreasing the distance b (as small as erection

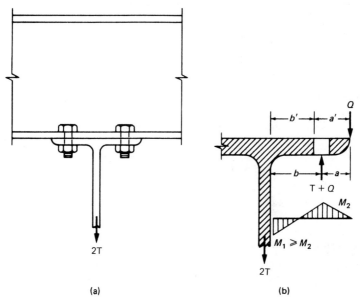

Fig. 7.15. Typical hanger-type connection: (a) applied tensile force to stem causes combined tension and bending in the flanges; (b) moment diagram for flange leg.

318 STEEL DESIGN FOR ENGINEERS AND ARCHITECTS

methods permit), the effects of the prying action will be reduced for a given load.

It is interesting to note that this prying action can be avoided by using thick enough washers between the surfaces in contact. Nothing in the AISCS seems to preclude this. From a construction point of view, however, this option may not be extremely desirable.

Exact modeling of the combined tension and prying forces is difficult to perform. A design and analysis procedure can be found on p. 4-91 of the AISCM. This procedure can also be used in the situations when prying and shear act together. The following example presents the design requirements recommend by the AISC for hanger-type connections.

Example 7.11. Complete the design of the tee section for the semirigid moment connection in Example 7.5. The tension flange force transferred by the tee is 67.5 k. Fasteners are to be $\frac{3}{4}$-in. A325-X bolts for the web and 1-in. A325-X bolts for the flange.

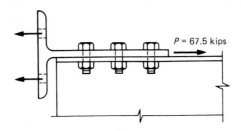

Solution. Number of tee flange bolts required

$$B = 34.6 \text{ k} \quad (\text{AISCM, Table I-A})$$

$$N_b = \frac{67.5 \text{ k}}{34.6 \text{ k}} = 2.0 \quad \text{four bolts minimum}$$

$$T = \frac{67.5 \text{ k}}{4} = 16.88 \text{ k}$$

Number of tee web bolts required (single shear)

$$R_v = 13.3 \text{ k} \quad (\text{AISCM, Table I-D})$$

$$N_b = \frac{67.5 \text{ k}}{13.3 \text{ k}} = 5.1 \quad \text{six bolts minimum}$$

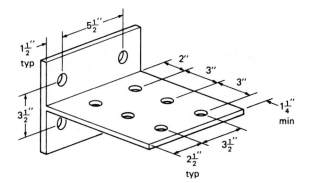

Try two flange bolts in each of two rows as shown.

$$\text{length tributary to bolt} = p = \frac{8.5 \text{ in.}}{2} = 4.25 \text{ in.}$$

Try three web bolts in each of two rows with $3\text{-}\frac{1}{2}$-in. gage. (tightening clearance is ok; p. 4-137)

Minimum tee length

$$d \geq 2.0 \text{ in.} + (2 \times 3.0 \text{ in.}) + 1.25 \text{ in.} = 9.25 \text{ in.}$$

Try WT 10.5×36.5, $8\frac{1}{2}$ in. long

$$d = 10.62 \text{ in.}, \quad f_b = 8.295 \text{ in.}, \quad t_f = 0.74 \text{ in.}, \quad t_w = 0.455 \text{ in.}$$

Check tee web for tension capacity

$$A_n \geq \frac{T}{0.5 F_u} = \frac{67.5 \text{ k}}{29 \text{ ksi}} = 2.33 \text{ in.}^2$$

$$A_g \geq \frac{T}{0.6 F_y} = \frac{67.5 \text{ k}}{22 \text{ ksi}} = 3.07 \text{ in.}^2$$

$A_g = 0.455 \text{ in.} \times 8.5 \text{ in.} = 3.87 \text{ in.}^2 > 3.07 \text{ in.}^2$ ok

$A_n = 3.87 \text{ in.}^2 - 2(1.0 \text{ in.} + 0.125 \text{ in.}) \times 0.455 \text{ in.}$

$\quad = 2.85 \text{ in.}^2 > 2.33 \text{ in.}^2$ ok

320 STEEL DESIGN FOR ENGINEERS AND ARCHITECTS

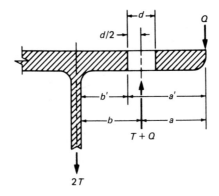

Check tee flange for tension and prying action:

$$b = \frac{3.5 - 0.455}{2} = 1.523 \text{ in.}$$

$$a = \frac{8.295 - 3.5}{2} = 2.40 \text{ in.}$$

$1.25b = 1.904$ in. < 2.40 in. Use $a = 1.904$ in.

$b' = 1.523 - \frac{1}{2} = 1.023$ in.

$a' = 1.904 + \frac{1}{2} = 2.404$ in.

$d' = 1 + 0.0625 = 1.0625$ in.

$p = 4.25$ in.

$\rho = 1.023/2.404 = 0.426$

$\delta = 1 - (1.0625/4.25) = 0.75$

$$t_c = \sqrt{\frac{8 \times 34.6 \text{ k} \times 1.023 \text{ in.}}{4.25 \text{ in.} \times 36 \text{ ksi}}} = 1.36 \text{ in.}$$

$$\alpha' = \frac{1}{0.75 \times (1 + 0.426)} \left[\left(\frac{1.36}{0.74}\right)^2 - 1 \right] = 2.22$$

$$T_{\text{all}} = 34.6 \text{ k} \times \left(\frac{0.74 \text{ in.}}{1.36 \text{ in.}}\right)^2 \times (1 + 0.75) = 17.93 \text{ k}$$

$T_{\text{all}} > T = 16.88$ k   ok

If the prying force is required:

$$\alpha = \frac{1}{0.75}\left[\frac{(16.88/34.6)}{(0.74/1.36)^2} - 1\right] = 0.864$$

$$Q = 34.6 \times 0.75 \times 0.864 \times 0.426 \times \left(\frac{0.74}{1.36}\right)^2 = 2.8 \text{ k per bolt}$$

Use WT 10.5×36.5, $8\frac{1}{2}$ in. long.

PROBLEMS TO BE SOLVED

7.1. Design the rigid connection shown to carry an end moment of 85 ft-kips and an end shear of 35 kips due to dead and live loads only. Use $\frac{3}{4}$-in. A325-N bearing type bolts. The connection plates are welded to the column and bolted to the beam.

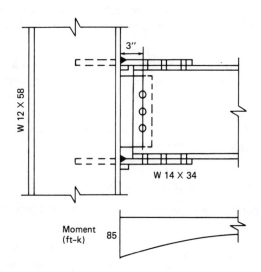

7.2. Design bolted flange plate moment connections for the beam and columns as shown. Design the connections to carry

a) full frame action negative moment of 125 ft-k with $\frac{3}{4}$-in. A325-SC bolts.

b) 75% of the full frame action moment with $\frac{3}{4}$-in. A325-N bolts.

c) 50% of the full frame action moment with $\frac{3}{4}$-in. A325-X bolts.

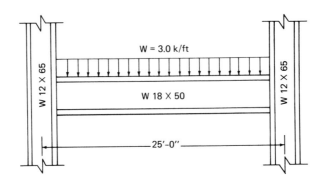

7.3. Rework Problem 7.2a for an end plate moment connection.

7.4. A W 14 × 43 beam frames into the web (weak axis) of a W 14 × 68 column. Design a semirigid moment connection of tee sections and shear angles to transfer a moment of 50 ft-kips and shear of 23 kips. Assume full lateral support and $\frac{7}{8}$-in. A325-N bolts.

7.5. Design a base plate for a W 10 × 88 column to resist a concentric axial load of 600 k. Assume that the base plate is on a very large caisson cap which has a compressive strength $f'_c = 5000$ psi. Using the approximate method, check the base plate for an axial load of 400 k and a bending moment about the major axis of 200 ft-k. Use two $1\frac{1}{2}$-in. A307 anchor bolts located 3 in. from the face of the flange on each side of the column.

7.6. Check Problem 7.5 using the method that assumes that the base plate acts similarly to a reinforced concrete beam. Use $n = 7$.

7.7. Design a bolted beam splice for a W 27 × 84 at a point where the shear is 150 kips and the moment is 150 ft-kips. Assume $\frac{3}{4}$-in. A325-X bolts.

7.8. Design a bolted beam splice to join a W 30 × 99 to a W 30 × 173 where the shear is 100 kips and the moment is 160 ft-kips. Use $\frac{3}{4}$-in. A325-X bolts. Sketch detail showing bolt spacings and necessary filler plates.

7.9. A hanger is supporting a 60-kip force from a W 27 × 146 beam as shown. Verify whether a WT 12 × 47 × $8\frac{1}{2}$ in. is adequate. Use $\frac{3}{4}$-in. A490-N bolts on $5\frac{1}{2}$-in. gage.

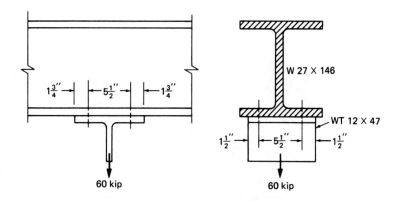

7.10. Select a double-angle hanger to support the loading shown in Problem 7.9. Use $\frac{7}{8}$-in. A325-X bolts.

8
Torsion

8.1 TORSION

Members that have loads applied away from their shear centers are subject to torsion.[1] These loads, when multiplied by their distances from the shear center of the member, cause deformations due to rotation that may also include warping. The rotation of the member is measured by the angle of the twist.

8.2 TORSION OF AXISYMMETRICAL MEMBERS

For pure torsion in the elastic range, a plane section perpendicular to the member's axis is assumed to remain plane after the torque has been applied. In this plane, shearing stress and strain vary linearly from the neutral axis. This behavior and distribution of stress resulting solely from pure torsion is also known as *St. Venant torsion*. When the ends of a member are restrained so that no deformation can occur, the stresses due to the warping of the surfaces govern. Essentially, this will reduce the St. Venant shear stresses and may slightly increase the longitudinal stresses. Hence, by considering only the St. Venant torsion, the design will be more conservative and the analysis will be greatly simplified. For a more in-depth analysis, see "Torsional Analysis of Steel Members," American Institute of Steel Construction (Chicago, 1983). Any shearing stress produced by bending would be added to the St. Venant shear stresses.

In a solid circular cylinder subjected to torsion (Fig. 8.1), the maximum shearing stress occurs at the perimeter. The stress at any point is obtained from

$$f_v = \frac{Tr}{J} \qquad (8.1)$$

---

[1]For sections with a center of symmetry, the shear center coincides with the center of symmetry. Refer to Sect. 8.5 for discussion of shear center.

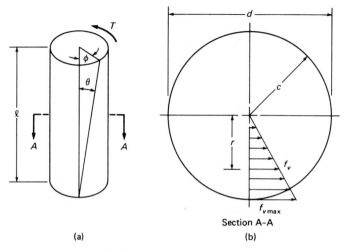

Fig. 8.1. Torsion on circular cross section.

and the angle of rotation is determined by

$$\phi = \frac{Tl}{GJ} \tag{8.2}$$

where T is the torque (in.-lb), r is the distance from the center of the member to the point under consideration (in.), l is the length of the cylinder (in.), G is the shear modulus $= E/2(1 + \mu)$ (psi), μ is Poisson's ratio, which for steel is between 0.25 and 0.30, and J is the polar moment of inertia (in.^{4}), which for a solid round section is expressed by

$$J = \frac{\pi d^4}{32} \tag{8.3}$$

When it is required to limit the torsional moment (torque), so that the maximum shear stress does not exceed its allowable value, eqn. (8.1) is transposed to

$$T = \frac{F_v J}{c} \tag{8.4}$$

where F_v is the allowable shear stress of the material, and c is the distance from the center of the member to the outer fibers.

Fig. 8.2. Sears Tower, Chicago.

Example 8.1. Determine the maximum allowable torque for a $1\frac{1}{2}$-in.-diameter steel shaft with an allowable stress $F_v = 14.5$ ksi.

Solution. The maximum torque is determined by the equation

$$T = F_v \frac{J}{c}$$

$$F_v = 14.5 \text{ ksi}$$

$$J = \frac{\pi d^4}{32} = 0.497 \text{ in.}^4$$

$$c = \frac{1.5 \text{ in.}}{2} = 0.75 \text{ in.}$$

$$T = \frac{14.5 \text{ ksi} \times 0.497 \text{ in.}^4}{0.75 \text{ in.}} = 9.6 \text{ in.-kips}$$

Converting to ft-kips

$$\frac{9.6 \text{ in.-kips}}{12 \text{ in./ft}} = 0.80 \text{ ft-kips}$$

Example 8.2. Determine the maximum torque for a hollow shaft with an outside diameter (d_o) of 9 in. and an inside diameter (d_i) of 8 in. Allowable stress is 8000 psi.

Solution. In a hollow, circular cylinder, J is given by $\pi/32(d_o^4 - d_i^4)$ where d_o and d_i are the outer and inner diameters, respectively.

$$T = F_v \frac{J}{c}$$

$$F_v = 8000 \text{ psi}$$

$$J = \frac{\pi(d_o^4 - d_i^4)}{32} = 242 \text{ in.}^4$$

$$c = 4.5 \text{ in.}$$

$$T = \frac{8000 \text{ psi} \times 242 \text{ in.}^4}{4.5 \text{ in.}} = 430{,}200 \text{ in.-lb}$$

Converting to ft-kips

$$\frac{430{,}200 \text{ in.-lb}}{12 \text{ in./ft} \times 1000 \text{ lb/k}} = 35.85 \text{ ft-kips}$$

Example 8.3. Determine the maximum torque a standard-weight 6-in.-diameter pipe can carry when allowable shear stress is 14.5 ksi.

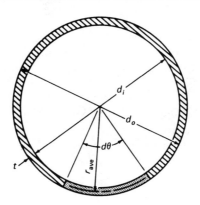

Solution.

a) Calculating J from AISC tables. (p. 1-93)

We know that $J_p = I_x + I_y$. In our case, $J_p = 2 \times 28.1 \text{ in.}^4 = 56.2 \text{ in.}^4$

b) Consider the pipe as a thin-walled element, such that its thickness is small with respect to the diameter. Hence, $d_o = d_i = d_{ave} = 6.345$ in.

$$J = \int r^2 \, dA$$

$$dA = r_{ave} \, d\theta \, t$$

$$J = \int_0^{2\pi} t r_{ave}^3 \, d\theta = 2\pi r^3 t$$

$$= 2\pi \left(\frac{6.345 \text{ in.}}{2}\right)^3 (0.28 \text{ in.}) = 56.17 \text{ in.}^4$$

c) Consider $J = J_o - J_i$

$$J = J_o - J_i = \frac{\pi}{32} [(6.625 \text{ in.})^4 - (6.065 \text{ in.})^4] = 56.28 \text{ in.}^4$$

We see that all three methods result in nearly identical values.

$$T = \frac{F_v J}{c} = \frac{14.5 \text{ ksi} \times 56.2 \text{ in.}^4}{3.313 \text{ in.}} = 245.97 \text{ in.-k} = 20.50 \text{ ft-k}$$

Example 8.4. What size solid steel circular bar is required to carry a torsional moment of 1 ft-kip if the allowable stress is 14.5 ksi?

Solution.

$$c = F_v \frac{J}{T}$$

$$J = \frac{\pi (2c)^4}{32} = \frac{\pi c^4}{2}$$

$T = 1$ ft-k $\times 12$ in./ft $\times 1000$ lb/k $= 12{,}000$ in.-lb

$$c = \frac{14{,}500 \text{ psi} \left(\dfrac{\pi c^4}{2}\right)}{12{,}000 \text{ in.-lb}}$$

$$c^3 = \frac{(2)\ 12{,}000\ \text{in.-lb}}{14{,}500\ \text{psi} \times \pi} = 0.527\ \text{in.}^3$$

$$c = \sqrt[3]{0.527\ \text{in.}^3} = 0.808\ \text{in.}$$

Use $1\text{-}\tfrac{5}{8}$-in.-diameter solid bar.

Example 8.5. A 6-ft-kip torque is applied to a 4-in. standard-weight pipe. Determine the stress at the inside face and outside face of the pipe and the angle of twist if the tube is 6 ft long. $E_s = 29 \times 10^6$ psi, and $\mu = 0.25$.

Solution. Member properties

$$d_o = 4.5\ \text{in.} \quad d_i = 4.026\ \text{in.}$$

$$f_v = \frac{Tr}{J}, \quad \text{where } r = 2.013\ \text{in. for the shear stress at inside face}$$
$$r = 2.25\ \text{in. for the shear stress at outside face}$$

$$\phi = \frac{Tl}{GJ}$$

$$J = \frac{\pi(d_o^4 - d_i^4)}{32} = 14.47\ \text{in.}^4$$

Also, from AISC tables, $J = 2 \times 7.23\ \text{in.}^4 = 14.46\ \text{in.}^4$

$$l = 6\ \text{ft} \times 12\ \text{in./ft} = 72\ \text{in.}$$

$$G = \frac{E}{2(1 + \mu)} = 11.6 \times 10^6\ \text{psi}$$

$$T = 6.0\ \text{ft-k} \times 12\ \text{in./ft} \times 1000\ \text{lb/k} = 72{,}000\ \text{in.-lb}$$

Stress at the inside face

$$f_v = \frac{72{,}000\ \text{in.-lb} \times 2.013\ \text{in.}}{14.47\ \text{in.}^4} = 10{,}000\ \text{psi}$$

Stress at the outside face

$$f_v = \frac{72{,}000\ \text{in.-lb} \times 2.25\ \text{in.}}{14.47\ \text{in.}^4} = 11{,}200\ \text{psi}$$

330 STEEL DESIGN FOR ENGINEERS AND ARCHITECTS

Angle of twist

$$\phi = \frac{72{,}000 \text{ in.-lb} \times 72 \text{ in.}}{11.6 \times 10^6 \text{ psi} \times 14.47 \text{ in.}^4} = 0.031 \text{ radians} = 1.77 \text{ degrees}$$

8.3 TORSION OF SOLID RECTANGULAR SECTIONS

When a torsional moment is applied to a noncircular section, the cross section rotates and deforms nonuniformly. As a result, a plane section perpendicular to the member's axis does not remain plane. This additional deformation is referred to as *warping*. In the case of solid rectangular sections, twisting will occur about the axis. No shear strain will occur at the corners, and consequently shear stresses at the corners are equal to zero.

In Fig. 8.3, the shear stress distribution indicates that the maximum stress occurs at the midpoint of the long side (see Example 8.6).

This nonuniform stress distribution is in direct contrast to the stress distribution of a circular section (Fig. 8.1b).

To determine the stress and rotation of a rectangular section, the use of two characteristic elements α and β becomes necessary.

The α factor is used in computing the torsional resistance for the stress formula. A solid rectangular section has a torsional constant of

$$J^* = \alpha b d^3 \tag{8.5}$$

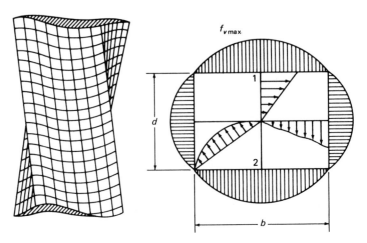

Fig. 8.3. Torsion on rectangular bars.

Table 8.1. Torsional Factors for Solid Rectangular Bars (Blodgett, *Design of Welded Structures*, reprinted with permission).

$b/d =$	1.00	1.50	1.75	2.00	2.50	3.00	4.00	6	8	10	∞
α	.208	.231	.239	.246	.258	.267	.282	.299	.307	.313	.333
β	.141	.196	.214	.229	.249	.263	.281	.299	.307	.313	.333

where b is the larger and d is the smaller dimension. The factor α is obtained from Table 8.1. The maximum shear stress due to a torque is then

$$f_v = \frac{Td}{J^*} \tag{8.6}$$

and is located at the middle of the long side. The angle of twist of the member is determined by

$$\phi = \frac{TL}{GJ^*} \tag{8.7}$$

where J^* is now given by

$$J^* = \beta b d^3 \tag{8.8}$$

It can be noted from Table 8.1 that α and β converge to the value $\frac{1}{3}$ as the ratio b/d becomes large. When the ratio is 4.0, the two coefficients are nearly the same, and the torsional resistance becomes the same for use in the stress and rotation formulas.

Example 8.6. Show that for a rectangular section under torsion, the shear stress at the corners is zero.

Solution. Consider a corner element as shown. Assuming that a shearing stress exists at the corner, two components can be shown parallel to the edges of the bar. However, as equal shear stress always occurs in pairs acting on mutually perpendicular planes, these components would have to be met by shear stresses lying in the plane of the outside surfaces. Because the surfaces are free surfaces, no stresses exist. Therefore, the shear stress at the corners must be zero.

332 STEEL DESIGN FOR ENGINEERS AND ARCHITECTS

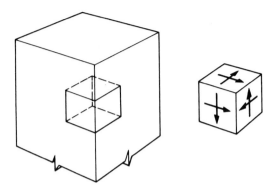

Example 8.7. Determine the maximum torque that can be applied to a 2-in. × 4-in. cantilevered steel bar. $F_v = 14.5$ ksi.

Solution.

$$F_v = \frac{Td}{J^*}, \quad T = \frac{F_v J^*}{d}, \quad J^* = \alpha b d^3$$

$d = 2$ in.

$b = 4$ in.

$b/d = 2$; From Table 8.1, $\alpha = 0.246$

$J^* = 0.246 \times 4 \text{ in.} \times (2 \text{ in.})^3 = 7.87 \text{ in.}^4$

$$T = \frac{14.5 \text{ ksi} \times 7.87 \text{ in.}^4}{2 \text{ in.}} = 57.06 \text{ in.-kip} \times \frac{1 \text{ ft}}{12 \text{ in.}} = 4.75 \text{ ft-kip}$$

Example 8.8. A 6.25-ft-kip torque is applied to a rectangular member. Determine the maximum stress when acting on a cantilevered 3-in. × 5-in. rectangular bar 4 ft long. Also find the rotation due to the applied torque if the shear modulus G is 11.6×10^6 psi.

Solution.

$$f_v = \frac{Td}{J^*}$$

$$\phi = \frac{Tl}{GJ^*}$$

$b = 5$ in.

$d = 3$ in.

$b/d = \dfrac{5}{3} = 1.67$; Interpolation must be used in Table 8.1 for b/d between 1.5 and 1.75.

$\alpha = 0.236$ J^* (stress) $= 0.236 \times 5$ in. $\times (3$ in.$)^3 = 31.86$ in.^{4}

$\beta = 0.208$ J^* (rotation) $= 0.208 \times 5$ in. $\times (3$ in.$)^3 = 28.08$ in.^{4}

$l = 4$ ft $\times 12$ in./ft $= 48$ in.

$T = 6.25$ ft-kip $\times 12{,}000 = 75{,}000$ in.-lb

$$f_v = \dfrac{75{,}000 \text{ in.-lb} \times 3 \text{ in.}}{31.86 \text{ in.}^4} = 7062 \text{ psi}$$

$$\phi = \dfrac{75{,}000 \text{ in.-lb} \times 48 \text{ in.}}{11.6 \times 10^6 \text{ psi} \times 28.08 \text{ in.}^4} = 0.011 \text{ radians} = 0.64 \text{ degrees}$$

Example 8.9. A $1\text{-}\tfrac{1}{2}$-in. $\times$ 3-in. bar 6 ft long is subject to a 1000 ft-lb torsional moment. Determine if the bar is safe with an allowable shear of 14.5 ksi and an angle of rotation limited to 1.5 degrees. $G = 11.6 \times 10^6$ psi.

Solution.

$b/d = 2$; $\alpha = 0.246$, $\beta = 0.229$

$f_v = \dfrac{Td}{J^*}$; J^* (stress) $= 0.246 \times 3$ in. $\times (1.5$ in.$)^3 = 2.49$ in.^{4}

$$f_v = \dfrac{12 \text{ in.-kip} \times 1.5 \text{ in.}}{2.49 \text{ in.}^4} = 7.23 \text{ ksi} < 14.5 \text{ ksi} \quad \text{ok}$$

$\phi = \dfrac{Tl}{GJ^*}$; J^* (rotation) $= 0.229 \times 3$ in. $\times (1.5$ in.$)^3 = 2.32$ in.^{4}

$$\phi = \dfrac{12 \text{ in.-kip} \times (6 \text{ ft} \times 12 \text{ in./ft})}{11.6 \times 10^3 \text{ ksi} \times 2.32 \text{ in.}^4} = 0.032 \text{ radian} = 1.84 \text{ degrees} \quad \text{N.G.}$$

The bar will rotate 0.34° greater than allowed. Therefore, the member is unsatisfactory due to excessive rotation.

Example 8.10. A 4-ft, simply supported solid steel bar is restrained against torsion at the ends. The bar will support a sign weighing 1.8 kips, which act at quarter points, 12 in. from the bar's shear center. Design the bar to carry the applied load. $F_y = 36$ ksi.

334 STEEL DESIGN FOR ENGINEERS AND ARCHITECTS

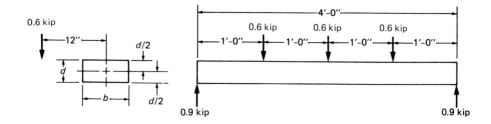

Solution. Rectangular sections bent about their weak axis develop greater bending and torsional strength due to their resistance to lateral-torsional buckling (see Commentary F2). Therefore, assume the bar to be b in. wide and $d = b/2$ in. deep.

Design the section for flexure, and check the increase in shear stress due to torsion.

$$\frac{M}{F_b} = S = \frac{bd^2}{6}$$

$M_{\text{max}} = (0.9 \text{ k} \times 2 \text{ ft}) - (0.6 \text{ k} \times 1 \text{ ft}) = 1.2 \text{ ft-kip} = 14,400 \text{ in.-lb}$

$F_b = 0.75 \, F_y = 27 \text{ ksi}$ (AISCS F2.1)

$$S = \frac{b}{6}\left(\frac{b}{2}\right)^2 = \frac{b^3}{24}$$

$$\frac{b^3}{24} = \frac{14,400 \text{ in.-lb}}{27,000 \text{ psi}}$$

$b^3 = 12.8 \text{ in.}^3$

$b = 2.34 \text{ in.}$ say $2\frac{1}{2}$ in.

$d = b/2 = 1\frac{1}{4}$ in.

Check shear stress

$F_v = 0.40 \, F_y = 14.5 \text{ ksi}$

$J^* = 0.246 \times 2.50 \text{ in.} \times (1.25 \text{ in.})^3 = 1.20 \text{ in.}^4$

$T = 0.9 \text{ k} \times 12 \text{ in.} = 10.8 \text{ in.-k}$

$$f_{v\text{torque}} = \frac{Tc}{J^*} = \frac{10.8 \text{ in.-k} \times 1.25 \text{ in.}}{1.20 \text{ in.}^4} = 11.25 \text{ ksi}$$

$$f_{v\text{bending}} = \frac{3}{2} \times \frac{0.9 \text{ k}}{2.50 \times 1.25 \text{ in.}} = 0.43 \text{ ksi}$$

$$f_{v\text{torque}} + f_{v\text{bending}} = 11.25 + 0.43 = 11.68 \text{ ksi} < F_v \quad \text{ok}$$

8.4 TORSION OF OPEN SECTIONS

Open structural steel members can often be considered as consisting of thin rectangular elements, where

$$J^* = \Sigma \frac{bd^3}{3} \tag{8.9}$$

and b is the largest dimension of each element, and d is the smallest.[2] For members with varying thicknesses, the average thickness is used as d. For commonly used rolled shapes, the AISCM has listed as J the torsional constant of each section. A slight variance can be observed by calculating J^* in a direct manner and comparing to the values for J listed in the properties table. These variations can be attributed to the effect of fillets, the rounded corners, and the ratio b/d not being infinite, which is implied by the factor 3 in the denominator of eqn. (8.9).

The open sections in Fig. 8.4 show their shear flow and the locations of maximum shear. It can be seen that each shape behaves as a series of rectangular elements with their shear stresses flowing continuously. An approximate value of the maximum shear occurring at the points indicated can be calculated from

$$f_v = \frac{Td}{J^*} \tag{8.10}$$

where d = thickness of the web or the flange as applicable.

Welds that may be used for a section subjected to torsional loading also must

---

[2]Some authors recommend the use of $J^* = \eta \Sigma \frac{1}{3} bd^3$, where η is a coefficient slightly larger than 1 depending on the type of section (i.e., η is equal to 1.0 for angles, 1.12 for channels, 1.15 for T sections and 1.2 for I and wide flange sections).

336 STEEL DESIGN FOR ENGINEERS AND ARCHITECTS

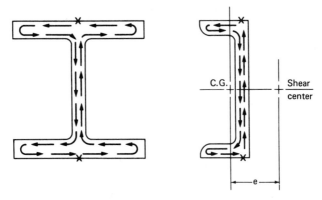

Fig. 8.4. Shear flow and maximum stress in open structural shapes in torsion.

be designed to carry the torsional stress. This stress is superimposed on the shear stress in the web. The weld is designed for the combined stresses.

Example 8.11. Determine the torsional constant J^* for a W 12 × 58, and compare the result with the value given in the AISCM.

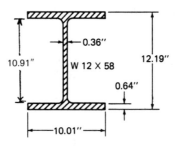

Solution. The value J^* can be determined from the formula

$$J^* = \Sigma \tfrac{1}{3} bd^3 \quad \text{where } b = \text{width of each element}$$
$$d = \text{thickness of each element}$$

$$J^* = \tfrac{1}{3} [2 \times (10.01 \text{ in.} \times (0.640 \text{ in.})^3) + (10.91 \text{ in.} \times (0.360 \text{ in.})^3)]$$

$$= 1.92 \text{ in.}^4$$

The value J from the AISCM is 2.10 (see p. 1-122). The variation can be explained by the approximation of the wide flange by rectangular elements. Note that J given in the AISCM is the same as J^* derived here.

Example 8.12. An externally applied force of 850 lb acts 18 in. to the right of the web centerline. Design the most economic wide flange for a 10-ft cantilever. $F_v = 13.5$ ksi, and $F_b = 20$ ksi. Initially, neglect the combination of shear stresses due to torsion and bending.

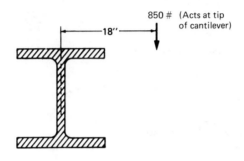

850 # (Acts at tip of cantilever)

Solution.

$$\text{From} \quad f_v = \frac{Tt_f}{J^*}, \quad \frac{t_f}{J^*} \leq \frac{F_v}{T}$$

The allowable stresses have been reduced to allow for the combined effect of the shear stress due to torsion and bending, resulting in a larger shear stress. This can be seen by the use of Mohr's circle.

Design the member first for torsion, then verify for bending and shear.

The member must be found by trial and error. Use the torsional constant in the AISCM.

$$\frac{t_f}{J} \leq \frac{13{,}500 \text{ psi}}{850 \text{ lb} \times 18 \text{ in.}} = .882$$

Try W 16 × 31, $t_f = 0.440$ in., $J = 0.46$ in.$^4 = J^*$

$$\frac{0.440}{0.46} = 0.957 > 0.882 \quad \text{N.G.}$$

338 STEEL DESIGN FOR ENGINEERS AND ARCHITECTS

Try W 10 × 30, $t_f = 0.51$ in., $J = 0.62$ in.^{4}

$$\frac{0.51}{0.62} = 0.822 < 0.882 \quad \text{ok}$$

Check for bending and shear

$$M = 0.850 \text{ k} \times 10 \text{ ft} \times 12 \text{ in./ft} = 102 \text{ in.-k}$$

$$S = 32.4 \text{ in.}^3$$

$$f_b = \frac{102}{32.4} = 3.15 \text{ ksi} < 20 \text{ ksi} \quad \text{ok}$$

$$f_{v\,\text{web}} = \frac{0.85}{10.47 \times 0.30} + \frac{0.85 \times 18 \times 0.30}{0.62} = 7.67 \text{ ksi} < 13.5 \text{ ksi} \quad \text{ok}$$

$$f_{v\,\text{flg}} = \frac{0.85 \times 18 \times 0.51}{0.62} = 12.59 \text{ ksi} < 13.5 \text{ ksi} \quad \text{ok}$$

For the reader familiar with Mohr's circle, the combination of stresses is indicated as follows.

For an element at the top flange near the point of fixity, we have $f_x = 3.15$ ksi and $f_y = 0$.

Considering Mohr's circle of center A,

$$OA = \frac{f_x + f_y}{2} = \frac{3.15 + 0}{2} = 1.58 \text{ ksi}$$

$$AB = \frac{f_x - f_y}{2} = \frac{3.15 - 0}{2} = 1.58 \text{ ksi}$$

$$BD = f_v = 12.59 \text{ ksi}$$

$$AD = \sqrt{AB^2 + BD^2} = 12.69 \text{ ksi} = AC = f_{v\,\text{max}} < 13.5 \text{ ksi} \quad \text{ok}$$

$$OE = f_{b\,\text{max}} = OA + AE = \frac{f_x + f_y}{2} + f_{v\,\text{max}} = 1.58 + 12.69$$

$$= 14.27 \text{ ksi} < 20 \text{ ksi} \quad \text{ok}$$

The orientation is at an angle $EAD/2$.

TORSION   339

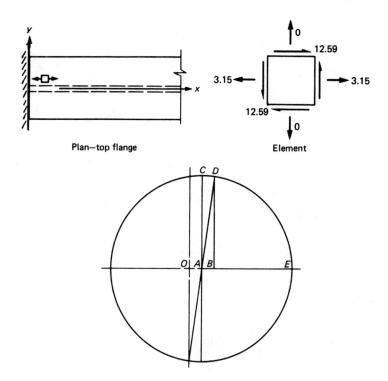

Plan—top flange   Element

Example 8.13. A wide-flange beam spanning 24 ft supports a 1-kip-per-foot brick wall, which acts 5 in. off center. Design the beam to withstand torsion and bending. Assuming the beam is simply supported, but restrained against torsion at the supports. $F_v = 14.5$ ksi.

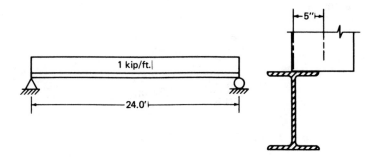

Solution. Determine the beam to carry the flexure (laterally unsupported); then check to see if it will be safe for the torsion.

340 STEEL DESIGN FOR ENGINEERS AND ARCHITECTS

$$S = \frac{M}{F_b}$$

$$F_v = \frac{Tc}{J}$$

$$M = \frac{wL^2}{8} = \frac{1 \text{ k/ft} \times (24 \text{ ft})^2}{8} = 72 \text{ ft-k}$$

Designing for flexure alone, the lightest available section is W 10 × 45 (AISCM p. 2-172).

$$S_x = 49.1 \text{ in.}^3$$

$$f_b = M/S_x = \frac{72 \text{ ft-k} \times 12 \text{ in./ft}}{49.1 \text{ in.}^3} = 17.60 \text{ ksi}$$

Check for torsion

$$T = \frac{5}{12} \text{ ft} \times \frac{24 \text{ ft}}{2} \times 1.0 \text{ k/ft} = 5.0 \text{ ft-k} = 60 \text{ in.-k}$$

$$J = 1.51 \text{ in.}^4$$

$$d = t_f = 0.62 \text{ in.}$$

$$f_v = \frac{60 \text{ in.-k} \times 0.62 \text{ in.}}{1.51 \text{ in.}^4} = 24.63 \text{ ksi} > 14.5 \text{ ksi} \quad \text{N.G.}$$

Calculating the ratio d/J,

$$\frac{d}{J} = \frac{F_v}{T}$$

$$\frac{d}{J} = \frac{14.5 \text{ ksi}}{60 \text{ in.-k}} = 0.24 \text{ in.}^{-3}$$

Try W 12 × 72, $d = t_f = 0.670$ in., $J = 2.93$ in.^{4}

$$\frac{0.670 \text{ in.}}{2.93 \text{ in.}^4} = 0.229 \text{ in.}^{-3} < 0.24 \text{ in.}^{-3} \quad \text{ok}$$

TORSION 341

Trying for a more economical section,

$$W\ 14 \times 68, \quad d = t_f = 0.720 \text{ in.}, \quad J = 3.02 \text{ in.}^4$$

$$\frac{0.720 \text{ in.}}{3.02 \text{ in.}^4} = 0.238 \text{ in.}^{-3} < 0.24 \text{ in.}^{-3} \quad \text{ok}$$

We see also by inspection (AISC, Part 2, "Allowable Moments in Beams") that the bending stress remains under control: $M_R = 185$ ft-k for $L = 24$ ft (p. 2-169).

Verify shear stresses:

$$f_{v\text{flg}} = \frac{60 \times 0.72}{3.02} = 14.3 \text{ ksi} < 14.5 \text{ ksi} \quad \text{ok}$$

$$f_{v\text{web}} = \frac{12}{14.04 \times 0.415} + \frac{60 \times 0.415}{3.02}$$

$$= 2.06 + 8.25 = 10.3 \text{ ksi} < 14.5 \text{ ksi} \quad \text{ok}$$

Use W 14 × 68.

Example 8.14. Determine the required weld length for the section shown at a location where the shear force is 10 kips and the torque is 300 in.-kips. Use A36 steel and $\frac{5}{16}$-in. E70 intermittent fillet welds.

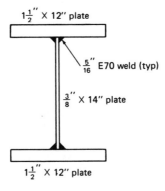

Solution. The force to be carried by the weld is produced at the interface of the flange and web. The force is the superposition of the shear flow due to vertical loading and the shear flow due to the torsional moment.

342 STEEL DESIGN FOR ENGINEERS AND ARCHITECTS

Shear stress due to direct shear force

$$f_v = \frac{V}{bd} = \frac{10}{\frac{3}{8} \times \left(14 + 2\left(1\frac{1}{2}\right)\right)} = 1.57 \text{ ksi}$$

$$J^* = \frac{1}{3} \times \left[2 \times 12 \times \left(1\frac{1}{2}\right)^3 + 14 \times \left(\frac{3}{8}\right)^3\right] = 27.25 \text{ in.}^4$$

Shear stress due to torsion

$$f_v = \frac{Tt_w}{J^*} = \frac{300 \times 0.375}{27.25} = 4.13 \text{ ksi}$$

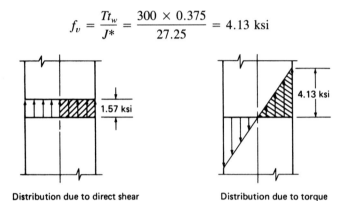

Distribution due to direct shear   Distribution due to torque

Shear stress distribution in web

The shear flow to be carried by the weld is the reaction at right.

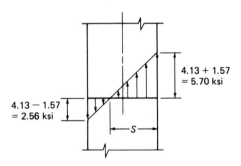

Location of zero shear flow on web:

$$(2.56 \text{ ksi} + 5.70 \text{ ksi}) \times S = 5.70 \text{ ksi} \times 0.375 \text{ in.}$$

$$S = 0.259 \text{ in.}$$

Force carried by weld (right side):

$$q = \frac{1}{(2 \times 0.375)} \left[(0.259 \times 5.70) \left(\frac{3}{8} - \frac{0.259}{3} \right) - \left(\frac{2.56}{3} \right) \left(\frac{3}{8} - 0.259 \right)^2 \right]$$
$$= 0.55 \text{ k/in.}$$

For E70 $\frac{5}{16}$ in. weld,

$$F_w = 5 \times 0.928 \text{ k/in./sixteenth} = 4.64 \text{ k/in.}$$

For 1 foot length,

$$q = 0.55 \text{ k/in.} \times 12 \text{ in./ft} = 6.6 \text{ k/ft}$$

Length of weld required[3]

$$L = \frac{6.6 \text{ k/ft}}{4.64 \text{ k/in./ft}} = 1.42 \text{ in.} \quad \text{use } 1\frac{1}{2} \text{ in. weld at 12 in. centers (min.).}$$

Example 8.15. Determine the required intermittent fillet weld length per foot of section for the plate section shown. Use E70 minimum size fillet weld. $V = 200$ kips; $T = 40$ ft-k.

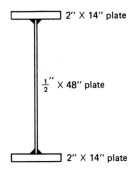

Solution.

$$V_{max} = 200 \text{ kips}; \quad T = 40 \text{ ft-kips}$$

---

[3]Welds in engineering practice are generally continuous. The design of intermittent welds is provided as an exercise.

Verify section, and design welds.

$$J = \frac{1}{3}\left[(14 \text{ in.} \times (2 \text{ in.})^3) \times 2 + \left(48 \text{ in.} \times \left(\frac{1}{2}\text{ in.}\right)^3\right)\right] = 76.7 \text{ in.}^4$$

$$f_{vV} = \frac{200 \text{ k}}{\frac{1}{2} \text{ in.} \times 52 \text{ in.}} = 7.69 \text{ ksi}$$

$$f_{vT_{fl}} = \frac{40 \text{ ft-k} \times 12 \text{ in./ft} \times 2 \text{ in.}}{76.7 \text{ in.}^4} = 12.52 \text{ ksi}$$

$$f_{vT_{web}} = \frac{40 \text{ ft-k} \times 12 \text{ in./ft} \times 0.5 \text{ in.}}{76.7 \text{ in.}^4} = 3.13 \text{ ksi}$$

$$f_{v\text{web}_{tot}} = 7.69 \text{ ksi} + 3.13 \text{ ksi} = 10.82 \text{ ksi}$$

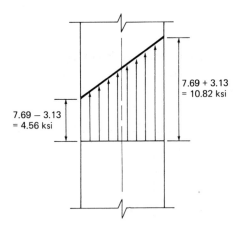

Shear flow at the right side of the web:

$$q = \tfrac{1}{2} \times [\tfrac{1}{3}(10.82 - 4.56) + \tfrac{1}{2}(4.56)] = 2.18 \text{ k/in.}$$

Minimum weld size = $\tfrac{5}{16}$ in.

$$F_w = 5 \times 0.928 \text{ k/in.} = 4.64 \text{ k/in.}$$

Length of weld required per foot

$$q = 2.18 \text{ k/in.} \times 12 \text{ in./ft} = 26.16 \text{ k/ft}$$

$$L = \frac{26.16 \text{ k/ft}}{4.64 \text{ k/in./ft}} = 5.6 \text{ in.} \quad \text{say 6 in.}$$

Use $\frac{5}{16}$-in. E70 weld 6 in. long, spaced 12 in. on center.

8.5 SHEAR CENTER

A prismatic member has a singular shear axis through which a force can be applied from any angle and no twisting moment will result. For any given cross section, the point on the axis is called the *shear center* of the section. If applied forces pass through the shear center, the member is subject only to flexural stresses. Forces applied away from the shear center develop a torsional moment that twists around the shear center.

Consider a section with an axis of symmetry transversely loaded in the plane of the axis, i.e., a wide flange loaded in the plane of the web (Fig. 8.5a). It can be seen that the external shear force and the internal shear forces due to the shear flow are in equilibrium. Consider now a channel loaded at the centroid (Fig. 8.5b). The forces are in equilibrium, but an unbalanced moment has developed due to the shear flow in the flanges. If the external force is applied at some distance away from the centroid, the moment can be brought to zero. This

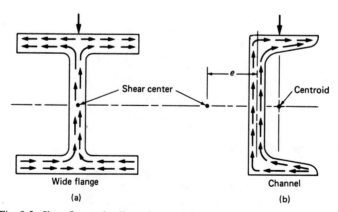

Fig. 8.5. Shear flow under direct shear and shear centers of common sections.

point of load application is the shear center. It can be proved that the distance between the shear center and the center of the web is given by

$$e = \frac{\Sigma I_x x}{\Sigma I_x} \tag{8.11}$$

In the properties section of the AISCM, the shear center location can be found for channel sections. It is important to note that, in general, the shear center must always lie on an axis of symmetry.

Example 8.16. Calculate the location of the shear center for the shape shown below.

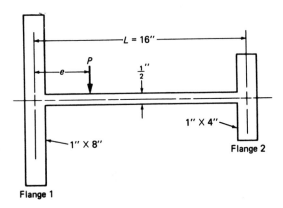

Solution. To locate the shear center, consider a force P applied to the section such that both flanges will deflect equally. Then

$$\frac{(L - e)P}{I_1} = \frac{eP}{I_2}$$

Solving for e,

$$e = \frac{LI_2}{I_1 + I_2}$$

Using the expression given

$$e = \frac{16 \text{ in.} \times 1 \text{ in.} \times \dfrac{(4 \text{ in.})^3}{12}}{1 \text{ in.} \times \dfrac{(8 \text{ in.})^3}{12} + 1 \text{ in.} \times \dfrac{(4 \text{ in.})^3}{12}} = 1.78 \text{ in.}$$

Example 8.17. Determine the approximate location of the shear center for a C 10 × 15.3. Compare the approximate location with the location given in the AISCM.

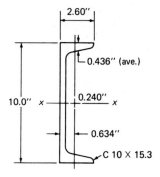

Solution. For symmetrical sections, the axis of symmetry is one shear axis. The other axis is found by

$$e = \frac{\Sigma I_x x}{\Sigma I_x}$$

where x is the distance from the Y axis to the centroid of the elements. Consider the Y axis passing through the center of the web. Assuming the flanges as separate parts, and neglecting their moments of inertia about their own centroidal axes,

$$e = \frac{I_{\text{web}} x + 2 I_{\text{flg}} x}{I_{\text{web}} + I_{\text{flg}}} = \frac{\left(\frac{t_w d^3}{12} \times 0\right) + 2(t_f b_f)\left(\frac{d}{2} - \frac{t_f}{2}\right)^2 \left(\frac{b_f}{2} - \frac{t_w}{2}\right)}{I_x}$$

$t_f = 0.436$ in., $t_w = 0.24$ in., $b_f = 2.60$ in., $d = 10.0$ in., $I_x = 67.4$ in.^{4}

$$e = \frac{2(0.436 \text{ in.} \times 2.60 \text{ in.})\left(\frac{10 \text{ in.}}{2} - \frac{0.436 \text{ in.}}{2}\right)^2 \times \left(\frac{2.60 \text{ in.}}{2} - \frac{0.24 \text{ in.}}{2}\right)}{67.4 \text{ in.}^4}$$

$= 0.908$ in.

The value for shear center location in the AISCM is measured from the back of the web. To determine e_0,

$$e_0 = 0.908 \text{ in.} - \frac{0.24 \text{ in.}}{2} = 0.788 \text{ in.}$$

The tabular value is $e_0 = 0.796$ in. The minor variation can be explained by the approximation of the flanges (which are tapered) by rectangular elements.

Example 8.18. Determine if a cantilevered C 12 × 30, 10 ft long will carry the applied force shown if $F_v = 14.5$ ksi and $F_b = 22.0$ ksi.

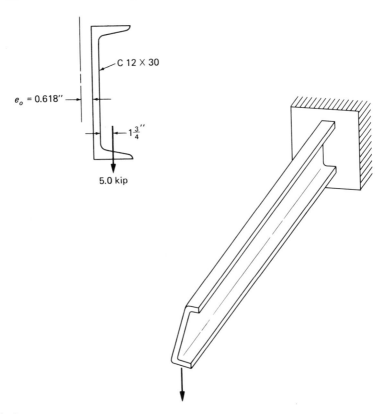

Solution.

$T = 5.0$ k × $(0.618$ in. $+ 1.75$ in.$) = 11.84$ in.-k

$M = 5.0$ k × 10 ft × 12 in./ft $= 600$ in.-k

$V = 5.0$ k

$J = 0.87$ in.^{4}

$S = 27.0$ in.^{3}

$$f_{v\text{flg}} = \frac{11.84 \text{ in.-k} \times 0.501 \text{ in.}}{0.87 \text{ in.}^4} = 6.82 \text{ ksi} < 14.5 \text{ ksi}$$

$$f_{v\text{web}} = \frac{11.84 \text{ in.-k} \times 0.510 \text{ in.}}{0.87 \text{ in.}^4} + \frac{5.0 \text{ k}}{12.0 \text{ in.} \times 0.510 \text{ in.}} = 6.94 \text{ ksi} + 0.82 \text{ ksi}$$

$$= 7.76 \text{ ksi} < 14.5 \text{ ksi}$$

$$f_b = \frac{600 \text{ in.-k}}{27 \text{ in.}^3} = 22.22 \text{ ksi} \approx 22 \text{ ksi} \quad \text{ok}$$

Investigating the angle of twist,

$$\phi = \frac{Tl}{GJ} = \frac{11.84 \text{ in.-k} \times 10 \text{ ft} \times 12 \text{ in.}/\text{ft}}{11.6 \times 10^3 \times 0.87 \text{ in.}^4} = 0.1408 \text{ radians} = 8.07°$$

Even though stress values are within allowable limits, the beam may be considered inadequate due to excessive deformation (i.e., rotation).

Example 8.19. Determine the shear center for the box beam shown. The wide flange is a W 14 × 90 with a 1-in. plate connected to the flanges as shown.

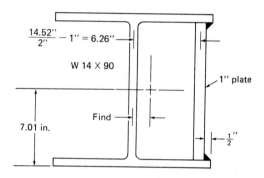

Solution. The shear center will be located somewhere along the X axis passing through the center of the members. Assume the Y axis passing through the centroid of the wide flange.

$$I_w = 999 \text{ in.}^4; \quad d \text{ (depth of plate)} = 14.02 - 2(0.710) = 12.60 \text{ in.}$$

$$e = \frac{\Sigma I_x x}{\Sigma I_x} = \frac{(999 \times 0) + \left(\frac{1 \times 12.60^3}{12} \times \left(\frac{14.52}{2} - 1\right)\right)}{999 + \left(\frac{1 \times 12.60^3}{12}\right)} = 0.895 \text{ in.}$$

350 STEEL DESIGN FOR ENGINEERS AND ARCHITECTS

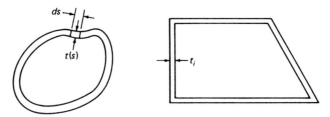

Fig. 8.6. Closed thin-walled sections.

8.6 TORSION OF CLOSED SECTIONS

A member subject to a torsional moment is considered to be a closed section if the cross section is a continuous thin-walled tube of any shape. With these conditions, a closed section can be analyzed using Bredt's formula (see Fig. 8.6).

For any thin-walled closed section, the shear flow is assumed to be constant throughout the plane. Describing this flow of shear through a continuous tube section, Bredt developed the formula

$$q = \frac{T}{2A} \tag{8.12}$$

where A is the area enclosed by the center line of the section's contour. The shearing stress at any point can then be found by

$$f_v = \frac{q}{t} = \frac{T}{2At} \tag{8.13}$$

where t is the thickness of the section.

The angle of twist of a thin-walled closed section is given by the same expression (8.2)

$$\phi = \frac{TL}{GJ} \tag{8.14}$$

where

$$J = \frac{4A^2}{\int \frac{ds}{t}} \tag{8.15}[4]$$

---

[4] Refer to Blodgett, *Design of Welded Structures*, 2.10-4.

ds is an element of length along the mid-thickness of the contour wall, and t is the wall thickness of that element.

Example 8.20. Determine the maximum torsion a thin wall shaft will carry if the outside shaft diameter is 9 in. and the inside diameter is 8 in. Allowable stress is 8000 psi. Compare the answer to Example 8.2.

Solution.

$$T = 2\,AtF_v$$

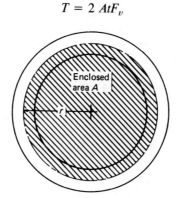

$$A = \pi \frac{d^2}{4} = 56.7 \text{ in.}^2 \ (d = d_{\text{ave}} = 8.5 \text{ in.})$$

$t = 0.5$ in.

$T = 2 \times (56.7 \text{ in.}^2) \times 0.5 \text{ in.} \times (8000 \text{ psi}) = 453{,}600$ in.-lb

Converting to ft-kips,

$$\frac{453{,}600 \text{ in.-lb}}{12 \text{ in./ft} \times 1000 \text{ lb/k}} = 37.8 \text{ ft-k}$$

The maximum torsion determined in Example 8.2 was 35.85 ft-kips. We see that, although the wall is somewhat thick, the result obtained by the approximate Bredt's formula is satisfactory.

Example 8.21. Determine the maximum torque and angle of rotation for

a) a closed round tube with an outside diameter of 8 in. and inside diameter of $7\text{-}\tfrac{1}{2}$ in.

b) the same thin-walled tube that has been slit along the wall.

$F_v = 14.5$ ksi, $G = 11.7 \times 10^3$ ksi, and $L = 10$ ft.

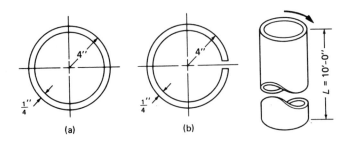

Solution.

a) Using thin-walled closed section theory,

$$T = 2\,AtF_v$$

$$A = \frac{\pi d^2}{4} = 47.17 \text{ in.}^2 \quad \left(d = d_{ave} = 7\frac{3}{4} \text{ in.}\right)$$

$t = 0.25$ in.

$T = 2 \times 47.17 \text{ in.}^2 \times 0.25 \text{ in.} \times 14.5 \text{ ksi} = 342 \text{ in.-kip} = 28.5 \text{ ft-kip}$

$$\phi = \frac{TL}{GJ}$$

$L = 10 \text{ ft} \times 12 \text{ in./ft} = 120$ in.

$$J = \frac{4A^2}{\int \frac{ds}{t}} = \frac{4 \times (47.17)^2}{\left(\dfrac{\pi \times 7.75}{0.25}\right)} = 91.39 \text{ in.}^4$$

$$\phi = \frac{342 \text{ in.-kip} \times 120 \text{ in.}}{11.7 \times 10^3 \text{ ksi} \times 91.39 \text{ in.}^4} = 0.0384 \text{ radians} = 2.20°$$

b) Consider the slit section as a thin rectangular section in torsion.

$$T = \frac{F_v J^*}{d}$$

$J^* = \alpha b d^3$; for a thin wall, $\alpha = 0.33$ (i.e., $b \gg d$; see Table 8.1)

$J^* = 0.33 \,(\pi \times 7.75 \text{ in.}) \times (0.25 \text{ in.})^3 = 0.1255 \text{ in.}^4$

$$T = \frac{14.5 \text{ ksi} \times 0.1255 \text{ in.}^4}{0.25 \text{ in.}} = 7.28 \text{ in.-kip} = 0.607 \text{ ft-kip}$$

$$\phi = \frac{TL}{GJ^*}$$

$L = 120$ in.

$$\phi = \frac{7.28 \text{ in.-kip} \times 120 \text{ in.}}{11.7 \times 10^3 \text{ ksi} \times 0.1255 \text{ in.}^4} = 0.595 \text{ radians} = 34.09°$$

The advantage, by having closed sections over open sections, can readily be seen.

$$\text{Ratio of maximum allowable torques} = \frac{28.50 \text{ ft-kip}}{0.607 \text{ ft-kip}} = 47$$

$$\text{Ratio of corresponding angles of twist} = \frac{0.0384 \text{ rad}}{0.595 \text{ rad}} = 0.0645$$

Example 8.22. For two C 10 × 15.3 channels with a length of 8 ft, determine the torsional capacity and the angle of twist for that torque

a) if connected as shown but do not act as a closed section.
b) if connected toe-to-toe, providing a closed section.
c) if connected back-to-back, providing an open section.

$F_v = 14.5$ ksi and $G = 11.7 \times 10^3$ ksi.

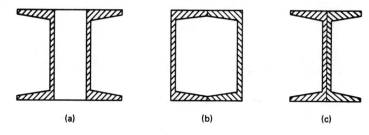

(a)　　　　(b)　　　　(c)

Solution.

a) $\quad T = \dfrac{F_v J^*}{t_f}$

354 STEEL DESIGN FOR ENGINEERS AND ARCHITECTS

$$J^* = \frac{1}{3}\left[(10 - 0.436) \times (0.24)^3 + 2\left(2.60 - \frac{0.24}{2}\right) \times (0.436)^3\right] \times 2$$

$$= 0.362 \text{ in.}^4$$

$t_f = 0.436$ in.

$$T = \frac{14.5 \text{ ksi} \times 0.362 \text{ in.}^4}{0.436 \text{ in.}} = 12.04 \text{ in.-k} = 1.0 \text{ ft-k}$$

$$\phi = \frac{Tl}{GJ^*}$$

$l = 8 \text{ ft} \times 12 \text{ in.}/\text{ft} = 96 \text{ in.}$

$$\phi = \frac{12.04 \text{ in.-kip} \times 96 \text{ in.}}{11.7 \times 10^3 \text{ ksi} \times 0.362 \text{ in.}^4} = 0.273 \text{ rad} = 15.64°$$

b) Treating the section as a thin-walled closed section,

$$T = 2 F_v A t$$

$A = (5.2 \text{ in.} - 0.24 \text{ in.}) \times (10 \text{ in.} - 0.436 \text{ in.}) = 47.44 \text{ in.}^2$

$t = 0.24$ in.

$T = 2 \times 14.5 \text{ ksi} \times 47.44 \text{ in.}^2 \times 0.24 \text{ in.} = 330.2 \text{ in.-k} = 27.5 \text{ ft-k}$

$$\phi = \frac{Tl}{GJ}$$

$l = 96$ in.

$$J = \frac{4A^2}{2 \times \left(\dfrac{d}{t_f} + \dfrac{b}{t_w}\right)} = \frac{4 \times (47.44)^2}{2 \times \left[\dfrac{5.20 - 0.24}{0.436} + \dfrac{10.0 - 0.436}{0.24}\right]} = 87.87 \text{ in.}^4$$

$$\phi = \frac{330.2 \text{ in.-k} \times 96 \text{ in.}}{11.7 \times 10^3 \text{ ksi} \times 87.87 \text{ in.}^4} = 0.0308 \text{ rad} = 1.767°$$

If we compare the rotation that would result from the torque of (a),

$$\phi_1 = 1.767° \left[\frac{12.04}{330.2}\right] = 0.064°$$

c) $$T = \frac{F_v J^*}{t_f}$$

$$J^* = 2\left(\frac{1}{3} \times 2\, b_f \times t_f^3\right) + \left(\frac{1}{3} \times (d - t_f) \times (2\, t_w)^3\right)$$

$$= 0.287 + 0.353 = 0.640 \text{ in.}^4$$

$$T = \frac{14.5 \text{ ksi} \times 0.640 \text{ in.}^4}{0.436 \text{ in.}} = 21.3 \text{ in.-k} = 1.77 \text{ ft-k}$$

$$\phi = \frac{Tl}{GJ^*}$$

$$l = 96 \text{ in.}$$

$$\phi = \frac{21.3 \text{ in.-k} \times 96 \text{ in.}}{11.7 \times 10^3 \text{ ksi} \times 0.640 \text{ in.}^4} = 0.273 \text{ rad} = 15.63°$$

If we compare the rotation that would result from the torque of (a),

$$\phi_1 = 15.63° \left[\frac{12.04}{21.3}\right] = 8.84°$$

Example 8.23. For the two 20-ft, simply supported beams below, determine

a) the allowable torque for $F_v = 14.5$ ksi.
b) the angle of twist for an applied torque of 150-in.-kip; $G = 11.7 \times 10^3$ ksi.
c) the required length of $\frac{5}{16}$-in. E70 intermittent fillet weld per foot; $V = 10$ kips and $T = 80$ in.-kip.

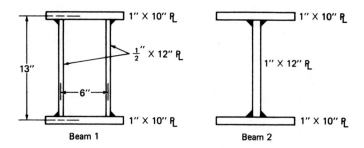

Beam 1   Beam 2

Solution.

a) Beam 1

$$T = F_v\, 2(At)$$

$A = 6$ in. $\times$ 13 in. $= 78.0$ in.^{2}

$t = \frac{1}{2}$ in.

$T = 14.5$ ksi $\times$ 2 $\times$ 78 in.^{2} $\times \frac{1}{2}$ in. $= 1131$ in.-kip $= 94.3$ ft-kip

Beam 2

$$T = F_v \frac{J^*}{t_f}$$

$J^* = 0.33\,[(10 \text{ in.} \times (1 \text{ in.})^3 \times 2) + (12 \text{ in.} \times (1 \text{ in.})^3)] = 10.67$ in.^{4}

$t_f = 1.0$ in.

$$T = \frac{14.5 \text{ ksi} \times 10.67 \text{ in.}^4}{1.0 \text{ in.}} = 154.7 \text{ in.-kip} = 12.9 \text{ ft-kip}$$

b) Beam 1

$$\phi = \frac{TL}{GJ}$$

$L = 20$ ft $\times$ 12 in./ft $= 240$ in.

$$J = \frac{4\,b^2\,d^2}{2 \times \left(\dfrac{b}{t_w} + \dfrac{d}{t_f}\right)} = \frac{4\,(13 \text{ in.})^2 (6 \text{ in.})^2}{2\left(\dfrac{13 \text{ in.}}{0.5 \text{ in.}} + \dfrac{6 \text{ in.}}{1 \text{ in.}}\right)} = 380.25 \text{ in.}^4$$

$$\phi = \frac{150 \text{ in.-kip} \times 240 \text{ in.}}{11.7 \times 10^3 \text{ ksi} \times 380.25 \text{ in.}^4} = 0.0081 \text{ rad} = 0.464 \text{ deg}$$

Beam 2

$$\phi = \frac{TL}{GJ^*}$$

$L = 240$ in.

$J^* = 10.67$ in.^{4}

$$\phi = \frac{150 \text{ in.-kip} \times 240 \text{ in.}}{11.7 \times 10^3 \text{ ksi} \times 10.67 \text{ in.}^4} = 0.288 \text{ rad} = 16.52 \text{ deg}$$

c) Beam 1

Consider web

$$\text{Shear} \quad f_v = \frac{10 \text{ k}}{2(0.5 \text{ in.}) \times 14 \text{ in.}} = 0.7 \text{ k/in.}^2$$

$$\text{Shear flow} \quad q_v = 0.7 \text{ k/in.}^2 \times 0.5 \text{ in.} = 0.35 \text{ k/in.}$$

$$\text{Torsion} \quad f_v = \frac{80.0 \text{ in.-kip}}{2 \times 78.0 \text{ in.}^2 \times 0.5 \text{ in.}} = 1.026 \text{ k/in.}^2$$

$$\text{Shear flow} \quad q_v = 1.026 \text{ k/in.}^2 \times 0.5 \text{ in.} = 0.51 \text{ k/in.}$$

$$\text{Total shear flow} \quad q_{\max} = 0.35 \text{ k/in.} + 0.51 \text{ k/in.} = 0.86 \text{ k/in.}$$

For $\frac{5}{16}$-in. E70 weld,

$$F_w = 5 \times 0.928 \text{ k/in.} = 4.64 \text{ k/in.}$$

For a 1-ft length,

$$\frac{0.86 \text{ k/in.} \times 12 \text{ in./ft}}{4.64 \text{ k/in.}} = 2.22 \text{ in./ft}$$

Use $2\frac{1}{4}$ in. intermittent welds at 12 in. centers.

Beam 2

The maximum shear flow is produced in the web and is the superposition of the shear flow due to vertical loading and the shear flow due to the torsional moment.

$$\text{Shear} \quad f_v = \frac{10 \text{ k}}{1 \text{ in.} \times 14 \text{ in.}} = 0.7 \text{ k/in.}^2$$

$$\frac{1}{2} \text{ shear flow} \quad q_v = 0.7 \text{ k/in.}^2 \times 1.0 \text{ in.} \times \frac{1}{2} = 0.35 \text{ k/in.}$$

$$\text{Torsion} \quad f_v = \frac{80 \text{ in.-k} \times 1.0 \text{ in.}}{10.67 \text{ in.}^4} = 7.5 \text{ k/in.}^2$$

Shear to be carried by weld on one side of web

$$f_v = 0.7 \text{ k/in.}^2 + 7.5 \text{ k/in.}^2 = 8.2 \text{ k/in.}^2$$
$$f_v = 0.7 \text{ k/in.}^2 - 7.5 \text{ k/in.}^2 = -6.8 \text{ k/in.}^2$$
$$q_{v\max} = 1.6 \text{ k/in. (right side)}$$

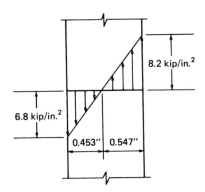

For $\tfrac{5}{16}$-in. E70 weld,

$$F_w = 5 \times 0.928 \text{ k/in.} = 4.64 \text{ k/in.}$$

For a 1-ft length,

$$\frac{1.6 \text{ k/in.} \times 12 \text{ in./ft}}{4.64 \text{ k/in.}} = 4.1 \text{ in. per 1-ft length}$$

Use $4\tfrac{1}{2}$-in. intermittent welds at 1-ft centers.

8.7 MEMBRANE ANALOGY

Prandtl showed that the governing equation for the torsion problem is the same for that of a thin membrane stretched across the cross section and subjected to uniform pressure.

If we have a tube conforming to the outline of the twisted section, cover the tube with a thin membrane glued to the tube, and apply pressure to the inside; the membrane will deflect, and we will note the following: the shear stresses due to the torsion correspond to the slopes of the membrane, and the lines of equal shear correspond to the contour lines, the direction of the shear stress to the tangent at the contour lines, and the torque to the volume under the membrane.

TORSION 359

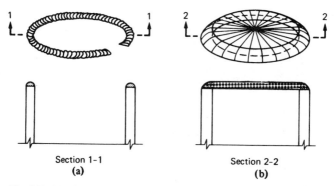

Fig. 8.7. Membrane analogy of (a) open sections and (b) closed sections.

In the case of a closed section, the boundaries of the tube will correspond to the limits of the material twisted, and the hole will be represented by a plate glued to the membrane (Fig. 8.7). From here, one can get a very good intuitive understanding of the stress conditions of a twisted member.

PROBLEMS TO BE SOLVED

8.1. For an applied torque of 15 in.-kips, determine the maximum shear stress for (a) a 2-in.-diameter steel bar and (b) a steel pipe having an outside diameter of $2\frac{1}{2}$ in. and an inside diameter of 2 in.

8.2. For the 2-in.-diameter steel bar and $2\frac{1}{2}$-in.-outside-diameter steel pipe mentioned in Problem 8.1, determine the maximum allowable torque, assuming A36 steel. Which member is more efficient, i.e., which carries more torque per unit weight?

8.3. For an applied torque of 18 in.-kips, determine the lightest circular bar required, assuming $F_y = 50$-ksi.

8.4. For an applied torque of 18 in.-kips, determine the lightest standard weight pipe, assuming A36 steel. For the pipe selected, calculate the angle of twist if $E_s = 29 \times 10^6$ psi; $\mu = 0.25$, and $L = 10$ ft.

8.5. For an applied torque of 15 in.-kips, determine the maximum shear stress for (a) a 2-in. by 2-in. square steel bar and (b) a 1-in. by 4-in. rectangular steel bar. Use A36 steel. Determine the angle of twist for both bars if $L = 7$ ft and $G = 11.6 \times 10^6$ psi.

8.6. For the two bars mentioned in Problem 8.5, determine the allowable torques and rotations corresponding to those torques if now the bars are made of aluminum with an allowable shear strength of 11.2 ksi and $G = 3.86 \times 10^3$ ksi. Assume linearly elastic behavior for aluminum.

8.7. For the bar shown, determine the maximum allowable load P if $F_y = 36$ ksi and $G = 11.6 \times 10^3$ ksi.

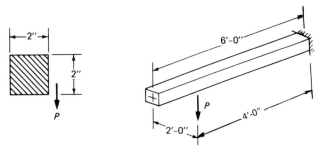

8.8. Design the most economical wide-flange section for an 8.0-ft cantilever on which a load of 5 k is applied at the free end, 12 in. to the right of the web centerline. Assume $F_v = 13.0$ ksi and $F_b = 22$ ksi. Neglect beam weight and combination of shear and bending stresses.

8.9. For the beam shown, determine the maximum allowable height for the concrete facia panel if lightweight concrete weighing 80 lb/ft² is used. The beam is simply supported and restrained against torsion at the supports. It is laterally supported at ends only.

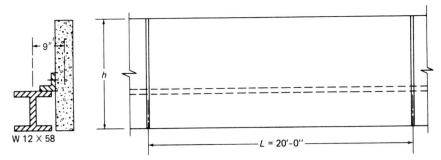

8.10. Design a square tube support column for the bus stop sign shown if wind pressure = 80 lb/ft², $L = 10$ ft, $F_v = 14.5$ ksi, $G = 11.6 \times 10^3$ ksi, and maximum allowable rotation = 4 degrees.

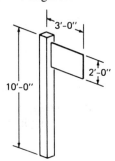

8.11. Compute the torsional constants J^* for the following open sections: (a) W 18 × 50, (b) W 16 × 31, and (c) C 12 × 20.7.

8.12. For a member subject to a torque of 3.5 ft-k, design the most economical W section and the most economical C section, if $F_v = 14.5$ ksi. For $L = 12$ ft, 0 in., which member has the greater resistance to twist (rotation) per unit weight?

8.13. For the fabricated edge beam shown, determine the maximum height of an 8-in. brick wall that can be supported if the ends of the beam are connected to very heavy columns. Assume A36 steel and that the brick weighs 120 psf. $L = 20$ ft.

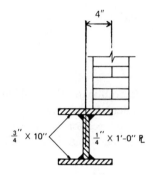

8.14. For the beam shown in Problem 8.13, design the welds, assuming E70 electrodes and the height of the brick wall to be 9 ft, 0 in.

8.15. Determine the maximum allowable load P that can be applied to a C 12 × 20.7 if the channel is simply supported over a span of 22 ft, 0 in. with the ends free to warp. The load P is applied at the centerline of the lower flange.

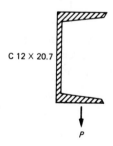

8.16. Determine the required weld for the section shown at a location where the shear force is 12 kips and the torque is 60 in.-kips. The member is constructed of 50-ksi steel. Use E70 weld.

362 STEEL DESIGN FOR ENGINEERS AND ARCHITECTS

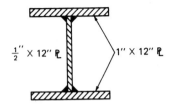

8.17. Locate the shear center for the following shapes:

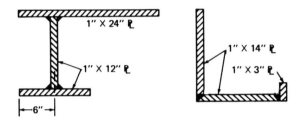

8.18. For the sections shown, determine the allowable torque if $F_v = 14.5$ ksi. Also, for each of the sections, calculate the rotation if $L = 10$ ft, $T = 30$ in.-kips, and $G = 11.7 \times 10^3$ ksi. Each section is made from four $\frac{1}{2}$-in. $\times$ 10-in. plates. For the left figure, assume the web to be one member, 1 in. thick.

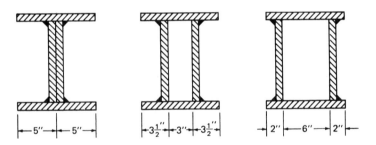

8.19. Calculate the required $\frac{1}{4}$-in. intermittent weld for the sections shown in the second and third sketches of Problem 8.19, if $V = 30$ kips, $T = 300$ in.-kips, and $F_v = 14.5$ ksi. Use E70 weld.

9
Composite Design

9.1 COMPOSITE DESIGN

For many years, reinforced concrete slabs on steel beams were used without considering any composite action. In the 1940s, it was shown that concrete slabs and steel beams act as a unit when joined together to resist horizontal shear. This resistance is provided by either encasing the beam in concrete or welding shear connectors to the beam (AISCS I1). By considering composite action, a significant increase in strength is achieved. When steel beams are totally encased in concrete cast integrally with the slab (see additional requirements in paragraphs 1 through 3 in AISCS I.1), the beams alone may be proportioned to resist, unassisted, both the dead and live load moments using an allowable bending stress of $0.76F_y$ (AISCS I2.1). In this case, temporary shoring is not required. When shear connectors are used, both materials are utilized to their fullest capacity, with the concrete slab acting in compression and the steel beam acting almost entirely in tension. Thus, in both cases, the composite member can carry larger loads (or similar loads on larger spans) than the same slab and beam acting individually. Also, since composite sections have greater stiffness, the deflections will be smaller.

Another basic advantage resulting from composite design is that shallower and lighter steel beams can be used. Consequently, both overall building weight and floor depth are reduced; for high-rise buildings, the ultimate savings in foundations, wiring, ductwork, walls, plumbing, etc. can be considerable.

If part of the concrete is in tension, that portion is considered to be nonexistent and is omitted in calculations, since the tensile capacity of concrete is very often assumed to be zero. Cover plates may be used on the bottom flange of the steel beams in composite sections to increase the efficiency of the member by lowering the neutral axis from within the concrete.[1] In bridge design, this

---
[1]Sharp increases in labor costs for welding have precluded most economic uses of cover plates in composite design in spite of large savings in steel poundage.

364 STEEL DESIGN FOR ENGINEERS AND ARCHITECTS

Fig. 9.1. A composite spandrel beam for a building. Note strands in conduits, perpendicular to the beam, for post-tensioning. By post-tensioning the slab, deflection and negative moment cracking are reduced in the direction perpendicular to the composite beam.

effect is achieved by using plate girders with heavy bottom flanges and relatively light top flanges. The following discussion is based on AISCS Chapter I for building design. Composite sections for bridge design will be briefly discussed in Appendix B.

Shores for composite construction are temporary formworks that support the steel beams at sufficiently close intervals to prevent the beam from sagging before the concrete hardens. When the steel beams are not shored during the casting and curing processes of the concrete, the steel section must be designed

Fig. 9.2. An interior girder of the building whose spandrel beam is shown in Fig. 9.1. Note how reinforcing bars and post-tensioning strands drape over the girder toward the top of the slab to provide negative moment reinforcement.

to support alone the wet concrete and all other construction loads. It is accepted practice to size the beam for dead plus live loads and full composite action when this action is obtained by the use of shear connectors (AISCS Commentary I2).

9.2 EFFECTIVE WIDTH OF CONCRETE SLAB

In AISCS I1, the width of the concrete slab that may participate in composite action is specified (see Fig. 9.3). For an interior beam (i.e., a beam which has beams on both sides of it), the maximum effective concrete flange width b may not exceed (1) one-fourth the beam span L nor (2) the sum of one-half of the

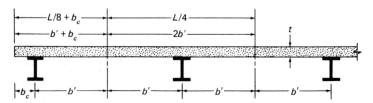

Fig. 9.3. Effective width of concrete flange in composite construction, as specified in AISCS I1.

distances to the centerlines of the adjacent beams on both sides. For an edge beam, the maximum effective width b may not exceed (1) one-eighth of the beam span L plus the distance from the beam centerline to the edge of the slab nor (2) one-half of the distance to the centerline of the adjacent beam plus the distance from the beam centerline to the edge of the slab.

Example 9.1. Determine the effective width b for the interior and edge beams shown when the span $L = 42$ ft, 0 in., slab thickness $t = 5\text{-}\tfrac{1}{2}$ in., and beam spacing is 8 ft, 0 in. on center for a W 18 × 35.

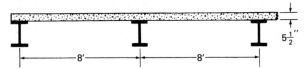

Solution. For a W 18 × 35, $b_f = 6$ in.

For interior beams:

1) $b \leq \dfrac{L}{4} = \dfrac{42.0 \text{ ft} \times 12 \text{ in.}/\text{ft}}{4} = 126$ in.

2) $b \leq (\tfrac{1}{2} \text{ centerline distance}) \times 2 = \dfrac{8 \text{ ft} \times 12 \text{ in.}/\text{ft}}{2} \times 2$

$= 96$ in.

Minimum $b = 96$ in. governs.

For edge beam:

1) $b \leq \dfrac{L}{8} + b_c = \dfrac{42.0 \text{ ft} \times 12 \text{ in.}/\text{ft}}{8} + \dfrac{6 \text{ in.}}{2} = 66$ in.

2) $b \leq (\frac{1}{2}$ centerline distance$) + b_c = \dfrac{8 \text{ ft} \times 12 \text{ in.}/\text{ft}}{2} + \dfrac{6 \text{ in.}}{2}$

$= 51$ in.

Minimum $b = 51$ in. governs.

Example 9.2. Rework Example 9.1 using a W 12 × 65 beam.

Solution. For a W 12 × 65, $b_f = 12$ in.

For interior beams, $b = 96$ in. (see Ex. 9.1)

For edge beam:

1) $b \leq \dfrac{42 \times 12}{8} + \dfrac{12}{2} = 69$ in.

2) $b \leq \dfrac{8 \times 12}{2} + \dfrac{12}{2} = 54$ in.

Minimum $b = 54$ in. governs.

9.3 STRESS CALCULATIONS

Stresses in composite sections are usually calculated by using the transformed area method, in which one of the two materials is transformed into an equivalent area of the other. Usually the available effective area of concrete is transformed into an equivalent area of steel. Assuming the strain of both materials to be the same at equal distances from the neutral axis, the unit stress in either material is then equal to its strain times its modulus of elasticity E. For steel, the modulus of elasticity $E_s = 29,000$ ksi, which is independent of the steel's yield stress F_y. For the concrete, the modulus of elasticity E_c is

$$E_c = w_c^{1.5} \, 33\sqrt{f'_c}$$

where w_c = unit weight of the concrete (pcf) and f'_c = 28-day compressive strength of the concrete (psi).[2] For normal-weight concrete (i.e., $w_c = 145$

---

[2] "Building Code Requirements for Reinforced Concrete (ACI 318-89) and Commentary-ACI 318R-89," American Concrete Institute (Detroit, 1989).

pcf), the above equation can be rewritten as

$$E_c = 57{,}000\sqrt{f'_c}$$

The unit stress in the steel is therefore E_s/E_c times the unit stress in the concrete. Defining the ratio of E_s/E_c as the modular ratio n (where n is usually taken as the closest integer value to the calculated value of E_s/E_c), the force resisted by 1 in.^{2} of concrete is equivalent to the force resisted by only $1/n$ in.^{2} of steel. Therefore, the effective area of concrete A_c (where $A_c = b \times t$) is replaced by the transformed area of concrete A_{ctr} where $A_{ctr} = A_c/n$ (see Fig. 9.4). Since the capacity of a composite steel-concrete beam designed for full composite action does not depend on the unit weight of the concrete (AISCS Commentary I2), stresses are typically calculated based on the transformed sec-

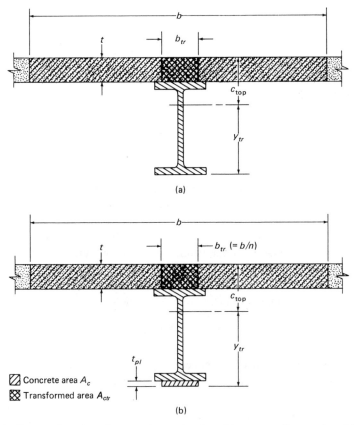

Fig. 9.4. Cross sections through composite construction: (a) noncover-plate section; (b) cover-plated section.

tion properties using normal-weight concrete (AISCS I2.2). However, for deflection calculations, the appropriate modular ratio n should be used when determining the transformed section properties (see Sect. 9.5). After the distance to the neutral axis y_{tr} is determined for the transformed section, the moment of inertia I_{tr} can then be calculated. After the transformed section properties are obtained, the stresses in the steel and concrete can be computed; the magnitudes of these stresses depend on whether or not temporary shoring is used during the construction of the floor system.

In the discussion which follows, all of the loads applied to the system before the concrete has cured (i.e., prior to composite action) will be defined as dead loads, and all of the loads applied after the concrete has cured (i.e., after the concrete has reached 75% of its required 28-day compressive strength) will be defined as live loads.

In the case when temporary shoring is used, the weights of the steel beam, the forms, and the wet concrete are all carried by the shores. After the concrete has cured, the shores are removed and the composite section must then carry the total dead plus live loads. The maximum bending stress for the steel is expressed by

$$f_{bs} = \frac{M y_{tr}}{I_{tr}} \qquad (9.1)$$

where M is the total moment due to both dead and live loads (i.e., $M = M_D + M_L$) and y_{tr} is the distance from the neutral axis to the extreme steel fibers. The maximum bending stress in the concrete is given by

$$f_{bc} = \frac{M c_{\text{top}}}{n I_{tr}} \qquad (9.2)$$

where c_{top} is the distance from the neutral axis to the extreme concrete fibers and n is the modular ratio. Note that eqn. (9.1) could be rewritten in terms of the transformed section modulus S_{tr}, which is defined by the following:

$$S_{tr} = \frac{I_{tr}}{y_{tr}} \qquad (9.3)$$

When temporary shoring is not used (i.e., unshored construction), the steel beam by itself must carry its own weight, plus the weights of the formwork and the wet concrete. The bending stress for the steel is expressed by

$$f_{bs} = \frac{M_D c}{I_s} \qquad (9.4)$$

where M_D is the moment due to dead loads only, c is the distance from the neutral axis of the steel alone to the extreme steel fibers, and I_s is the total moment of inertia of the steel, including cover plates if used. After the concrete has cured and the forms have been removed, the composite section will resist all of the loads placed on the system after the curing of the concrete. Thus, the additional stress in the steel is

$$f_{bs} = \frac{M_L y_{tr}}{I_{tr}} \tag{9.5}$$

where M_L is the moment due to the live loads (i.e., the loads placed on the system after the concrete has cured). The stress in the concrete is expressed by

$$f_{bc} = \frac{M_L c_{top}}{n I_{tr}} \tag{9.6}$$

In general, the composite section should be proportioned to carry both the dead and live loads without exceeding an allowable stress of $0.66F_y$ in the steel even when the steel section is not shored (AISCS I2.2). However, if the steel beam is unshored, the stress in the steel due to the dead loads (eqn. (9.4)) plus the stress due to the live loads (eqn. (9.5)) should not exceed $0.90F_y$ (AISCS I2.2). This provision ensures that no permanent deformation under service loads will occur.

The stress in the concrete, which is given by eqn. (9.2) or (9.6) as appropriate, should not exceed $0.45f'_c$, where f'_c is the compressive strength of the concrete (AISCS I2.2).

Example 9.3. For Example 9.1, determine the transformed section of the concrete when $E_{steel} = 29 \times 10^6$ psi and $E_{concrete} = 2.9 \times 10^6$ psi.

Solution.

$$n = \frac{E_{steel}}{E_{conc}} = \frac{29 \times 10^6 \text{ psi}}{2.9 \times 10^6 \text{ psi}} = 10$$

a) For Example 9.1 interior beams, $t = 5\text{-}\frac{1}{2}$ in., $b = 96$ in.

Effective area of concrete

$$A_c = 5.50 \text{ in.} \times 96 \text{ in.} = 528 \text{ in.}^2$$

COMPOSITE DESIGN 371

$$A_{ctr} = \frac{A_c}{n} = \frac{528 \text{ in.}^2}{10} = 52.8 \text{ in.}^2$$

b) For Example 9.1 edge beam, $t = 5\text{-}\frac{1}{2}$ in., $b = 51$ in.

Effective area of concrete

$$A_c = 5.50 \text{ in.} \times 51 \text{ in.} = 280.5 \text{ in.}^2$$

$$A_{ctr} = \frac{A_c}{n} = \frac{280.5 \text{ in.}^2}{10} = 28.1 \text{ in.}^2$$

Example 9.4. For the case shown, determine the transformed section properties and the stresses at points 1 and 2 given $M = 160$ ft-k. Use a W 18 × 35 with $b = 70$ in., $t = 4$ in., $n = 10$, and shored construction.

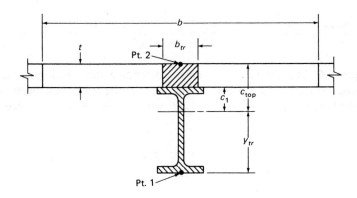

Solution. For W 18 × 35,

$$A = 10.3 \text{ in.}^2, \quad d = 17.70 \text{ in.}, \quad I = 510 \text{ in.}^4$$

Determine transformed properties:

$$b_{tr} = \frac{b}{n} = \frac{70 \text{ in.}}{10} = 7 \text{ in.}$$

$$A_{ctr} = 4 \text{ in.} \times 7 \text{ in.} = 28 \text{ in.}^2$$

$$y_{tr} = \frac{10.3 \text{ in.}^2 \times (17.70 \text{ in.}/2) + 28 \text{ in.}^2 \times (17.70 \text{ in.} + 4 \text{ in.}/2)}{10.3 \text{ in.}^2 + 28 \text{ in.}^2}$$

$$= 16.78 \text{ in.}$$

372 STEEL DESIGN FOR ENGINEERS AND ARCHITECTS

$$I_{tr} = I_0 + \Sigma(Ad^2)$$

$$= 510 + 10.3 \times \left(\frac{17.70}{2} - 16.78\right)^2 + \left(\frac{7 \times (4)^3}{12}\right)$$

$$+ 28.0 \times (17.70 + 2.0 - 16.78)^2$$

$$= 1434 \text{ in.}^4$$

Stresses at extreme fibers:

Point 1

$$f_{bs} = \frac{My_{tr}}{I_{tr}} = \frac{160 \text{ ft-k} \times 12 \text{ in./ft} \times 16.78 \text{ in.}}{1434 \text{ in.}^4} = 22.47 \text{ ksi}$$

Point 2

$$f_{bc} = \frac{Mc_{top}}{nI_{tr}} = \frac{160 \text{ ft-k} \times 12 \text{ in./ft} \times (21.70 \text{ in.} - 16.78 \text{ in.})}{10 \times 1434 \text{ in.}^4} = 0.66 \text{ ksi}$$

Example 9.5. For the situation shown, determine the transformed section properties and the stresses in the extreme fibers of the steel and concrete given $M = 560$ ft-k. Use $b = 88$ in., $t = 5$ in., $n = 10$, and shored construction.

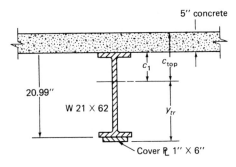

Solution. For a W 21 × 62,

$$A = 18.3 \text{ in.}^2$$

$$d = 20.99 \text{ in.}$$

$$I = 1330 \text{ in.}^4$$

$$b_{tr} = \frac{88}{10} = 8.8 \text{ in.}$$

$$A_{ctr} = 8.8 \text{ in.} \times 5 \text{ in.} = 44 \text{ in.}^2$$

COMPOSITE DESIGN 373

Determine centroid of transformed section:

$$y_{tr} = \frac{18.3 \text{ in.}^2 \times [1 \text{ in.} + (20.99 \text{ in.}/2)] + 6.0 \text{ in.}^2 \times (1 \text{ in.}/2) + 44.0 \text{ in.}^2 \times [21.99 \text{ in.} + (5 \text{ in.}/2)]}{18.3 \text{ in.}^2 + 6.0 \text{ in.}^2 + 44.0 \text{ in.}^2}$$

$$= 18.90 \text{ in.}$$

$$I_{tr} = I_0 + \Sigma A d^2 = 5831.9 \text{ in.}^4$$

Stresses at extreme fibers:

$$f_{bs} = \frac{M y_{tr}}{I_{tr}} = \frac{560 \text{ ft-k} \times 12 \text{ in./ft} \times 18.90 \text{ in.}}{5831.9 \text{ in.}^4} = 21.78 \text{ ksi}$$

$$f_{bc} = \frac{M c_{top}}{n I_{tr}} = \frac{560 \text{ ft-k} \times 12 \text{ in./ft} \times (26.99 \text{ in.} - 18.90 \text{ in.})}{10 \times 5831.9 \text{ in.}^4} = 0.93 \text{ ksi}$$

Example 9.6. Determine the moment-resisting capacity of the composite section shown. Assume A36 steel, $f'_c = 3000$ psi, $n = 9$, $I_{tr} = 8636$ in.^{4}, $y_{tr} = 17.99$ in. Shored construction.

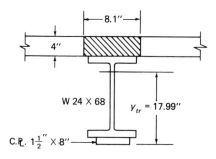

Solution.

$$F_y = 36 \text{ ksi}, \quad F_b = 24 \text{ ksi} \quad \text{(AISCS I2.2)}$$

Allowable concrete stress $= f_c = 0.45 f'_c$

$$= 0.45 \times 3000 \text{ psi} = 1.35 \text{ ksi} \quad \text{(AISCS I2.2)}$$

$$M_{\max} = \frac{F_b I_{tr}}{y_{tr}} = \frac{24 \text{ ksi} \times 8636 \text{ in.}^4}{17.99 \text{ in.}} = 11520 \text{ in.-k}$$

$$= 960.0 \text{ ft-k}$$

374 STEEL DESIGN FOR ENGINEERS AND ARCHITECTS

The maximum concrete stress can be transposed to determine the maximum moment capacity considering the concrete.

$$f_{bc} = \frac{Mc_{top}}{nI_{tr}}$$

$$M_{max} = \frac{nf_c I_{tr}}{c_{top}} = \frac{9 \times 1.35 \text{ ksi} \times 8636 \text{ in.}^4}{1.5 \text{ in.} + 23.73 \text{ in.} + 4 \text{ in.} - 17.99 \text{ in.}} = 9340 \text{ in.-k}$$

$$= 778.0 \text{ ft-k}$$

Since the moment capacity of the concrete is less than that of the steel, the concrete will fail first. Therefore,

$$M_{max} = 778 \text{ ft-k}$$

Note that for unshored construction, the moment capacity of the concrete would be based on the live load moment only.

Example 9.7. Assuming that no shoring will be used, determine the dead load moment capacity of the composite section of Example 9.6.

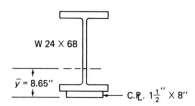

Solution. Since no shoring will be used, the steel alone will carry the dead load.

For a W 24 × 68,

$$A = 20.1 \text{ in.}^2$$

$$d = 23.73 \text{ in.}$$

$$I = 1830 \text{ in.}^4$$

$$\bar{y}_{steel} = \frac{(1.5 \text{ in.} \times 8 \text{ in.})(1.5 \text{ in.}/2) + 20.1 \text{ in.}^2 (23.73 \text{ in.}/2 + 1.5 \text{ in.})}{(1.5 \text{ in.} \times 8 \text{ in.}) + 20.1 \text{ in.}^2}$$

$$= 8.65 \text{ in. from bottom}$$

$$I_{\text{steel}} = I_0 + \Sigma(Ad^2) = \frac{8 \text{ in. } (1.5 \text{ in.})^3}{12} + 12 \text{ in.}^2 (8.65 \text{ in.} - 0.75 \text{ in.})^2$$

$$+ 1830 \text{ in.}^4 + 20.1 \text{ in.}^2 \left(\frac{23.73 \text{ in.}}{2} + 1.5 \text{ in.} - 8.65 \text{ in.}\right)^2$$

$$= 3028 \text{ in.}^4$$

Note that the quantity $8 \times (1.5)^3/12$, which is the moment of inertia of the plate with respect to its weak axis, is usually neglected.

Maximum moment capacity of steel section

$$M_{\text{top}} = \frac{F_b I_s}{c_{\text{top}}} = \frac{24 \text{ ksi} \times 3028 \text{ in.}^4}{23.73 \text{ in.} + 1.5 \text{ in.} - 8.65 \text{ in.}} = 4383 \text{ in.-k} = 365 \text{ ft-k}$$

$$M_{\text{bot}} = \frac{F_b I_s}{\bar{y}_{\text{steel}}} = \frac{24 \text{ ksi} \times 3028 \text{ in.}^4}{8.65 \text{ in.}} = 8401 \text{ in.-k} = 700 \text{ ft-k}$$

Moment capacity of the top governs

$$M_{\text{max}} = 365 \text{ ft-k}$$

Example 9.8. Check the adequacy of the composite section shown below. Span = 48 ft, 0 in. Beam spacing = 10 ft, 0 in. Slab is 6 in. normal-weight concrete with $f'_c = 4000$ psi ($n = 8$). Assume 25 psf for mechanical, ceiling, and lightweight partitions. Live load = 200 psf. Unshored construction.

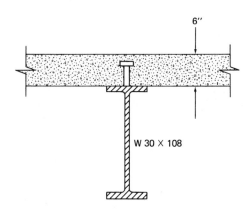

Solution.

$$b = \begin{cases} \dfrac{L}{4} = \dfrac{48 \times 12}{4} = 144 \text{ in.} \\ \text{center-to-center spacing} = 10 \times 12 = 120 \text{ in. (governs)} \end{cases}$$

$$b_{tr} = \frac{120}{8} = 15 \text{ in.}$$

$$A_{ctr} = 15 \times 6 = 90 \text{ in.}^2$$

$$y_{tr} = \frac{\left(31.7 \times \dfrac{29.83}{2}\right) + [90 \times (29.83 + 3)]}{31.7 + 90} = 28.16 \text{ in.}$$

$$I_{tr} = 4470 + 31.7 \left(\frac{29.83}{2} - 28.16\right)^2 + \frac{15 \times 6^3}{12} + 90(32.83 - 28.16)^2$$

$$= 4470 + 5561 + 270 + 1963 = 12{,}264 \text{ in.}^4$$

$$S_{tr} = \frac{12{,}264}{28.16} = 435.5 \text{ in.}^3$$

$$S_{\text{top}} = \frac{12{,}264}{29.83 + 6 - 28.16} = 1599 \text{ in.}^3$$

Maximum applied moments:

$$w_D = 150 \, \frac{\text{lb}}{\text{ft}^3} \times \left(\frac{6 \text{ in.}}{12 \text{ in.}/\text{ft}}\right) \times 10 \text{ ft} + 108 \text{ lb/ft} = 858 \text{ lb/ft}$$

$$M_D = 0.858 \times \frac{48^2}{8} = 247 \text{ ft-k}$$

$$w_L = (25 \text{ psf} + 200 \text{ psf}) \times 10 \text{ ft} = 2250 \text{ lb/ft}$$

$$M_L = 2.25 \times \frac{48^2}{8} = 648 \text{ ft-k}$$

Maximum applied stresses:

In steel alone:

$$f_{bs} = \frac{247 \times 12}{299} = 9.9 \text{ ksi} < 24 \text{ ksi} \quad \text{ok} \qquad (9.4)$$

Live load only:

$$f_{bs} = \frac{648 \times 12}{435.5} = 17.9 \text{ ksi} \qquad (9.5)$$

$$f_{bc} = \frac{648 \times 12}{8 \times 1599} = 0.61 \text{ ksi} \qquad (9.6)$$

$$0.45 f'_c = 1.8 \text{ ksi} > 0.61 \text{ ksi} \quad \text{ok}$$

Dead load + live load:

$$f_{bs} = \frac{(648 + 247) \times 12}{435.5} = 24.7 \text{ ksi}$$

Since 24.7 ksi > 0.66 × 36 = 24 ksi, beam is slightly overstressed. Say ok.

Also, since the beam is unshored,

f_{bs} (dead load only on steel alone)

$+ f_{bs}$ (live load only on transformed section) $\leq 0.90 F_y$

9.9 ksi + 17.9 ksi = 27.8 ksi < 0.9 × 36 ksi = 32.4 ksi   ok

9.4 SHEAR CONNECTORS

For a steel beam and a concrete slab to act compositely, the two materials must be thoroughly connected to each other so that longitudinal shear may be transferred between them. When the steel beam is fully encased in the slab, mechanical connectors are not necessary, since longitudinal shear can be fully transferred by the bond between the steel and concrete. Otherwise, mechanical shear connectors are needed.

Round studs or channels, welded to the top flange of beams, are the type of shear connectors most commonly used (see Fig. 9.5). The studs are round bars with one end upset to prevent vertical separation and the other welded to the beam. Because welded studs are often damaged when beams are shipped, field welding of the studs by stud welding guns is preferable. The allowable horizontal shear loads for channels and round studs of various weights or diameters are tabulated in Table I4.1 of AISCS (reproduced as Table 9.1 here).

According to the AISC, the number of shear connectors required for full composite action is determined by dividing the total shear force V_h to be resisted, between the points of maximum positive moment and zero moment, by the shear capacity of one connector. For a simply supported beam, this number is

378 STEEL DESIGN FOR ENGINEERS AND ARCHITECTS

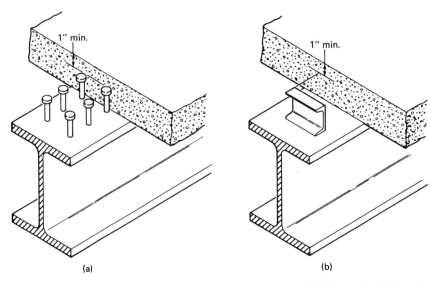

Fig. 9.5. Shear connectors for composite construction: (a) round-headed shear studs welded to the beam; (b) short length of a channel section welded to the beam.

doubled to obtain the total number of connectors required for the entire span. The shear force V_h is the smaller of the two values as determined by the following formulas ((I4-1) and (I4-2), respectively):

$$V_h = \frac{0.85 f'_c A_c}{2} \tag{9.7}$$

$$V_h = \frac{A_s F_y}{2} \tag{9.8}$$

The AISC permits uniform spacing of the connectors between the points of maximum positive moment and the point of zero moment. For concentrated loads, the number of shear connectors required between any concentrated load and the nearest point of zero moment must be determined by (I4-5).

Note that when the longitudinal reinforcing bars in the slab are included in the calculation of the composite section properties (AISCS I4), the footnote on p. 5-58 is applicable. However, in practice, including the reinforcing bars is seldom done. Also, minimum and maximum shear connector spacings are given in AISCS I4.

When the unit weight of the concrete falls between 90 and 120 lb/ft^3 (i.e., lightweight concrete), the allowable shear load for one connector is obtained

Table 9.1. Allowable Horizontal Shear Load for One Connector (q), kips^a (AISC Table I4.1, reprinted with permission).

Connector^b	Specified Compressive Strength of Concrete (f'_c), ksi		
	3.0	3.5	>4.0
$\frac{1}{2}$ in. diam. × 2 in. hooked or headed stud	5.1	5.5	5.9
$\frac{5}{8}$ in. diam. × $2\frac{1}{2}$ in. hooked or headed stud	8.0	8.6	9.2
$\frac{3}{4}$ in. diam. × 3 in. hooked or headed stud	11.5	12.5	13.3
$\frac{7}{8}$ in. diam. × $3\frac{1}{2}$ in. hooked or headed stud	15.6	16.8	18.0
Channel C3 × 4.1	4.3 w^c	4.7 w^c	5.0 w^c
Channel C4 × 5.4	4.6 w^c	5.0 w^c	5.3 w^c
Channel C5 × 6.7	4.9 w^c	5.3 w^c	5.6 w^c

^{a}Applicable only to concrete made with ASTM C33 aggregates.
^{b}The allowable horizontal loads tabulated may also be used for studs longer than shown.
$^c w$ = length of channel, inches.

Table 9.2. Coefficients for Use with Concrete Made with C330 Aggregates (AISC Table I4.2, reprinted with permission).

Specified Compressive Strength of Concrete (f'_c)	Air Dry Unit Weight of Concrete, pcf						
	90	95	100	105	110	115	120
≤4.0 ksi	0.73	0.76	0.78	0.81	0.83	0.86	0.88
≥5.0 ksi	0.82	0.85	0.87	0.91	0.93	0.96	0.99

by multiplying the value given in Table 9.1 by the appropriate coefficient given in AISCS Table I4.2, p. 5-59 (see Table 9.2).

Example 9.9. Determine the number of $\frac{7}{8}$-in.-diameter studs required when $f'_c = 3000$ psi and $F_y = 36$ ksi are used for the interior beams in

a) Example 9.1.
b) Example 9.2.

Solution. Shear force V_h to be resisted by connectors

$$V_h \text{ for concrete} = \frac{0.85 f'_c A_c}{2}$$

380 STEEL DESIGN FOR ENGINEERS AND ARCHITECTS

$$V_h \text{ for steel} = \frac{A_s F_y}{2}$$

Capacity of one $\frac{7}{8}$ in. stud = 15.6 k (AISCS Table I4.1)

a) From Example 9.1, $b = 96$ in.

$$A_c = b \times t = 96 \text{ in.} \times 5.50 \text{ in.} = 528 \text{ in.}^2$$

$$A_s = 10.3 \text{ in.}^2$$

$$V_{h \, conc} = \frac{0.85 \times 3.0 \text{ ksi} \times 528 \text{ in.}^2}{2} = 673.2 \text{ k}$$

$$V_{h \, stl} = \frac{10.3 \text{ in.}^2 \times 36 \text{ ksi}}{2} = 185.4 \text{ k}$$

V_h for steel governs.

Total number of shear connectors required

$$\frac{2 \times 185.4 \text{ k}}{15.6 \text{ k}} = 23.76$$

Use 24 $\frac{7}{8}$-in. studs

b) From Example 9.2, $b = 96$ in.

$$A_c = b \times t = 96 \text{ in.} \times 5.50 \text{ in.} = 528 \text{ in.}^2$$

$$A_s = 19.1 \text{ in.}^2$$

$$V_{h \, conc} = \frac{0.85 \times 3.0 \text{ ksi} \times 528 \text{ in.}^2}{2} = 673.2 \text{ k}$$

$$V_{h \, stl} = \frac{19.1 \text{ in.}^2 \times 36 \text{ ksi}}{2} = 343.8 \text{ k}$$

V_h for steel governs

Total number of shear connectors required

$$\frac{2 \times 343.8 \text{ k}}{15.6 \text{ k}} = 44.08$$

Use 45 $\frac{7}{8}$-in. studs

Example 9.10. Determine the number of 4-in. channels needed for shear transfer when $f'_c = 3.0$ ksi and $F_y = 36$ ksi for the composite section shown. Assume the length of each channel to be 3 in.

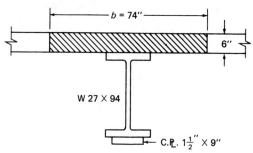

Solution.

$$A_c = tb = 6 \text{ in.} \times 74 \text{ in.} = 444 \text{ in.}^2$$

$$A_s = A_{WF} + A_{pl} = 27.7 \text{ in.}^2 + \left(1\frac{1}{2} \text{ in.} \times 9 \text{ in.}\right) = 41.2 \text{ in.}^2$$

$$V_{h\,\text{conc}} = \frac{0.85 f'_c A_c}{2} = \frac{0.85 \times 3.0 \text{ ksi} \times 444 \text{ in.}^2}{2} = 566.1 \text{ k} \quad \text{(governs)}$$

$$V_{h\,\text{stl}} = \frac{A_s F_y}{2} = \frac{41.2 \text{ in.}^2 \times 36 \text{ ksi}}{2} = 741.6 \text{ k}$$

Total number of shear connectors required

$$\frac{2 \times 566.1 \text{ k}}{13.8 \text{ k}} = 82 \quad \text{(capacity of one shear connector} = 4.6 \times 3 = 13.8 \text{ k)}$$

Use 82 4-in. channels 3 in. long.

Example 9.11. Verify the adequacy of the composite beam shown and determine for full composite action the number of $\frac{3}{4}$-in.-diameter studs required between A and B and between B and C. Beams are spaced 8 ft, 0 in. on center; $f'_c = 4000$ psi (normal weight concrete). Shored construction.

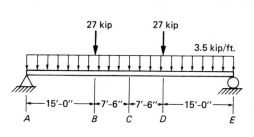
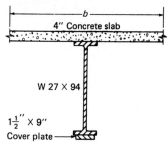

Solution.

$$R = \frac{wL}{2} + P = 105.75 \text{ k}$$

$$M_C = \frac{105.75 \text{ k} \times 45 \text{ ft}}{2} - \frac{27 \text{ k} \times 15 \text{ ft}}{2} - \frac{3.5 \text{ k/ft} \times (45 \text{ ft})^2}{8}$$

$$= 1292 \text{ ft-k}$$

$$M_B = 105.75 \text{ k} \times 15 \text{ ft} - \frac{3.5 \text{ k/ft} \times (15)^2}{2}$$

$$= 1193 \text{ ft-k}$$

Determine transformed section properties:

$$b \leq \frac{L}{4} = \frac{45 \text{ ft}}{4} = 11.25 \text{ ft} = 135 \text{ in.}$$

$$b \leq 8.0 \text{ ft} \times 12 \text{ in./ft} = 96 \text{ in.} \quad \text{(governs)}$$

$$n = \frac{29{,}000{,}000}{57{,}000 \sqrt{4000}} = 8.04 \quad \text{Use } n = 8$$

$$b_{tr} = \frac{b}{n} = \frac{96 \text{ in.}}{8} = 12.0 \text{ in.}$$

$$y_{tr} = \frac{(1.5 \text{ in.} \times 9 \text{ in.}) \times 0.75 \text{ in.} + 27.7 \text{ in.}^2 \times 14.96 \text{ in.} + [(4 \text{ in.} \times 12.0 \text{ in.}) \times 30.42 \text{ in.}]}{(1.5 \text{ in.} \times 9 \text{ in.}) + 27.7 \text{ in.}^2 + (4 \text{ in.} \times 12.0 \text{ in.})}$$

$$= 21.1 \text{ in.}$$

$$I_{tr} = \Sigma(I_0 + Ad^2) = (1.5 \times 9)(21.1 - 0.75)^2 + 3270$$

$$+ 27.7 \left(\frac{26.92}{2} + 1.5 - 21.1\right)^2 + \frac{12.0 \times (4.0)^3}{12}$$

$$+ (4 \times 12.0)(30.42 - 21.1)^2$$

$$= 14{,}138 \text{ in.}^4$$

$$S_{tr} = \frac{I_{tr}}{y_{tr}} = \frac{14{,}138 \text{ in.}^4}{21.1} = 670 \text{ in.}^3$$

$$M_{all} = 670 \text{ in.}^3 \times \frac{24 \text{ ksi}}{12 \text{ in.}/\text{ft}} = 1340 \text{ ft-k} > 1292 \text{ ft-k} \quad \text{ok}$$

$$S_{top} = \frac{I_{tr}}{y_{top}} = 1249 \text{ in.}^3$$

$$f_c = 0.45 f'_c = 1.8 \text{ ksi}$$

$$M_{all} = n \times S_{top} \times f_c = 1499 \text{ ft-k} > 1292 \text{ ft-k} \quad \text{ok}$$

$$y_s = \frac{(1.5 \text{ in.} \times 9 \text{ in.}) \times 0.75 \text{ in.} + 27.7 \text{ in.}^2 \times \left(\dfrac{26.92 \text{ in.}}{2} + 1.5 \text{ in.}\right)}{(1.5 \text{ in.} \times 9 \text{ in.}) + 27.7 \text{ in.}^2}$$

$$= 10.3 \text{ in.}$$

$$I_s = \Sigma(I_0 + Ad^2) = (1.5 \times 9)(10.3 - 0.75)^2 + 3270$$

$$+ 27.7 \left(\frac{26.92}{2} + 1.5 - 10.3\right)^2$$

$$= 5103 \text{ in.}^4$$

$$S_s = \frac{I_s}{y_s} = \frac{5103 \text{ in.}^4}{10.3 \text{ in.}} = 496 \text{ in.}^3$$

$$\beta = \frac{S_{tr}}{S_s} = \frac{670 \text{ in.}^3}{496 \text{ in.}^3} = 1.35$$

$$V_{h\,conc} = \frac{0.85 f'_c A_c}{2} = \frac{0.85 \times 4.0 \text{ ksi} \times (96 \text{ in.} \times 4 \text{ in.})}{2} = 652.8 \text{ k}$$

$$V_{h\,steel} = \frac{A_s F_y}{2} = \frac{41.2 \text{ in.}^2 \times 36 \text{ ksi}}{2} = 741.6 \text{ k}$$

Governing V_h is due to concrete, $V_h = 652.8$ k.

Shear capacity of $\frac{3}{4}$-in. stud $= 13.3$ k

Number of studs required between M_{max} and $M = 0$

$$N_1 = \frac{V_h}{q} = \frac{652.8 \text{ k}}{13.3 \text{ k}} = 49.1$$

Use 50 studs.

384 STEEL DESIGN FOR ENGINEERS AND ARCHITECTS

Minimum number of studs N_2 required between A and B

$$N_2 = \frac{N_1(M\beta/M_{\max} - 1)}{\beta - 1} = \frac{49.1\left(\dfrac{1193 \times 1.35}{1292} - 1\right)}{1.35 - 1} \quad \text{(AISCS I4)}$$

$$= 34.6 \quad \text{say 35 studs}$$

Number of studs between A and $B = 35$

Number of studs between B and $C = 50 - 35 = 15$

When the required number of shear connectors for full composite action is not used, the AISC calls for an effective section modulus to be used as determined by eqn. (I2-1):

$$S_{\text{eff}} = S_s + \sqrt{\frac{V'_h}{V_h}}\,(S_{tr} - S_s) \tag{9.9}$$

where S_s is the section modulus of the steel alone (referred to the bottom flange) and V'_h is the shear capacity provided by the connectors used and is obtained by multiplying the number of connectors used and the shear capacity of one connector (AISCS I.4). Conversely, when the required S_{tr} is less than the available S_{tr}, transposition of eqn. (9.9) into

$$V'_h = V_h \left(\frac{S_{tr\,\text{req}} - S_s}{S_{tr} - S_s}\right)^2 \tag{9.10}$$

will give the horizontal force needed to be carried by shear connectors to accommodate the required section modulus. The effective moment of inertia, I_{eff} is determined by (I4-4):

$$I_{\text{eff}} = I_s + \sqrt{\frac{V'_h}{V_h}}\,(I_{tr} - I_s) \tag{9.11}$$

where I_s is the moment of inertia of the steel beam and I_{tr} is the moment of inertia of the transformed composite section. In all cases, V'_h shall not be less than $\frac{1}{4}\,V_h$ (AISCS I4).

Example 9.12. Referring to Example 9.11, determine S_{eff} if only 28 3-in. channels 3 in. long are used on half of the beam.

Solution.

$$S_{\text{eff}} = S_s + \sqrt{\frac{V_h'}{V_h}}(S_{tr} - S_s)$$

$$S_{tr} = 670 \text{ in.}^3$$

$$S_s = 496 \text{ in.}^3$$

$$V_h = 652.8 \text{ k}$$

Shear capacity of each channel = $5.0\, w$

w = length of channel = 3 in.

5.0 k/in. × 3 in. = 15.0 k

$$\frac{652.8 \text{ k}}{15.0 \text{ k}} = 43.5 \quad \text{Say 44 are needed for full capacity.}$$

Since only 28 channels are used,

$$V_h' = 28 \times 15.0 \text{ k} = 420.0 \text{ k}$$

$$S_{\text{eff}} = 496 + \sqrt{\frac{420.0}{652.8}}(670 - 496) = 635.6 \text{ in.}^3$$

Note that $M_{\text{all}} = 635.6 \times \dfrac{24}{12} = 1271$ ft-k < 1292 ft-k   N.G.

9.5 FORMED STEEL DECK

When formed steel deck (FSD) is used as permanent formwork for the concrete, additional requirements governing the composite construction are given in AISCS I5. A typical composite floor system utilizing FSD is shown in Fig. 9.6. In general, AISCS I5.1 requires that the nominal rib height h_r should not exceed 3 in., and that the average width of the concrete rib w_r should not be less than

386 STEEL DESIGN FOR ENGINEERS AND ARCHITECTS

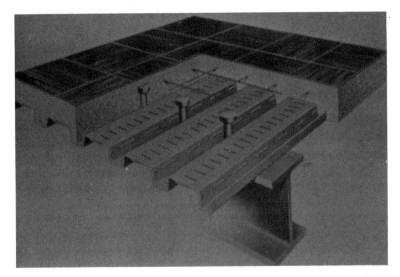

Fig. 9.6. Cutaway view of a typical composite floor system using formed steel deck (FSD).

2 in. Also, the slab thickness above the deck should not be less than 2 in. The welded stud shear connectors shall be $\frac{3}{4}$ in. or less in diameter and shall extend at least $1\frac{1}{2}$ in. over the top of the steel deck after installation. A schematic presentation of these requirements is given in Fig. 9.7.

When the deck ribs are oriented perpendicular to the steel beams or girders, the main requirements of the AISCS are (see AISCS I5.2 and Fig. 9.8a):

1. Concrete below the top of steel decking shall be neglected in computations of section properties and in calculating A_c for eqn. (9.7).
2. The spacing of stud shear connectors shall not exceed 36 in. along the length of the beam.
3. The allowable horizontal shear capacity of stud connectors shall be reduced from the values given in AISCS Tables I4.1 and I4.2 by the reduction factor given in (I5-1).

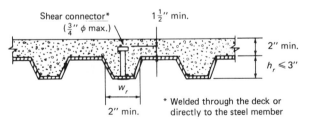

Fig. 9.7. General requirements of formed steel deck as given in AISCS I5.1.

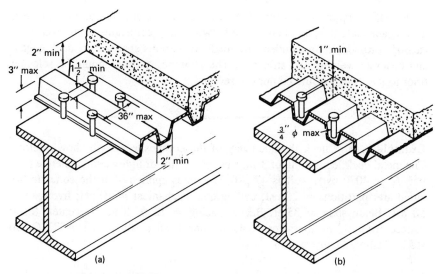

Fig. 9.8. Formed steel deck (FSD) used as permanent formwork for composite construction: (a) deck ribs oriented perpendicular to the steel beam; (b) deck ribs oriented parallel to the steel beam. Dimension and clearance restrictions shown in either (a) or (b) apply to both.

4. The steel deck shall be anchored to the beam or girder by welding or anchoring devices, whose spacing shall not exceed 16 in.

For composite sections with the metal deck ribs oriented parallel to the steel beam or girder (Fig. 9.8b), reference should be made to AISCS I5.3. The major difference to note between perpendicular and parallel orientation of deck ribs is that for the latter, the concrete below the top of the steel deck may be included when determining the transformed section properties and must be included when calculating A_c in eqn. (9.7). In general, A_c may be calculated from the following equation:

$$A_c = b\left[t_c + \left(\frac{w_r h_r}{s}\right)\right] \tag{9.12}$$

where b = effective width of the concrete
t_c = thickness of the concrete above the steel decking
s = rib spacing (center-to-center of rib).

When the ratio of the average rib width to the rib height w_r/h_r is less than 1.5, the allowable shear load per stud must be multiplied by the reduction factor given in (I5-2).

388 STEEL DESIGN FOR ENGINEERS AND ARCHITECTS

The AISCS requirements for FSD are based on the assumption that the steel deck is adequate to span the distance between the steel beams. Manufacturers' catalogs provide recommended safe loads for various spans, steel deck gages, and concrete weights and strengths. The appropriate deck should be selected prior to the design of the composite section.

Example 9.13. Check the adequacy of the composite system shown below. The metal deck, consisting of 3-in. ribs with $2\frac{1}{2}$ in. of lightweight concrete (110 pcf, f'_c = 4000 psi), weighs 47 psf. The loads applied after the concrete has cured are: partitions = 20 psf; ceiling and mechanical = 10 psf; live load = 50 psf. Beam span = 35 ft. Beam spacing = 12 ft, 0 in. on center. Shear connectors are $\frac{3}{4}$-in.-diameter × $4\frac{7}{8}$-in.-long headed studs. Use A572 Gr 50 steel. Unshored construction.

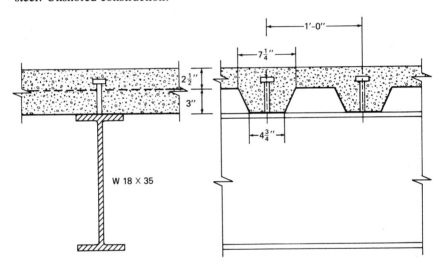

Solution.

$$w_D = 0.047 \text{ k/ft}^2 \times 12 \text{ ft} + 0.035 = 0.6 \text{ k/ft}$$

$$w_L = (0.020 + 0.010 + 0.050) \times 12 = 0.96 \text{ k/ft}$$

$$M_D = \frac{0.6 \times (35)^2}{8} = 91.9 \text{ ft-k}$$

$$M_L = \frac{0.96 \times (35)^2}{8} = 147 \text{ ft-k}$$

Effective flange width:

$$b = \begin{cases} 35 \times \dfrac{12}{4} = 105 \text{ in.} \quad \text{(governs)} \\ 12 \times 12 = 144 \text{ in.} \end{cases}$$

$$n = \frac{29{,}000{,}000}{57{,}000 \sqrt{4000}} = 8.04 \quad \text{Use } n = 8 \quad \text{(base transformed section properties on normal-weight concrete for stress computations per AISCS I2.2)}$$

$$b_{tr} = \frac{105}{8} = 13.13 \text{ in.}$$

$$A_{ctr} = 13.13 \text{ in.} \times 2.5 \text{ in.} = 32.83 \text{ in.}^2$$

$$y_{tr} = \frac{10.3 \text{ in.}^2 \times \left(\dfrac{17.70 \text{ in.}}{2}\right) + 32.83 \text{ in.}^2 \times \left(17.7 \text{ in.} + 3.0 \text{ in.} + \dfrac{2.5 \text{ in.}}{2}\right)}{10.3 \text{ in.}^2 + 32.83 \text{ in.}^2}$$

$$= 18.82 \text{ in.}$$

$$I_{tr} = 510 \text{ in.}^4 + 10.3 \text{ in.}^2 \times \left(18.82 \text{ in.} - \frac{17.70 \text{ in.}}{2}\right)^2$$

$$+ \frac{1}{12} \times (13.13 \text{ in.}) \times (2.5 \text{ in.})^3$$

$$+ 32.83 \text{ in.}^2 \times \left(17.7 \text{ in.} + 3 \text{ in.} + \frac{2.5 \text{ in.}}{2} - 18.82 \text{ in.}\right)^2$$

$$= 510 + 1023.8 + 17.1 + 321.6 = 1872.5 \text{ in.}^4$$

$$S_{tr} = \frac{1872.5 \text{ in.}^4}{18.82 \text{ in.}} = 99.5 \text{ in.}^3$$

$$S_{top} = \frac{1872.5 \text{ in.}^4}{23.2 \text{ in.} - 18.82 \text{ in.}} = 427.5 \text{ in.}^3$$

Check maximum stresses for full composite action:

In steel alone:

$$f_{bs} = \frac{91.9 \times 12}{57.6} = 19.2 \text{ ksi} < 33 \text{ ksi} \quad \text{ok}$$

390 STEEL DESIGN FOR ENGINEERS AND ARCHITECTS

Live load only:

$$f_{bs} = \frac{147 \times 12}{99.5} = 17.7 \text{ ksi}$$

$$f_{bc} = \frac{147 \times 12}{8 \times 427.5} = 0.52 \text{ ksi} < 0.45 f'_c = 1.8 \text{ ksi} \quad \text{ok}$$

Dead load + Live load:

$$f_{bs} = \frac{(91.9 + 147) \times 12}{99.5} = 28.8 \text{ ksi} < 33 \text{ ksi} \quad \text{ok}$$

For unshored construction:

$$19.2 \text{ ksi} + 17.7 \text{ ksi} = 36.9 \text{ ksi} < 0.9 \times 50 = 45 \text{ ksi} \quad \text{ok}$$

Therefore, the W 18 × 35 is adequate when using full composite action. However, a more economical solution is to use the same wide flange with partial composite action since, in this case, the actual stresses are much lower than the allowable ones. We will first compute the required number of studs for full composite action, and then recompute the stresses when less than this required number is provided. Typically, one stud per rib is usually provided along the span length.

Determine the number of shear connectors required for full composite action.

The allowable horizontal shear load for the $\frac{3}{4}$-in.-diameter stud is 13.3 kips, as given in Table 9.1; however, this value must be modified, since we have lightweight concrete and a metal deck spanning perpendicular to the beam.

- From Table 9.2, the required coefficient = 0.83 (110 pcf concrete with a compressive strength of 4000 psi).
- From (I5-1), the following reduction factor is obtained:

$$\left(\frac{0.85}{\sqrt{N_r}}\right)\left(\frac{w_r}{h_r}\right)\left(\frac{H_s}{h_r} - 1.0\right) \leq 1.0$$

where N_r = number of studs per rib and H_s = length of the stud after welding.

In our case:

$$N_r = 1$$

$w_r = (4.75 \text{ in.} + 7.25 \text{ in.})/2 = 6.0 \text{ in.}$

$h_r = 3.0 \text{ in.}$

$H_s = 4.875 \text{ in.} - 0.1875 \text{ in.} = 4.6875 \text{ in.}$ (Note: see manufacturers' catalogs for typical losses in total stud length after welding.)

Thus, the reduction factor is

$$\left(\frac{0.85}{\sqrt{1}}\right)\left(\frac{6.0}{3.0}\right)\left(\frac{4.6875}{3.0} - 1\right) = 0.96$$

Reduced stud capacity = 0.83 × 0.96 × 13.3 k = 10.6 k

$$V_{h\,\text{conc}} = \frac{0.85 \times 4 \text{ ksi} \times (105 \text{ in.} \times 2.5 \text{ in.})}{2} = 446.3 \text{ k}$$

$$V_{h\,\text{stl}} = \frac{50 \text{ ksi} \times 10.3 \text{ in.}^2}{2} = 257.5 \text{ k} \quad \text{(governs)}$$

Total no. of studs required for full composite action $= \dfrac{2 \times 257.5}{10.6}$

$= 48.6 \quad \text{(say 49)}$

Try using 35 studs total (one stud per rib).

Since we have only 35 studs on this beam, we have less than full composite action, and the effective section modulus must be determined from eqn. (9.9):

$$S_{\text{eff}} = S_s + \sqrt{\frac{V_h'}{V_h}}(S_{tr} - S_s)$$

$S_s = 57.6 \text{ in.}^3$

$V_h' = 35 \text{ studs} \times 10.6 \text{ k} = 371 \text{ k} > \dfrac{2 \times 257.5}{4} = 128.8 \text{ k} \quad \text{ok}$

$$S_{\text{eff}} = 57.6 + \sqrt{\frac{371}{2 \times 257.5}}(99.5 - 57.6) = 93.2 \text{ in.}^3$$

(Note: using 35 studs corresponds to $[371/(2 \times 257.5)] \times 100\% = 72\%$ of full composite action.)

Recheck the stresses:

$$f_{bs} = \frac{(91.9 + 147) \times 12}{93.2} = 30.8 \text{ ksi} < 33 \text{ ksi} \quad \text{ok}$$

$$19.2 + \frac{147 \times 12}{93.2} = 38.1 \text{ ksi} < 0.9 \times 50 = 45 \text{ ksi} \quad \text{ok}$$

Note: The stress in the concrete does not need to be rechecked, since it is maximum when there is full composite action. If the magnitude of this stress were required, it could be obtained by using the procedure given in the AISCM on p. 2-243. Therefore, W 18 × 35 with $\frac{3}{4}$-in.-diameter studs spaced 12 in. o.c. (one stud/rib) is adequate.

9.6 COVER PLATES

When cover plates are used in composite construction, it is not generally required to extend the cover plate to the ends of the beam where the moments are small. To determine the theoretical cutoff point for cover plates, the easiest method is to draw the moment diagram of the span and locate the parts of the beam that will require the extra moment capacity provided by the cover plate.

It is generally good practice to use cover plates that are at least 1 in. narrower than the bottom flange of a shop-fabricated beam to leave sufficient space along the edge for welding. When reinforcing existing beams, the bottom plates are usually wider than the bottom flange so as to avoid the difficulty of overhead welding. The AISCS requires that partial-length cover plates be extended beyond the theoretical cut-off point and welded or high-strength-bolted (SC) to the beam flange to transfer the flexural stresses that the cover plate would have received had it been extended the full length of the beam (see AISCS B10 and Sect. 3.1 of this book). The force to be developed in the cover plate extension is equal to MQ/I, where M is the moment at the cutoff point, Q is the statical moment of the cover plate about the neutral axis of the composite cover-plated section, and I is the moment of inertia of the composite cover-plated section (AISCS Commentary B10). For further details, see Chapters 3–5.

Example 9.14. Determine the theoretical cutoff point and the connection of the cover plates for loading as shown for the beam in Example 9.11. Use f'_c of 4000 psi, and neglect the weight of the beam.

S]MPOSITE DESIGN 393

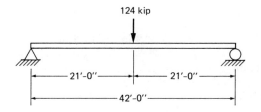

Solution.

$$M_{max} = \frac{PL}{4} = \frac{124 \text{ k} \times 42 \text{ ft}}{4} = 1302 \text{ ft-k}$$

S_{tr} with cover plates = 670 in.³ (see Example 9.11)

S_{tr} without cover plates = 324 in.³

Moment capacity of beam without cover plate

$$M = S_{tr} \times F_b = 324 \text{ in.}^3 \times 24.0 \text{ ksi} = 7776 \text{ in.-k}$$
$$= 648 \text{ ft-k}$$

Moment capacity of beam with cover plate

$$M_{max} = 670 \text{ ft-k} \times 24 \text{ ksi} = 16,080 \text{ in.-k} = 1340 \text{ ft-k}$$

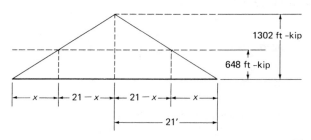

To determine x,

$$\frac{x}{648} = \frac{21}{1302}, \quad x = 10.45 \text{ ft}$$

$$21.0 - x = 21.0 - 10.45 = 10.55 \text{ ft}$$

A $1\frac{1}{2}$-in. × 9-in. cover plate, 2 × 10.55 ft = 21.1 ft long is theoretically required 10.45 ft from either end. See Chapter 3 for development length necessary according to AISC requirements (AISCS B10).

394 STEEL DESIGN FOR ENGINEERS AND ARCHITECTS

Extension length

$$L = 1.5 \times 9 \text{ in.} = 13.5 \text{ in.}$$

Shear at cutoff point

$$V = 62.0 \text{ k}$$

$$Q = A_{pl} \times y = 13.5 \text{ in.}^2 \times (21.1 \text{ in.} - 0.75 \text{ in.}) = 274.7 \text{ in.}^3$$

Shear flow

$$q = \frac{VQ}{I} = \frac{62.0 \text{ k} \times 274.7 \text{ in.}^3}{14{,}138 \text{ in.}^4} = 1.20 \text{ k/in.}$$

Using minimum E70 intermittent fillet welds at 12 in. on center,

$$\text{Weld size} = \frac{5}{16} \text{ in.} \quad \text{(AISCS J2.2b)}$$

$$F_w = 5 \times 0.928 \text{ k/in./sixteenth} = 4.64 \text{ k/in.}$$

$$l_w = \frac{12 \text{ in./ft} \times 1.20 \text{ k/in.}}{4.64 \text{ k/in.} \times 2 \text{ sides}} = 1.55 \text{ in./ft length}$$

Say $l_w = 1\frac{3}{4}$ in. per 1 ft length E70 $\frac{5}{16}$ in. intermittent fillet weld.

Termination welds (AISCS B10)

$$H(\text{force needed}) = \frac{648 \text{ ft-k} \times 12 \text{ in./ft} \times 274.7 \text{ in.}^3}{14{,}138 \text{ in.}^4} = 151 \text{ k}$$

Total weld length

$$l_w = 1.5 \times 2 \times 9 \text{ in.} + 9 \text{ in. (end)} = 36 \text{ in.}$$

Force developed

$$36 \text{ in.} \times 4.64 \text{ k/in.} = 167.0 \text{ k} > 151.0 \text{ k} \quad \text{ok}$$

Example 9.15. Rework Example 9.14 with distributed loading of 5.8 kips per foot.

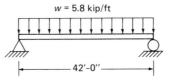

COMPOSITE DESIGN

Solution.

$$M_{max} = \frac{wL^2}{8} = \frac{5.8 \text{ k/ft} \times (42 \text{ ft})^2}{8}$$

$$= 1279 \text{ ft-k}$$

From Example 9.14, moment capacity of the section without cover plate is 648 ft-k.

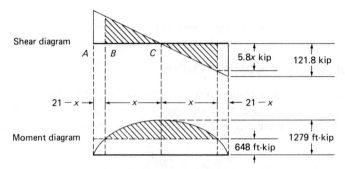

From statics, the area under the shear diagram is equal to the change in moment.

Change in moment between B and C = $1279 - 648 = 631$ ft-k

Area under shear diagram between B and C = 631 ft-k

Area under shear diagram between B and C = $(x)(5.8\,x) \times \frac{1}{2}$

$$= 2.90\,x^2$$

$$2.90\,x^2 = 631 \text{ ft-k}$$

$$x^2 = 217.6; \quad x = 14.75 \text{ ft}$$

$$21 - x = 6.25 \text{ ft}$$

The theoretical length of cover plate is $(2 \times 15 \text{ ft}) = 30$ ft long, 6 ft from either end.

Shear at termination point

$$V_B = 5.8 \text{ k/ft} \times 14.75 \text{ ft} = 85.6 \text{ k}$$

$$Q = 274.7 \text{ in.}^3 \quad (\text{Example 9.14})$$

$$q = \frac{VQ}{I} = \frac{85.6 \text{ k} \times 274.7 \text{ in.}^3}{14{,}138 \text{ in.}^4} = 1.66 \text{ k/in.}$$

396 STEEL DESIGN FOR ENGINEERS AND ARCHITECTS

Use $\frac{5}{16}$-in. E70 intermittent fillet weld at 12 in. on center.

$$l_w = \frac{12 \text{ in.}/\text{ft} \times 1.66 \text{ k/in.}}{4.64 \text{ k/in.} \times 2} = 2.15 \text{ in.}$$

Use $l_w = 2\frac{1}{2}$ in. E70 $\frac{5}{16}$ in. intermittent fillet weld.

Termination welds—same as in Example 9.14.

9.7 DEFLECTION COMPUTATIONS

To simplify deflection computations of simply supported beams under uniformly distributed loading, the following formulas may be used. They are derived from the expression $\Delta = 5\, wl^4/384\, EI$. For unshored construction

$$\Delta_D = \frac{M_D L^2}{161\, I_s} \tag{9.13}$$

$$\Delta_L = \frac{M_L L^2}{161\, I_{tr}} \tag{9.14}$$

and for shored construction

$$\Delta_D = \frac{M_D L^2}{161\, I_{tr}} \tag{9.15}$$

$$\Delta_L = \frac{M_L L^2}{161\, I_{tr}} \tag{9.16}$$

where M is the moment in ft-kips, L is the span in feet, and I is the moment of inertia in inch^4, and where subscript D is for dead load, subscript L is for live load, subscript s is for steel alone, and subscript tr is for transformed section.

Example 9.16. Show that the deflection formula commonly used for composite construction for uniformly distributed loading is derived from simple beam deflection.

COMPOSITE DESIGN

Solution. The equation for deflection of simply supported beams is

$$\Delta = \frac{5\,wl^4}{384\,EI}, \quad \text{in inches}$$

When the moment due to uniform loading is introduced,

$$\Delta = \frac{5\,l^2}{48\,EI} \times \frac{wl^2}{8} = \frac{5\,Ml^2}{48\,EI}$$

Inserting the value for E_{steel} yields

$$\Delta = \frac{5\,Ml^2}{48 \times 29{,}000 \times I} = \frac{Ml^2}{278{,}400\,I}$$

Converting units,

$$\Delta \text{ in.} = \frac{M \text{ ft-k } L \text{ ft}^2}{278{,}400 \text{ ksi } I \text{ in.}^4} (12 \text{ in./ft})^3$$

$$\Delta = \frac{ML^2}{161\,I}, \quad \text{where } \Delta \text{ is in inches, } M \text{ is in foot-kips, } I \text{ is in inches}^4, \text{ and } L \text{ is in feet.}$$

Example 9.17. For the composite section in Example 9.11, determine dead load deflection, live load deflection, and total deflection when the dead load moment is 748 ft-k and the live load moment is 554 ft-k. Assume unshored construction. Moments are due to uniform loads. Span = 42 ft.

Solution. From Example 9.11,

$$I_{tr} = 14{,}138 \text{ in.}^4$$

$$I_s = 5105 \text{ in.}^4$$

$$\Delta_D = \frac{M_D L^2}{161\,I_s} = \frac{554 \text{ ft-k} \times (42 \text{ ft})^2}{161 \times 5105 \text{ in.}^4} = 1.19 \text{ in.}$$

$$\Delta_L = \frac{M_L L^2}{161\,I_{tr}} = \frac{748 \text{ ft-k} \times (42 \text{ ft})^2}{161 \times 14{,}138 \text{ in.}^4} = 0.58 \text{ in.}$$

$$\Delta_T = \Delta_D + \Delta_L = 1.19 \text{ in.} + 0.58 \text{ in.} = 1.77 \text{ in.}$$

398 STEEL DESIGN FOR ENGINEERS AND ARCHITECTS

Example 9.18. Determine the dead load deflection, live load deflection, and total deflection of Example 9.17 if shored construction is used.

Solution.

$$\Delta_D = \frac{M_D L^2}{161\, I_{tr}} = \frac{554 \text{ ft-k} \times (42 \text{ ft})^2}{161 \times 14{,}138 \text{ in.}^4} = 0.43 \text{ in.}$$

$$\Delta_L = \frac{M_L L^2}{161\, I_{tr}} = \frac{748 \text{ ft-k} \times (42 \text{ ft})^2}{161 \times 14{,}138 \text{ in.}^4} = 0.58 \text{ in.}$$

$$\Delta_T = \Delta_D + \Delta_L = 0.43 \text{ in.} + 0.58 \text{ in.} = 1.01 \text{ in.}$$

Example 9.19. Determine the dead load deflection, live load deflection, and total deflection for the composite section in Example 9.13.

Solution. From Example 9.13:

$$M_D = 91.9 \text{ ft-k}, \quad M_L = 147 \text{ ft-k}, \quad I_s = 510 \text{ in.}^4$$

$$\Delta_D = \frac{M_D L^2}{161\, I_s} = \frac{91.9 \text{ ft-k} \times (35 \text{ ft})^2}{161 \times 510 \text{ in.}^4} = 1.37 \text{ in.}$$

In order to compute the live load deflection, the effective moment of inertia I_{eff} must be computed based on the transformed section properties of the lightweight concrete.

$$I_{\text{eff}} = I_s + \sqrt{\frac{V_h'}{V_h}}\,(I_{tr} - I_s)$$

$$b = 105 \text{ in.} \quad \text{(see Example 9.13)} \tag{9.11}$$

In general, $E_c = w_c^{1.5}\, 33\, \sqrt{f_c'}$ where w_c is the unit weight of concrete, lb/ft^3

$$E_c = (110)^{1.5} \times 33 \times \sqrt{4000} = 2{,}407{,}870 \text{ psi}$$

$$n = \frac{29{,}000{,}000}{2{,}407{,}870} = 12.04 \quad \text{Use } n = 12$$

$$b_{tr} = \frac{105}{12} = 8.75 \text{ in.}$$

$A_{ctr} = 8.75 \times 2.5 = 21.88$ in.^{2}

$y_{tr} = 17.76$ in.

$I_{tr} = 1723.1$ in.^{4}

$I_{eff} = 510 + \sqrt{0.72}\,(1723.1 - 510) = 1539.3$ in.^{4}

$$\Delta_L = \frac{M_L L^2}{161\, I_{eff}} = \frac{147 \text{ ft-k} \times (35 \text{ ft})^2}{161 \times 1539.3 \text{ in.}^4} = 0.73 \text{ in.} < \frac{35 \times 12}{360} = 1.17 \text{ in.} \quad \text{ok}$$

$\Delta_T = \Delta_D + \Delta_L = 1.37 + 0.73 = 2.10$ in.

(Note: In general, some or all of the dead load deflection can be taken out by cambering the steel beam.)

9.8 AISCM COMPOSITE DESIGN TABLES

Since the compression flange of the steel beam has continuous lateral support in standard composite construction, the allowable bending stress in the steel F_b equals $0.66 F_y$. To aid in the selection of an economical beam size, Composite Beam Selection Tables have been provided in the AISCM starting on p. 2-259. The discussion on the use of these tables starts on p. 2-243, and examples begin on p. 2-249. Note that to simplify the presentation, some approximations were introduced which make the values in the tables slightly conservative. The following examples illustrate the use of these design aids.

Example 9.20. Design a composite section for dead load of 90 lb/ft^2 (including beam weight) and live load of 100 lb/ft^2 when W 24 beams are placed 10 ft on center and no cover plates are to be used. The beam length is 45 ft, concrete thickness 5 in., $f'_c = 3000$ psi and $n = 9$. Use $\frac{7}{8}$-in.-diameter headed studs. Unshored construction.

Solution. Loads on beam:

$$w_D = 0.09 \text{ k/ft}^2 \times 10 \text{ ft} = 0.9 \text{ k/ft}$$

$$w_L = 0.10 \text{ k/ft}^2 \times 10 \text{ ft} = 1.0 \text{ k/ft}$$

$$M_D = \frac{0.9 \text{ k/ft} \times (45 \text{ ft})^2}{8} = 227.8 \text{ ft-k}$$

$$M_L = \frac{1.0 \text{ k/ft} \times (45 \text{ ft})^2}{8} = 253.1 \text{ ft-k}$$

$$M_T = 227.8 + 253.1 = 480.9 \text{ ft-k} \quad \text{say } 481 \text{ ft-k}$$

$$S_{tr \text{ req}} = \frac{M}{F_b} = \frac{481 \text{ ft-k} \times 12 \text{ in./ft}}{24 \text{ ksi}} = 240.5 \text{ in.}^3$$

$$b = \text{lesser of} \begin{cases} \dfrac{L}{4} = \dfrac{45 \times 12}{4} = 135 \text{ in.} \\ \text{center-to-center spacing} = 10 \times 12 = 120 \text{ in.} \end{cases}$$

$$b = 120 \text{ in.}$$

$$b_{tr} = \frac{120}{9} = 13.3 \text{ in.}$$

$$A_{ctr} = 5 \text{ in.} \times 13.3 \text{ in.} = 66.7 \text{ in.}^2$$

From the Composite Beam Table on p. 2-271, try a W 24 × 76 which has a $\overline{S}_{tr} \simeq S_{tr} = 246.3 \text{ in.}^3$ for $Y2 = 5.0/2 = 2.5$ in. (value of $\overline{S}_{tr}$, obtained from double interpolation, is conservative). Note that if the values of S_{tr} fall within the gray shaded area in the tables, the neutral axis of the composite section is within the concrete.

Also from the Composite Beam Table:

$$I_s = 2100 \text{ in.}^4$$

$$S_s = 176 \text{ in.}^3$$

$$A_s = 22.4 \text{ in.}^2$$

$$d = 23.9 \text{ in.}$$

$$\overline{I}_{tr} \simeq I_{tr} = 5590 \text{ in.}^4 \quad \text{(from double interpolation)}$$

$$y_{tr} = \frac{5590 \text{ in.}^4}{246.3 \text{ in.}^3} = 22.7 \text{ in.}$$

$$c_{top} = 23.9 \text{ in.} + 5.0 \text{ in.} - 22.7 \text{ in.} = 6.2 \text{ in.}$$

Check stresses:

Dead load:

$$f_{bs} = \frac{227.8 \text{ ft-k} \times 12 \text{ in./ft}}{176 \text{ in.}^3} = 15.5 \text{ ksi} < 24 \text{ ksi} \quad \text{ok}$$

Live load:

$$f_{bs} = \frac{253.1 \text{ ft-k} \times 12 \text{ in./ft}}{246.3 \text{ in.}^3} = 12.3 \text{ ksi}$$

$$f_{bc} = \frac{253.1 \text{ ft-k} \times 12 \text{ in./ft} \times 6.2 \text{ in.}}{9 \times 5590 \text{ in.}^4}$$

$$= 0.37 \text{ ksi} < 0.45 \times 3.0 \text{ ksi} = 1.35 \text{ ksi} \quad \text{ok}$$

Dead load + live load:

$$f_{bs} = \frac{481 \text{ ft-k} \times 12 \text{ in./ft}}{246.3 \text{ in.}^3} = 23.4 \text{ ksi} < 24 \text{ ksi} \quad \text{ok}$$

For unshored construction:

15.5 ksi + 12.3 ksi = 27.8 ksi < 0.9 × 36 ksi = 32.4 ksi   ok

$$V_h = \text{lesser of} \begin{cases} 403 \text{ k (from Composite Beam Table for steel)} \quad \text{(governs)} \\ 0.85 \times 3 \text{ ksi} \times 120 \text{ in.} \times 5 \text{ in.}/2 = 765 \text{ k} \end{cases}$$

Total number of studs required:

$$N = \frac{2 \times 403 \text{ k}}{15.6 \text{ k/stud}} = 51.6 \quad \text{say 52 studs for full composite action}$$

Determine the total number of studs required for partial composite action:
Solve (I2-1) for V'_h:

$$V'_h = V_h \left[\frac{S_{tr\,req} - S_s}{S_{tr} - S_s} \right]^2$$

$$= (2 \times 403) \times \left[\frac{240.5 - 176.0}{246.3 - 176.0} \right]^2 = 678.5 \text{ k} > \frac{2 \times 403}{4} = 201.5 \text{ k} \quad \text{ok}$$

$$N = \frac{678.5}{15.6} = 43.5 \quad \text{say 45 studs for the entire length.}$$

Check deflection:

$$I_{\text{eff}} = 2100 + \sqrt{\frac{678.5}{806.0}} (5590 - 2100) = 5302 \text{ in.}^4$$

$$\Delta_L = \frac{253.1 \text{ ft-k} \times (45 \text{ ft})^2}{161 \times 5302.0 \text{ in.}^4} = 0.60 \text{ in.} < \frac{45 \times 12}{360} = 1.5 \text{ in.} \quad \text{ok}$$

$$\Delta_D = \frac{227.8 \text{ ft-k} \times (45 \text{ ft})^2}{161 \times 2100.0 \text{ in.}^4} = 1.36 \text{ in.}$$

$$\Delta_T = 0.60 + 1.36 = 1.96 \text{ in.}$$

Example 9.21. Determine the required length of the cover plate for the composite beam shown below. Use E70 welds and $\frac{3}{4}$-in.-diameter headed studs. The beam span is 50 ft, and the beams are spaced 10 ft on center. Total load on a beam is 4.1 k/ft (DL + LL). Use $f'_c = 3000$ psi concrete ($n = 9$). Shored construction.

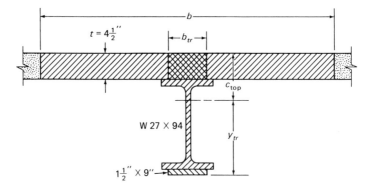

Solution.

$$M_{\text{max}} = \frac{wL^2}{8} = \frac{4.1 \text{ k/ft} \times (50 \text{ ft})^2}{8} = 1281 \text{ ft-k}$$

$$F_b = 0.66 \times 36 \text{ ksi} = 24.0 \text{ ksi}$$

$$S_{tr\,req} = \frac{M}{F_b} = \frac{1281 \text{ ft-k} \times 12 \text{ in./ft}}{24.0 \text{ ksi}} = 640.5 \text{ in.}^3$$

$$b = \begin{cases} \dfrac{50 \text{ ft} \times 12 \text{ in./ft}}{4} = 150 \text{ in.} \\ 10 \text{ ft.} \times 12 \text{ in./ft} = 120 \text{ in.} \quad \text{(governs)} \end{cases}$$

$$b_{tr} = \frac{120 \text{ in.}}{9} = 13.3 \text{ in.}$$

$$A_{ctr} = 13.3 \text{ in.} \times 4.5 \text{ in.} = 60 \text{ in.}^2$$

$$y_{tr} = \frac{(9 \text{ in.} \times 1.5 \text{ in.}) \times 0.75 \text{ in.} + 27.7 \text{ in.}^2 \times \left(\dfrac{26.92}{2} + 1.50\right) + (60 \text{ in.}^2) \times \left(1.50 + 26.92 + \dfrac{4.5}{2}\right)}{(9 \text{ in.} \times 1.5 \text{ in.}) + 27.7 \text{ in.}^2 + 60 \text{ in.}^2}$$

$$= 22.4 \text{ in.}$$

$$I_{tr} = (9 \times 1.5)(22.4 - 0.75)^2 + 3270 + 27.7\left(22.4 - \frac{26.92}{2} - 1.5\right)^2$$

$$+ \frac{13.3 \times (4.5)^3}{12} + (60)\left(\frac{4.5}{2} + 26.92 + 1.5 - 22.4\right)^2$$

$$= 15{,}336 \text{ in.}^4$$

$$S_{tr} = \frac{15{,}336 \text{ in.}^4}{22.4 \text{ in.}} = 684.6 \text{ in.}^3 > 640.5 \text{ in.}^3 \text{ required} \quad \text{ok}$$

Find theoretical length of cover plate:

Transformed section modulus of W 27 × 94 without cover plate:

Enter Composite Beam Selection Table with $A_{ctr} = 60$ in.^{2} and $Y2 = 4.50$ in./2 = 2.25 in. to obtain (from interpolation):

$$S_{tr} = 329 \text{ in.}^3$$

$$I_{tr} = 7956 \text{ in.}^4$$

Moment capacity of section without cover plate

$$M = S_{tr}F_b = \frac{329 \text{ in.}^3 \times 24 \text{ ksi}}{12 \text{ in./ft}} = 658 \text{ ft-k}$$

404 STEEL DESIGN FOR ENGINEERS AND ARCHITECTS

Change in moment between maximum and theoretical cutoff point

$$1281 \text{ ft-k} - 658 \text{ ft-k} = 623 \text{ ft-k}$$

Area under the shear diagram between the same two points

$$623 \text{ ft-k} = (X)(4.1\ X) \times \tfrac{1}{2} = 2.05\ X^2$$

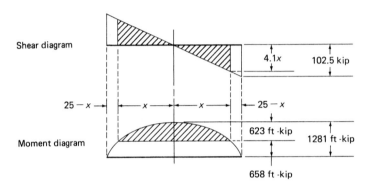

$$2.05\ X^2 = 623 \text{ ft-k}$$
$$X = 17.43 \text{ ft} \quad \text{say} \quad 17.50 \text{ ft}$$

Theoretical length of cover plate

$$2 \times 17.50 \text{ ft} = 35 \text{ ft-0 in.}$$

Theoretical end point of cover plate

$$\frac{50 \text{ ft} - 35 \text{ ft}}{2} = 7.5 \text{ ft from beam ends}$$

Minimum size of weld $= \tfrac{5}{16}$ in.

Capacity of $\tfrac{5}{16}$-in. weld for both sides of cover plate

$$F_w = 2 \times (5 \times 0.928) = 9.28 \text{ k/in.}$$

Shear at cutoff points

$$V = \frac{35.0 \text{ ft} \times 4.1 \text{ k/ft}}{2} = 71.8 \text{ k}$$

Shear flow

$$q = \frac{VQ}{I}$$

$$Q = (1.5 \text{ in.} \times 9 \text{ in.}) \times \left(22.4 \text{ in.} - \frac{1.5 \text{ in.}}{2}\right) = 292.3 \text{ in.}^3$$

$$q = \frac{71.8 \text{ k} \times 292.3 \text{ in.}^3}{15,336 \text{ in.}^4} = 1.37 \text{ k/in.}$$

Assuming intermittent fillet welds spaced 12 in. o.c.,

$$\text{Required weld length} = \frac{1.37 \text{ k/in.} \times 12 \text{ in.}}{9.28 \text{ k/in.}} = 1.77 \text{ in.}$$

Use 2-in. weld at 12-in. spacing center-to-center.

Force to be resisted by end weld

$$\frac{MQ}{I} = \frac{658 \text{ ft-k} \times 12 \text{ in./ft} \times 292.3 \text{ k/in.}}{15,336 \text{ in.}^4} = 150.5 \text{ k}$$

Using AISCS B10, extend plate $13\frac{1}{2}$ in. beyond the theoretical cutoff point with continuous fillet weld.

Force available

$$L = (2 \times 13.5 \text{ in.}) + 9 \text{ in.} = 36 \text{ in.}$$

$$H = 36 \text{ in.} \times 4.64 \text{ k/in.} = 167.0 \text{ k} > 150.5 \text{ k} \quad \text{ok}$$

Number of studs required for steel for one-half span

$$V_{h\,\text{steel}} = \frac{A_s F_y}{2} = \frac{(27.7 + 13.5) \times 36}{2} = 741.6 \text{ k}$$

$$A_c = t \times b = 4.5 \text{ in.} \times 120 \text{ in.} = 540 \text{ in.}^2$$

Number of studs required for concrete

$$V_{h\,\text{conc}} = \frac{0.85 f'_c A_c}{2} = \frac{0.85 \times 3 \times 540}{2} = 688.5 \text{ k} \quad \text{(governs)}$$

406 STEEL DESIGN FOR ENGINEERS AND ARCHITECTS

Capacity of $\tfrac{3}{4}$-in. studs = 11.5 k

$$N = \frac{688.5 \text{ k}}{11.5 \text{ k}} = 59.9 \quad \text{say 60 studs for one-half span}$$

Use total of 120 $\tfrac{3}{4}$-in. ϕ studs.

Note: This is the number of studs needed for full moment capacity. As the required moment is less than the available, the number of shear connectors may be reduced by

$$V'_h = V_h \left(\frac{S_{tr\,req} - S_s}{S_{tr} - S_s}\right)^2$$

Calculating S_s (section modulus of steel beam referred to its bottom flange),

$$y_s = \frac{(9 \times 1.5 \times 0.75) + 27.7 \times \left(\frac{26.92}{2} + 1.50\right)}{9 \times 1.5 + 27.7} = 10.30 \text{ in.}$$

$$I_s = 9 \times 1.5 \times (10.30 - 0.75)^2 + 3270 + 27.7 \times \left(10.30 - \frac{26.92}{2} - 1.5\right)^2$$

$$= 5103 \text{ in.}^4$$

$$S_s = \frac{5103 \text{ in.}^4}{10.30 \text{ in.}} = 495.4 \text{ in.}^3$$

$$V'_h = 688.5 \times \left(\frac{640.5 - 495.4}{684.6 - 495.4}\right)^2 = 405 \text{ k}$$

$$N = \frac{405 \text{ k}}{688.5 \text{ k}} \times 59.9 \times 2 = 70.5 \quad \text{say 72 $\tfrac{3}{4}$-in. ϕ studs (total)}$$

Check shear

$$V = 102.5 \text{ k}$$

$$f_v = \frac{102.5 \text{ k}}{26.92 \text{ in.} \times 0.490 \text{ in.}} = 7.77 \text{ ksi} < 14.5 \text{ ksi} \quad \text{ok}$$

As the beam is not to be coped, there is no need to verify safety in tearing of web (block shear).

COMPOSITE DESIGN 407

Example 9.22. Design beam A for the situation shown. Because of clearance limitations, beam A cannot exceed 17 in. in depth. Also, verify beam B, which is a W 27 × 102 with a $1\frac{1}{2}$ × 9 in. cover plate on its bottom flange. Note that beam A will be coped where it frames into beam B, so that both top flanges will be at the same elevation. The slab is $4\frac{1}{2}$ in. normal-weight concrete with $f'_c = 3$ ksi ($n = 9$). Shear connectors are $\frac{3}{4}$-in.-diameter headed studs. Unshored construction.

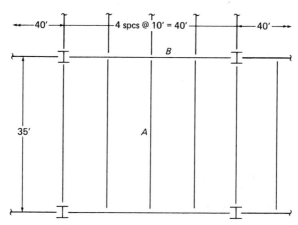

Solution.

Beam A

DL $\qquad \dfrac{4.5 \text{ in.}}{12 \text{ in.}/\text{ft}} \times 0.15 \text{ kcf} \times 10 \text{ ft} = 0.563 \text{ k/ft}$

$\qquad\qquad\qquad\qquad\qquad$ Beam (estimated) $= \underline{0.060 \text{ k/ft}}$

$\qquad\qquad\qquad\qquad\qquad\qquad\qquad\qquad w_D = 0.623 \text{ k/ft}$

LL $\qquad$ Ceiling, flooring, partition $\quad 0.030 \text{ ksf} \times 10 \text{ ft} = 0.30 \text{ k/ft}$

$\qquad\qquad\qquad\qquad\qquad\qquad 0.100 \text{ ksf} \times 10 \text{ ft} = \underline{1.00 \text{ k/ft}}$

$\qquad\qquad\qquad\qquad\qquad\qquad\qquad\qquad w_L = 1.30 \text{ k/ft}$

$$M_D = \frac{w_D L^2}{8} = \frac{0.623 \text{ k/ft} \times (35 \text{ ft})^2}{8} = 95.4 \text{ ft-k}$$

$$M_L = \frac{w_L L^2}{8} = \frac{1.30 \text{ k/ft} \times (35 \text{ ft})^2}{8} = 199.1 \text{ ft-k}$$

$$M_{\text{tot}} = M_D + M_L = 294.5 \text{ ft-k}$$

$$S_{tr\,\text{req}} = \frac{12 \text{ in.}/\text{ft} \times M_{\text{tot}}}{F_b} = \frac{12 \text{ in.}/\text{ft} \times 294.5 \text{ ft-k}}{24 \text{ ksi}} = 147.3 \text{ in.}^3$$

408 STEEL DESIGN FOR ENGINEERS AND ARCHITECTS

Since the Composite Beam Tables do not provide values for heavy W 16 sections, a trial and error procedure must be performed:

- Try W 16 × 57:

$$A = 16.8 \text{ in.}^2, \quad d = 16.43 \text{ in.}, \quad I_s = 758 \text{ in.}^4, \quad S_s = 92.2 \text{ in.}^3$$

$$b = \begin{cases} \dfrac{35 \times 12}{4} = 105 \text{ in.} \quad \text{(governs)} \\ 10 \times 12 = 120 \text{ in.} \end{cases}$$

$$b_{tr} = \frac{105 \text{ in.}}{9} = 11.67 \text{ in.}$$

Transformed section properties are as follows:

$$y_{tr} = 16.14 \text{ in.}$$
$$I_{tr} = 2241 \text{ in.}^4$$
$$S_{tr} = 138.8 \text{ in.}^3 < 147.3 \text{ in.}^3 \quad \text{N.G.}$$

- Try the next heaviest section, which is W 16 × 67:

$$A = 19.7 \text{ in.}^2, \quad d = 16.33 \text{ in.}, \quad I_s = 954 \text{ in.}^4, \quad S_s = 117 \text{ in.}^3$$

$$b_{tr} = 11.67 \text{ in.} \quad \text{(as before)}$$

Transformed section properties:

$$y_{tr} = 15.74 \text{ in.}$$
$$I_{tr} = 2597 \text{ in.}^4$$
$$S_{tr} = 165.0 \text{ in.}^3 > 147.3 \text{ in.}^3 \quad \text{ok}$$

Check stresses:

Dead load:

$$f_{bs} = \frac{95.4 \times 12}{117} = 9.78 \text{ ksi} < 24 \text{ ksi} \quad \text{ok}$$

Live load:

$$f_{bs} = \frac{199.1 \times 12}{165} = 14.48 \text{ ksi}$$

$$f_{bc} = \frac{199.1 \times 12 \times 5.09}{9 \times 2597.0} = 0.52 \text{ ksi}$$

$$0.52 \text{ ksi} < 0.45 \times 3 = 1.35 \text{ ksi} \quad \text{ok}$$

Dead load + live load:

$$\text{Section is ok since } S_{tr} > S_{tr\,req} \quad (\text{see above})$$

For unshored construction:

$$9.78 \text{ ksi} + 14.48 \text{ ksi} = 24.26 \text{ ksi} < 32.4 \text{ ksi} \quad \text{ok}$$

W 16 × 67 is adequate for the case of full composite action.

$$V_h = 0.85 f'_c A_c / 2 = \frac{0.85 \times 3.0 \text{ ksi} \times 105 \text{ in.} \times 4.5 \text{ in.}}{2} = 602.4 \text{ k}$$

$$V_h = A_s F_y / 2 = \frac{19.70 \text{ in.}^2 \times 36 \text{ ksi}}{2} = 354.5 \text{ k} \quad (\text{governs})$$

Use $V_h = 354.5$ k

Considering that $S_{tr\,req} < S_{tr}$, reduce V_h to V'_h:

$$V'_h = V_h \times \left[\frac{S_{tr\,req} - S_s}{S_{tr} - S_s}\right]^2 = 354.5 \times \left(\frac{147.3 - 117.0}{165.0 - 117.0}\right)^2 = 141.3 \text{ k}$$

$$N = \frac{141.3 \text{ k}}{11.5 \text{ k}} \times 2 = 24.6$$

Use 26 connectors total.

Beam end shear

$$V = 1.923 \text{ k/ft} \times \frac{35 \text{ ft}}{2} = 33.7 \text{ k}$$

$$f_v = \frac{33.7 \text{ k}}{16.33 \text{ in.} \times 0.395 \text{ in.}} = 5.22 \text{ ksi} < 14.5 \text{ ksi} \quad \text{ok}$$

410 STEEL DESIGN FOR ENGINEERS AND ARCHITECTS

Framed connection

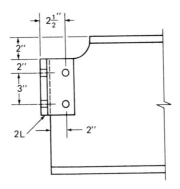

W 16 × 67 beam will be coped 2 in. deep with reaction of 33.7 k. Bolting will be with two $\frac{3}{4}$-in. ϕ A325-N high-strength bolts.

From Table I-D (AISCM, p. 4-5),

$$\text{Shear capacity} = 2 \times 18.6 \text{ k} = 37.2 \text{ k} > 33.7 \text{ k}$$

From Table I-F, for bearing on the $\frac{3}{8}$-in. beam web:

$$l_v = 2 \text{ in.} > 1.5 \times 0.75 = 1.125 \text{ in.}$$

$$\text{Bearing capacity} = \tfrac{3}{8} \text{ in.} \times (2 \times 52.2 \text{ k/in.}) = 39.2 \text{ k} > 33.7 \text{ k} \quad \text{ok}$$

From Table I-F, for bearing on the $\frac{5}{16}$-in. framing angles:

$$\text{Bearing capacity} = \tfrac{5}{16} \text{ in.} \times (2 \times 52.2 \text{ k/in.}) = 32.6 \text{ k} > \frac{33.7 \text{ k}}{2} \quad \text{ok}$$

From Table I-G, for web tearout (block shear):

$$l_v = 2 \text{ in.}, \quad l_h = 2 \text{ in.}$$

$$R_{BS} = (1.60 + 0.33) \times 58 \times \tfrac{3}{8} = 42.0 \text{ k} > 33.7 \text{ k} \quad \text{ok}$$

Use 2 L 4 × 4 × $\tfrac{5}{16}$ × 7 in. long for the framing angles.

Deflection

$$I_{\text{eff}} = I_s + \sqrt{\frac{V'_h}{V_h}}\,(I_{tr} - I_s)$$

$$= 954 + \sqrt{\frac{141.3}{354.5}}\,(2597 - 954) = 1991 \text{ in.}^4$$

$$\Delta_D = \frac{95.4 \times (35)^2}{161 \times 954} = 0.76 \text{ in.}$$

$$\Delta_L = \frac{199.1 \times (35)^2}{161 \times 1991} = 0.76 \text{ in.}$$

$$\frac{l}{360} = \frac{35 \text{ ft} \times 12 \text{ in./ft}}{360} = 1.17 > 0.76 \text{ in.} \quad \text{ok}$$

Beam B

Weight of beam + plate = 102 + 46 = 148 lb/ft   say 150 lb/ft
Allowance for concentrated load

$$\frac{0.15 \text{ k/ft} \times 40 \text{ ft}}{3} = 2 \text{ k}$$

Concentrated DL (0.623 k/ft × 35 ft/2 × 2) + 2.0 k = 23.8 k

Concentrated LL 1.30 k/ft × 35 ft/2 × 2 = 45.5 k

Total load = 69.3 k

$P_D = 23.8$ k $R_D = 35.7$ k

$P_L = 45.5$ k $R_L = 68.3$ k

$M_D = 35.7$ k × 20 ft − 23.8 k × 10 ft = 476 ft-k

$M_L = 68.3$ k × 20 ft − 45.5 k × 10 ft = 911 ft-k

412 STEEL DESIGN FOR ENGINEERS AND ARCHITECTS

Using W 27 × 102 + 1½-in. × 9-in. cover plate, calculate properties with and without concrete.

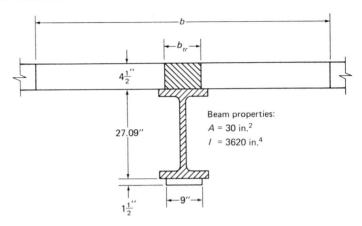

Beam properties:
$A = 30$ in.^{2}
$I = 3620$ in.^{4}

Steel properties

$$y_s = \frac{9 \text{ in.} \times 1.5 \text{ in.} \times 0.75 \text{ in.} + 30 \text{ in.}^2 \times \left(\frac{27.09}{2} + 1.5 \text{ in.}\right)}{9 \text{ in.} \times 1.5 \text{ in.} + 30 \text{ in.}^2}$$

$= 10.61$ in.

$I_s = 13.5 \text{ in.}^2 \times (.75 \text{ in.} - 10.61 \text{ in.})^2$

$+ 30 \text{ in.}^2 \times \left(\frac{27.09 \text{ in.}}{2} + 1.5 \text{ in.} - 10.61 \text{ in.}\right)^2 + 3620 \text{ in.}^4$

$= 5522$ in.^{4}

$$S_{s\,top} = \frac{5522 \text{ in.}^4}{27.09 \text{ in.} + 1.5 \text{ in.} - 10.61 \text{ in.}} = 307 \text{ in.}^3$$

$$S_{s\,bot} = \frac{5522 \text{ in.}^4}{10.61 \text{ in.}} = 520 \text{ in.}^3$$

$$M_{all} = \frac{307 \text{ in.}^3 \times 24 \text{ ksi}}{12 \text{ in.}/\text{ft}} = 614 \text{ ft-k} > 476 \text{ ft-k} \quad \text{ok}$$

Transformed section properties

$$b = \begin{cases} \dfrac{40 \times 12}{4} = 120 \text{ in.} \quad \text{(governs)} \\ 35 \times 12 = 420 \text{ in.} \end{cases}$$

$$b_{tr} = \frac{120 \text{ in.}}{9} = 13.3 \text{ in.}$$

$$y_{tr} = \frac{9 \text{ in.} \times 1.5 \text{ in.} \times 0.75 \text{ in.} + 30 \text{ in.}^2 \times (27.09 \text{ in.}/2 + 1.5 \text{ in.}) + 4.5 \text{ in.} \times 13.3 \text{ in.} \times (27.09 \text{ in.} + 1.5 \text{ in.} + 2.25 \text{ in.})}{9 \text{ in.} \times 1.5 \text{ in.} + 30 \text{ in.}^2 + 13.3 \text{ in.} \times 4.5 \text{ in.}}$$

$$= 22.3 \text{ in.}$$

$$I_{tr} = 13.5 \text{ in.}^2 \times (0.75 \text{ in.} - 22.3 \text{ in.})^2 + 30 \text{ in.}^2$$
$$\times \left(\frac{27.09 \text{ in.}}{2} + 1.5 \text{ in.} - 22.3 \text{ in.}\right)^2$$
$$+ 3620 \text{ in.}^4 + 13.3 \text{ in.} \times \frac{(4.5 \text{ in.})^3}{12}$$
$$+ 4.5 \text{ in.} \times 13.3 \text{ in.} \times (27.09 \text{ in.} + 1.5 \text{ in.}$$
$$+ 2.25 \text{ in.} - 22.3 \text{ in.})^2$$
$$= 15{,}934 \text{ in.}^4$$

$$S_{top} = \frac{15{,}934 \text{ in.}^4}{27.09 \text{ in.} + 1.5 \text{ in.} + 4.5 \text{ in.} - 22.3 \text{ in.}} = 1477 \text{ in.}^3$$

$$S_{tr} = \frac{15{,}934 \text{ in.}^4}{22.3 \text{ in.}} = 715 \text{ in.}^3$$

$$M_{max} = \frac{715 \text{ in.}^3 \times 24 \text{ ksi}}{12 \text{ in./ft}} = 1430 \text{ ft-k} > 1387 \text{ ft-k} \quad \text{ok}$$

$$S_{tr\,req} = 1387 \text{ ft-k} \times \frac{12 \text{ in./ft}}{24 \text{ ksi}} = 694 \text{ in.}^3$$

Check concrete stress:

$$f_{bc} = \frac{911 \text{ ft-k} \times 12 \text{ in./ft}}{9 \times 1477 \text{ in.}^3} = 0.82 \text{ ksi} < 1.35 \text{ ksi} \quad \text{ok}$$

Verify steel stresses for unshored construction:

Dead load:

Top: $$f_{bs} = \frac{476 \text{ ft-k} \times 12 \text{ in./ft}}{307 \text{ in.}^3} = 18.6 \text{ ksi} < 24 \text{ ksi} \quad \text{ok}$$

414 STEEL DESIGN FOR ENGINEERS AND ARCHITECTS

$$\text{Bottom:} \quad f_{bs} = \frac{476 \text{ ft-k} \times 12 \text{ in./ft}}{520 \text{ in.}^3} = 11.0 \text{ ksi}$$

Live load:

$$\text{Top:} \quad f_{bs} = \frac{911 \text{ ft-k} \times 12 \text{ in./ft} \times (28.59 \text{ in.} - 22.3 \text{ in.})}{15{,}934 \text{ in.}^4} = 4.3 \text{ ksi}$$

$$\text{Bottom:} \quad f_{bs} = \frac{911 \text{ ft-k} \times 12 \text{ in./ft}}{715 \text{ in.}^3} = 15.3 \text{ ksi}$$

Totals:

Top: 18.6 ksi + 4.3 ksi = 22.9 ksi < 0.9 × 36 ksi = 32.4 ksi ok

Bottom: 11.0 ksi + 15.3 ksi = 26.3 ksi < 32.4 ksi ok

Shear connectors

$$V_h = (30 \text{ in.}^2 + 13.5 \text{ in.}^2) \times \frac{36 \text{ ksi}}{2} = 783 \text{ k}$$

$$V_h = 0.85 \times 3.0 \text{ ksi} \times \frac{120 \text{ in.} \times 4.5 \text{ in.}}{2} = 689 \text{ k} \quad \text{(governs)}$$

$$V'_h = 689 \text{ k} \times \left(\frac{694 - 520}{715 - 520}\right)^2 = 549 \text{ k}$$

$$N = 2 \times \frac{549 \text{ k}}{11.5 \text{ k}} = 96 \ \tfrac{3}{4}\text{-in. studs}$$

Use 48 shear connectors in each half of beam B

Connectors needed between load at quarter points and support

$$N_2 = \frac{N_1 \left(\dfrac{M\beta}{M_{\max}} - 1\right)}{\beta - 1} \quad \text{(AISCS I4)}$$

$$\beta = \frac{694 \text{ in.}^3}{520 \text{ in.}^3} = 1.335$$

$$M = (35.7 \text{ k} + 68.3 \text{ k}) \times 10 \text{ ft} = 1040 \text{ ft-k}$$

$$M_{max} = 1387 \text{ ft-k}$$

$$N_2 = \frac{48 \times \left(\frac{1040}{1387} \times 1.335 - 1\right)}{1.335 - 1} \approx 0$$

Studs may be uniformly spaced.

Cover plate cutoff

Calculating transformed properties for W 27 × 102 without cover plate:

$$y_{tr} = 24.1 \text{ in.}$$

$$I_{tr} = 8706 \text{ in.}^4$$

$$S_{tr} = \frac{8706 \text{ in.}^4}{24.1 \text{ in.}} = 361 \text{ in.}^3$$

$$S_{top} = \frac{8706 \text{ in.}^4}{27.09 \text{ in.} + 4.5 \text{ in.} - 24.1 \text{ in.}} = 1162 \text{ in.}^3$$

Allowable moment without cover plate

$$M_{all} = 361 \text{ in.}^3 \times \frac{24 \text{ ksi}}{12 \text{ in./ft}} = 722 \text{ ft-k}$$

Distance of cutoff point to end beam

$$104 \text{ k} \times X = 722 \text{ ft-k}, \quad X = 6.94 \text{ ft}$$

Welding

$$Q = 13.5 \text{ in.}^2 \times (22.3 \text{ in.} - 0.75 \text{ in.}) = 290.9 \text{ in.}^3$$

Shear flow between concentrated load at quarter point and cutoff

$$V = 104 \text{ k}$$

$$q = \frac{104 \text{ k} \times 290.9 \text{ in.}^3}{15{,}934 \text{ in.}^4} = 1.90 \text{ k/in.}$$

416 STEEL DESIGN FOR ENGINEERS AND ARCHITECTS

Shear flow between two concentrated loads

$$V = 104 \text{ k} - 69.3 \text{ k} = 34.7 \text{ k}$$

$$q = \frac{34.7 \text{ k} \times 290.9 \text{ in.}^3}{15{,}934 \text{ in.}^4} = 0.63 \text{ k/in.}$$

Use $\tfrac{5}{16}$-in. E70 intermittent fillet welds

$$F_w = 5 \times 0.928 \text{ k/in.} = 4.64 \text{ k/in.}$$

Minimum length = 1.5 in.

Spacing of welds between concentrated load to cutoff

$$s = \frac{4.64 \text{ k/in.} \times 1.5 \text{ in.} \times 2}{1.90 \text{ k/in.}} = 7.33 \text{ in.}$$

Use $\tfrac{5}{16}$-in. weld $1\tfrac{1}{2}$-in. long at 7-in. centers

If 12-in. spacing is desired,

$$l_w = \frac{1.90 \text{ k/in.} \times 12 \text{ in./ft}}{4.64 \text{ k/in.} \times 2} = 2.46 \text{ in.}$$

Use $\tfrac{5}{16}$-in. weld $2\tfrac{1}{2}$ in. long at 12-in. centers

Spacing of welds between two concentrated loads

$$\frac{4.64 \text{ k/in.} \times 1.5 \text{ in.} \times 2}{0.63 \text{ k/in.}} = 22.09 \text{ in.}$$

Use $\tfrac{5}{16}$-in. weld $1\tfrac{1}{2}$ in. long at 12-in. centers

Termination welds

$$l_w = (9 \text{ in.} \times 1.5 \text{ in.} \times 2) + 9 \text{ in.} = 36 \text{ in.}$$

Weld capacity

$$F_w = 36 \times 4.64 \text{ k/in.} = 167 \text{ k}$$

Moment at cutoff point

$M = 722$ ft-k

$$H = \frac{MQ}{I} = \frac{722 \text{ ft-k} \times 12 \text{ in./ft} \times 290.9 \text{ in.}^3}{15{,}934 \text{ in.}^4} = 158 \text{ k} < 167 \text{ k} \quad \text{ok}$$

PROBLEMS TO BE SOLVED

9.1. Determine the effective width b for the interior and edge beams for W 16 × 40 beams with 5-in. concrete slab. Assume that the beams are spaced 8 ft, 6 in. on center and span 35 ft.

9.2. Determine the transformed area for the concrete in Problem 9.1 for a) $f'_c = 3000$ psi, b) $f'_c = 4000$ psi, and c) $f'_c = 5000$ psi.

9.3. For the interior composite beam described in Problem 9.1, determine the section properties and stresses if $M_D = 60$ ft-k and $M_L = 100$ ft-k. Unshored construction and $n = 9$. Check to see if allowable stresses are satisfied.

9.4. For the composite beam shown, determine the section properties and stresses for the steel and the composite section when $M_D = 80$ ft-k and $M_L = 340$ ft-k. Unshored construction, $n = 8$. Also, check to see if the allowable stresses are satisfied.

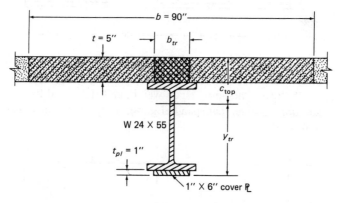

9.5. For the composite beam shown in Problem 9.4, determine the dead load and live load deflections if $w_{DL} = 400$ lb/lin. ft, $w_{LL} = 1700$ lb/lin. ft, and $L = 40$ ft.

9.6. Determine the properties and the allowable moment, considering both maximum steel and concrete stresses for a composite section made from a W 18 × 35 beam, $1\frac{1}{2}$-in. × 5-in. cover plate and 5-in. slab. Assume $b = 86$ in., $f'_c = 3000$ psi, and $n = 9$. Shored construction.

9.7. Rework Problem 9.6, but for a W 16 × 40 beam and 1 × 5-in. cover plate, both of A588 Gr 50 steel.

9.8. Determine the number of $\frac{3}{4}$-in.-diameter studs required for the composite section of Problem 9.4 to develop full composite action. Assume $f'_c = 3000$ psi.

9.9. Rework Problem 9.8, but for C3 × 4.1 channel shear connectors, each 4 in. long. Use a minimum number of connectors to develop the moment resistance.

9.10. Determine the number of $\frac{7}{8}$-in.-diameter stud shear connectors required to develop the full composite capacity of the section described in Problem 9.6.

9.11. Show that a W 24 × 76 beam with a 2 × 7-in. cover plate is satisfactory for the given loads. Assume A36 steel, $\frac{5}{16}$-in. E70 welds, 5-in. slab, $f'_c = 3500$ psi, $n = 9$, and $b_{tr} = 7$ in. Also, calculate the theoretical cutoff point of cover plate, connections of the cover plate to the flange, termination welds, and the required number of $\frac{3}{4}$-in.-diameter stud shear connectors. Unshored construction.

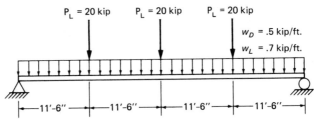

9.12., 9.13., and 9.14. For the following problems, assume A36 steel, E70 welds, $\frac{3}{4}$-in.-diameter stud shear connectors, 5-in. slab, $f'_c = 4000$ psi, total dead load = 75 lb/ft² (including beam weight and slab), and live load = 125 lb/ft². In addition, assume that spandrel girder 3 supports a facia panel weighing 400 lb/in. ft.

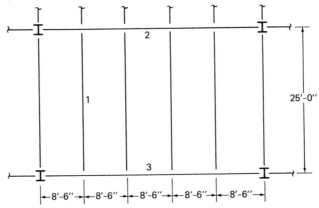

9.12. Design beam 1. Due to clearance limitations, depth cannot exceed 20 in. (including slab). Do no use cover-plated beam.

9.13. Design girder 2. Due to clearance limitations, depth, including slab, cannot exceed 36 in. Use 2-in.-thick cover plate.

9.14. Design girder 3. There are no clearance limitations.

9.15. Design a composite beam to span 25 ft, with a FSD consisting of 3-in. ribs + 3.25-in. lightweight concrete ($f'_c = 4000$ psi). Dimensions of FSD are the same as those given in Example 9.13. Beam spacing = 9 ft, 0 in. Dead load = 46 psf. Live load = 110 psf. Use $\frac{3}{4}$-in.-diameter × $4\frac{7}{8}$-in.-long headed studs, A572 Gr 50 steel, and unshored construction.

10
Plastic Design of Steel Beams

10.1 PLASTIC DESIGN OF STEEL BEAMS

Plastic design is a method by which structural elements are selected considering the system's overall ultimate capacity. For safety, the applied loads are increased by load factors dictated by appropriate codes. This design is based on the yield property of the steel. Because not all steels have properties suitable for plastic design, only those appropriate are specified in AISCS N2.

Consider the stress-strain diagram of a steel appropriate for plastic design, (Fig. 10.1).[1] It can be seen that until the steel reaches point 1, corresponding to a stress F_y, stress and strain are proportional, and the material is linearly elastic in the range 0-1. Beyond point 1, the material begins to exhibit some yield, so that if the stress is released at a point between 1 and 3, the material will experience some permanent deformation. For example, if the stress is released at point A, the member will undergo a permanent deformation ϵ_A. We note that between points 2 and 3 the material seems to flow, i.e., to increase its strain without any appreciable increase in stress. The stress corresponding to point 2 is called the *yield stress* and is indicated by F_y. Point 3 is the beginning of the strain-hardening state, where the material needs additional stress to increase its strain. Finally, the material's ultimate capacity is reached at point 4 with the corresponding stress F_u.

In working stress (elastic) design, the beam is considered to fail when the stress in the steel reaches F_y. In view of the safety factor imposed on the stresses, the working stress is approximately one-third below the yield stress. Plastic design is based on the behavior of steel in the range 2-3. In Fig. 10.1b, the stress is idealized, showing one stress range below yield (linearly elastic) and one range at yield. In practice, as a result of the introduction of load factors,

---

[1] See the Introduction for stress-strain diagrams drawn to scale.

PLASTIC DESIGN OF STEEL BEAMS 421

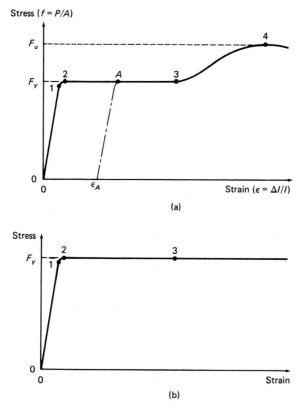

Fig. 10.1. Stress-strain diagram for mild structural steel: (a) normal diagram; (b) idealized diagram for plastic design.

the stress F_y is very seldom reached in plastic design; only the additional load-carrying capacity of the structure is considered.

To clarify the behavior of a beam, consider a rectangular member under flexure. The relationship between stress and moment is given by

$$f = \frac{M}{S} \tag{10.1}$$

where $S = bh^2/6$. When the moment increases, the stress f also increases and reaches F_y, the yield stress. We can recover the expression (10.1) by considering the equilibrium of moments in Fig. 10.2c

$$M_y = \frac{bh}{2}\frac{1}{2}\frac{2}{3}\frac{h}{}F_y = \frac{bh^2}{6}F_y \tag{10.2}$$

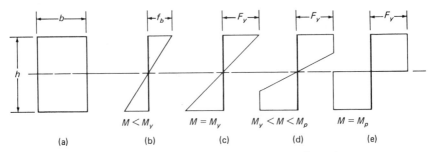

Fig. 10.2. Plastic hinge development in a rectangular section.

From this point on, fibers stressed at F_y cannot increase their stress and hence begin to yield. If the moment M is increased, fibers closer to the neutral axis start to yield, and a stress distribution, illustrated in Fig. 10.2d, occurs. When the stress distribution in Fig. 10.2e is reached, the member will have exhausted its carrying capacity, and all the fibers will have yielded, i.e., have plastified. The moment corresponding to this stage is called the *plastic moment* and is represented by M_p.

Obtaining the relationship between the plastic moment and the yield stress, we have

$$M_p = b \frac{h}{2} \frac{h}{2} F_y = \frac{bh^2}{4} F_y \tag{10.3}$$

Fig. 10.3. Bank of Villa Park, Villa Park, Illinois.

PLASTIC DESIGN OF STEEL BEAMS

The quantity $bh^2/4$ is the plastic section modulus and is represented by Z. The ratio $M_p/M_y = Z/S$ is called the *shape factor*. This shape factor, which is 1.5 for a rectangle, varies considerably with the shape of the section. Common values for wide flanges range from 1.08 to 1.20. For the case when a cross section is not symmetrical with respect to the tension and compression fibers, Z should be calculated by using the critical section modulus S.

The location where there is no stress, and consequently no strain, is the *plastic neutral axis*. In plastic design, for tension to equal compression, the tension area must be equal to the compression area, as both the tensile and compressive stresses are equal to F_y. Therefore, the plastic neutral axis must divide the cross section into two parts, both having equal areas. In a section that is symmetrical with respect to the horizontal axis, the elastic neutral axis (center of gravity) and the plastic neutral axis coincide.

Example 10.1. Calculate the values of S, Z, and the shape factor about the X axis for the figure shown.

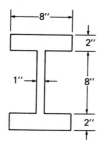

Solution. By inspection, the neutral axis (NA) is at midheight, 6 in. from the bottom.

$$I = \Sigma \left(\frac{bh^3}{12} + Ad^2\right) = 2 \times \left[\frac{8 \text{ in. } (2 \text{ in.})^3}{12} + (2 \text{ in.} \times 8 \text{ in.})(5 \text{ in.})^2\right]$$

$$+ \frac{1 \text{ in. } (8 \text{ in.})^3}{12}$$

$$I = 853 \text{ in.}^4$$

$$S = \frac{I}{c} = \frac{853 \text{ in.}^4}{6 \text{ in.}} = 142.2 \text{ in.}^3$$

The plastic section modulus can be determined by summing the areas of the stress blocks times their distance to the neutral axis.

$$Z = \Sigma\, Ad = 2 \times [(2 \text{ in.} \times 8 \text{ in.}) \times 5 \text{ in.}$$
$$+ (1 \text{ in.} \times 4 \text{ in.}) \times 2 \text{ in.}] = 176 \text{ in.}^3$$

Shape factor

$$\frac{Z}{S} = \frac{176 \text{ in.}^3}{142.2 \text{ in.}^3} = 1.24$$

Example 10.2. Calculate the values of S, Z, and the shape factor about the X axis for the figure shown.

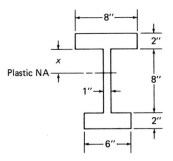

Solution. The elastic properties are:

$$\bar{y} = \frac{(6 \text{ in.} \times 2 \text{ in.}) \times 1 \text{ in.} + (1 \text{ in.} \times 8 \text{ in.}) \times 6 \text{ in.} + (8 \text{ in.} \times 2 \text{ in.}) \times 11 \text{ in.}}{(6 \text{ in.} \times 2 \text{ in.}) + (8 \text{ in.} \times 1 \text{ in.}) + (8 \text{ in.} \times 2 \text{ in.})}$$

$$= 6.56 \text{ in.}$$

$$I = \frac{8 \text{ in.} (2 \text{ in.})^3}{12} + 16 \text{ in.}^2 (11 \text{ in.} - 6.56 \text{ in.})^2 + \frac{1 \text{ in.} (8 \text{ in.})^3}{12}$$

$$+ 8 \text{ in.}^2 (6 \text{ in.} - 6.56 \text{ in.})^2 + \frac{6 \text{ in.} (2 \text{ in.})^2}{12}$$

$$+ 12 \text{ in.}^2 (1 \text{ in.} - 6.56 \text{ in.})^2$$

$$= 740.9 \text{ in.}^4$$

$$S = \frac{I}{c} = \frac{740.9 \text{ in.}^4}{6.56 \text{ in.}} = 112.9 \text{ in.}^3 \quad \text{(critical)}$$

To locate the plastic NA, find the distance X, such that the area above the neutral axis is equal to the area below.

$$(8 \text{ in.} \times 2 \text{ in.}) + (1 \text{ in.} \times (X)) = (1 \text{ in.} \times (8 \text{ in.} - X)) + (6 \text{ in.} \times 2 \text{ in.})$$

$$X = 2 \text{ in.}$$
$$Z = (8 \text{ in.} \times 2 \text{ in.}) \times 3 \text{ in.} + (2 \text{ in.} \times 1 \text{ in.})$$
$$\times 1 \text{ in.} + (1 \text{ in.} \times 6 \text{ in.})$$
$$\times 3 \text{ in.} + (6 \text{ in.} \times 2 \text{ in.}) \times 7 \text{ in.} = 152 \text{ in.}^3$$

$$\text{Shape factor} = \frac{Z}{S} = \frac{152 \text{ in.}^3}{112.9 \text{ in.}^3} = 1.35$$

Examples 10.3 and 10.4. Find the values of S, Z, and the shape factor for the shapes shown.

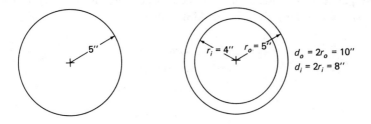

Solution. For a solid circular section, the elastic and plastic section moduli are found to be

$$S = \frac{\pi r^3}{4} = \frac{\pi d^3}{32} \quad \text{(AISCM, p. 6-20 Properties of Geometric Sections)}$$

$$Z = \frac{d^3}{6}$$

$$S = \frac{\pi (10 \text{ in.})^3}{32} = 98.2 \text{ in.}^3$$

$$Z = \frac{(10 \text{ in.})^3}{6} = 166.7 \text{ in.}^3$$

$$\text{Shape factor} \frac{Z}{S} = \frac{166.7 \text{ in.}^3}{98.2 \text{ in.}^3} = 1.70$$

For a hollow circular section, the elastic and plastic section moduli are found to be

$$S = \frac{\pi(d_o^4 - d_i^4)}{32 \, d_o} \quad \text{(AISCM, p. 6-20)}$$

$$Z = \frac{d_o^3 - d_i^3}{6}$$

$$S = \frac{\pi \times [(10 \text{ in.})^4 - (8 \text{ in.})^4]}{32 \times 10 \text{ in.}} = 57.96 \text{ in.}^3$$

$$Z = \frac{(10 \text{ in.})^3 - (8 \text{ in.})^3}{6} = 81.33 \text{ in.}^3$$

$$\text{Shape factor } \frac{Z}{S} = \frac{81.33 \text{ in.}^3}{57.96 \text{ in.}^3} = 1.40$$

Example 10.5. Calculate the S, Z, and shape factor for a W 24 × 104, and compare with values given in the AISC tables.

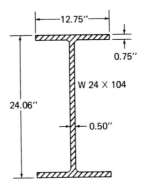

Solution.

$d = 24.06$ in.; $t_w = 0.50$ in.

$b_f = 12.75$ in.; $t_f = 0.75$ in.

$$I = \frac{0.5 \text{ in.} \times (24.06 \text{ in.} - 2(0.75 \text{ in.}))^3}{12} + 2 \times \left(\frac{12.75 \text{ in.} \times (0.75 \text{ in.})^3}{12}\right.$$

$$\left. + (0.75 \text{ in.} \times 12.75 \text{ in.}) \times \left(\frac{24.06 \text{ in.}}{2} - 0.375 \text{ in.}\right)^2\right) = 3077 \text{ in.}^4$$

$$S = \frac{I}{c} = \frac{3077 \text{ in.}^4}{12.03 \text{ in.}} = 255.8 \text{ in.}^3$$

$$Z = 2 \times \left((0.75 \text{ in} \times 12.75 \text{ in.}) \times \frac{24.06 \text{ in.} - 0.75 \text{ in.}}{2}\right.$$

$$\left. + \frac{24.06 \text{ in.} - 1.5 \text{ in.}}{2} \times 0.50 \text{ in.} \times \frac{24.06 \text{ in.} - 1.5 \text{ in.}}{4}\right) = 286.5 \text{ in.}^3$$

The AISCM lists a value of 258 in.^{3} for the member section modulus, and 289 in.^{3} for the member plastic modulus.[2]

$$\text{Shape factor } \frac{Z}{S} = \frac{286.5 \text{ in.}^3}{255.8 \text{ in.}^3} = 1.12 \quad \text{(shape factor same by AISC)}$$

10.2 PLASTIC HINGES

When a flexural member is subjected to loading, so as to cause certain parts of the member to acquire its plastic moment, the member will begin to yield at those points and transfer any moment caused by further loading, which could have been resisted at that location had it not yielded, to other parts of the beam still below the plastic moment. Where the moment reaches the plastic moment, the steel section yields and the structure, for any additional load, behaves as if a hinge were located there.

Consider a beam with fixed supports uniformly loaded with a load w (Fig. 10.4a). If the load w is increased until the plastic moment is reached at the supports, plastic hinges will form at A and B. The load corresponding to this stage is $w_1 = 12 M_p/L^2$. If the load is further increased, no more moment will be carried at points A and B because they have reached the plastic hinge conditions at M_p. For increased loads, the beam will behave as a simply supported beam (Fig. 10.4b). The collapse will occur when the third hinge forms at the center of the beam where the moment M_p is attained here. The total load applied to the beam is then $w = w_1 + w_2$. At collapse, the moment M at the center is

$$\frac{w_1 L^2}{24} + \frac{w_2 L^2}{8} = M_p \tag{10.4}$$

---

[2]The differences are due to the fillets at the web/flange corners.

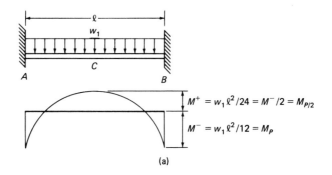
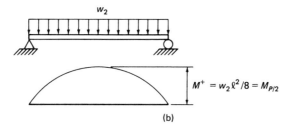
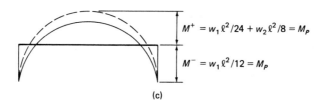

Fig. 10.4. Development of plastic hinges in a fixed beam.

which yields

$$w_2 = \frac{4 M_p}{L^2} \tag{10.5}$$

The total capacity of the beam at collapse is $12 M_p/L^2 + 4 M_p/L^2 = 16 M_p/L^2$ instead of only $12 M_p/L^2$ for the elastic approach. This behavior is due to redistribution of moments.[3] (Here safety and load factors have not been considered.)

---

[3]This fact is recognized by the AISC code in the working stress design approach in F1.1.

Example 10.6. Calculate P, considering elastic and plastic design approaches. Omit load and safety factors.

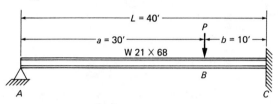

Solution. Elastic Design Approach

Formulas are from AISCM beam diagram and formulas No. 14

$$M_C = \frac{Pab}{2L^2}(a + L)$$

$$a = 30 \text{ ft}; \quad b = 10 \text{ ft}; \quad L = 40 \text{ ft}$$

$$M_C = \frac{30 \times 10}{2(40)^2}(30 + 40)P = 6.56\,P$$

$$M_B = R_A a$$

$$R_A = \frac{Pb^2}{2L^3}(a + 2L) = 0.086\,P$$

$$M_B = 0.086\,P \times 30 = 2.58\,P$$

$$F_y = \frac{M_{max}}{S}; \quad M_{max} = F_y S$$

$$F_y = 36.0 \text{ ksi}$$

$$S = 140 \text{ in.}^3$$

$$M_{max} = 6.56\,P = \frac{36.0 \text{ ksi} \times 140 \text{ in.}^3}{12 \text{ in./ft}} = 420 \text{ ft-k}$$

$$P = 64.0 \text{ k}$$

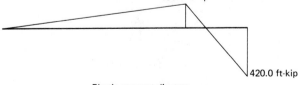

Elastic moment diagram

430 STEEL DESIGN FOR ENGINEERS AND ARCHITECTS

Plastic Design Approach

$$F_y = \frac{M_p}{Z}; \quad M_p = ZF_y$$

$$F_y = 36 \text{ ksi}$$

$$Z = 160 \text{ in.}^3$$

$$M_p = 160 \text{ in.}^3 \times \frac{36 \text{ ksi}}{12 \text{ in./ft}} = 480 \text{ ft-k}$$

The load P_1 is the load that will cause a plastic hinge at C.

$$M_p = 6.56 \, P_1$$

$$P_1 = \frac{480}{6.56} = 73.17 \text{ k}$$

To plastify B (i.e., to obtain the third hinge, the second being already at A), we need an additional moment at B of

$$M = 480 - (2.58 \times 73.17) = 291.22 \text{ ft-k}$$

This moment will be obtained by an additional load P_2 acting on a simply supported beam. As A is a normal hinge and C a plastic one,

$$P_2 \times \frac{10 \times 30}{40} = 291.22$$

$$P_2 = 38.83 \text{ k}$$

$$P = P_1 + P_2 = 73.17 \text{ k} + 38.83 \text{ k} = 112.0 \text{ k}$$

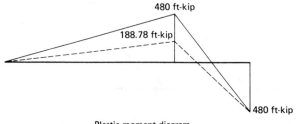

Plastic moment diagram

10.3 BEAM ANALYSIS BY VIRTUAL WORK

In mechanics, energy is defined as the capacity to do work. Work is the product of a force and the distance traveled by the force in its own direction or the product of a moment and the angle of its rotation. From conservation of energy, internal work U_i is equal to the external work U_o. By introducing a small fictitious (virtual) displacement in a structure, and thereby causing the forces and moments to do work, the plastic moment in the structure may be determined.

In plastic design, because hinges form at the locations of plastic moments, the internal virtual work is the sum of the products of the rotations of the member at the hinge locations and the plastic moment of the member.

External virtual work of a flexural member is the summation of the externally applied loads and the respective deformations of the member at the locations of the loads.

In the process of analysis by virtual work for a one-span beam, hinges occur at the supports and under the concentrated load. In case of more than one concentrated load, the location of the hinge in the span cannot be readily determined.

The location of the hinge in the span is assumed, the plastic moment M_p is calculated, and the corresponding moment diagram is shown. If at no point there is a moment larger in absolute value than M_p, then the assumption is correct, and M_p is the plastic moment. If at some point the moment is larger than M_p, then the hinge is erroneously located and has to be moved to that point. It can be proved that the plastic moment corresponding to the proper hinge location is the maximum M_p.

Example 10.7. Calculate the plastic moment M_p for the beam shown using a) the method of virtual work and b) the method of superposition.

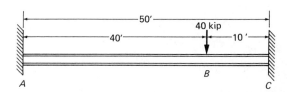

Solution. a) We assume that plastic hinges form at points A, B, and C.

Consider a virtual displacement given to the system, so that AB rotates by θ. The virtual displacement at B becomes $40\,\theta$ ft, and the rotation at C, $40\,\theta/10 = 4\,\theta$.

432 STEEL DESIGN FOR ENGINEERS AND ARCHITECTS

The angle at B is the sum of the angles at A and C, which equals 5θ.

The virtual work of the external force is $40\theta \text{ ft} \times 40 \text{ k} = 1600\theta \text{ ft-k}$

The virtual work of the internal effects is

$$(M_p \times \theta) + (M_p \times 5\theta) + (M_p \times 4\theta) = 10\theta M_p$$

Internal work is equal to the external work

$$10 M_p \theta = 1600 \theta \text{ (ft-k)}$$

$$M_p = 160 \text{ ft-k}$$

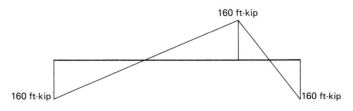

b) A quick way to determine the plastic moment of a beam section involves the superposition of moment diagrams due to hinged supports and fixed supports. The three necessary plastic hinges are formed at the two ends and the point of maximum positive bending. Using these concepts, the plastic moment can be expressed as the moment diagram of a statically determinate span superimposed on the moment diagram due to given end conditions.

In this case, assume simple supports (hinges) at points A and C, so that the moment at point B is

$$M_B = \frac{Pab}{L} = \frac{40 \times 40 \times 10}{50} = 320 \text{ ft-kips}$$

Considering fixed supports, the moment at B will be M_p. Consequently, superimposing the statically determinate moment diagram on the moment diagram due to the actual end conditions, and knowing that the final moment must be equal to M_p, we can write the following equation for the moments at B:

$$320 - M_p = M_p$$

or
$$M_p = 160 \text{ ft-k}$$

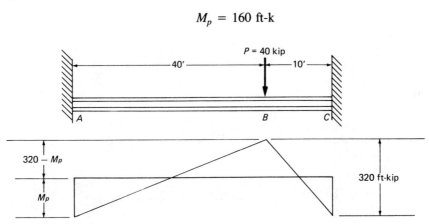

Example 10.8. Calculate the plastic moment for the uniformly loaded beam shown, using a) the principle of virtual work. b) the method of superposition.

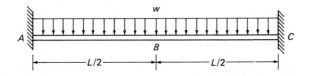

Solution. a) Hinges will form at A, B, and C (by inspection).

Virtual work of internal effects:

$$(M_p \times \theta) + (M_p \times 2\theta) + (M_p \times \theta) = 4 M_p \theta$$

Virtual work of external forces:

We can consider that the virtual work in member AB is equal to the load in that member times the displacement of its center. Hence, it is equal to

$$\frac{wL}{2} \times \frac{\theta L}{2} \times \frac{1}{2} \times 2 \text{ sections}$$

$$4 M_p \theta = \frac{wL^2 \theta}{4}$$

$$M_p = \frac{wL^2}{16}$$

b)

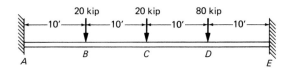

$$\frac{wL^2}{8} - M_p = M_p$$

$$M_p = \frac{wL^2}{16}$$

Example 10.8. Calculate the plastic moment for the uniformly loaded beam shown, using a) the principle of virtual work. b) the method of superposition.

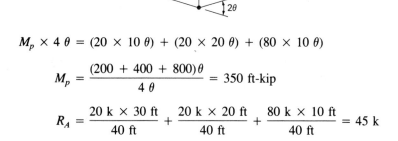

Solution. a) Assume plastic hinges at A, C, and E.

$$M_p \times 4\theta = (20 \times 10\,\theta) + (20 \times 20\,\theta) + (80 \times 10\,\theta)$$

$$M_p = \frac{(200 + 400 + 800)\theta}{4\,\theta} = 350 \text{ ft-kip}$$

$$R_A = \frac{20 \text{ k} \times 30 \text{ ft}}{40 \text{ ft}} + \frac{20 \text{ k} \times 20 \text{ ft}}{40 \text{ ft}} + \frac{80 \text{ k} \times 10 \text{ ft}}{40 \text{ ft}} = 45 \text{ k}$$

$R_B = 120 \text{ k} - 45 \text{ k} = 75 \text{ k}$

$M_A = -350 \text{ ft-k}$

$M_B = -350 \text{ ft-k} + (45 \text{ k} \times 10 \text{ ft}) = 100 \text{ ft-k}$

$M_C = -350 \text{ ft-k} + (45 \text{ k} \times 20 \text{ ft}) - (20 \text{ k} \times 10 \text{ ft}) = 350 \text{ ft-k}$

$M_D = (75 \text{ k} \times 10 \text{ ft}) - 350 \text{ ft-k} = 400 \text{ ft-k} > M_p = 350 \text{ ft-k}$

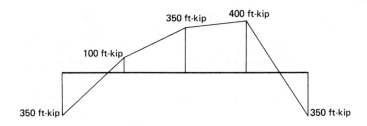

We see that the hinge selection is wrong and that the hinge located at C must be moved to D because $M_D > M_p$.

Selecting the new location for hinges A, D, E,

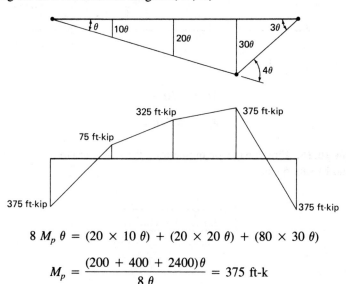

$$8 M_p \theta = (20 \times 10 \theta) + (20 \times 20 \theta) + (80 \times 30 \theta)$$

$$M_p = \frac{(200 + 400 + 2400)\theta}{8 \theta} = 375 \text{ ft-k}$$

Recalculating,

$R_A = 45 \text{ k} \quad R_B = 75 \text{ k}$

$M_A = -375$ ft-k

$M_B = -375$ ft-k $+ (45$ k $\times 10$ ft$) = 75$ ft-k

$M_C = -375$ ft-k $+ (45$ k $\times 20$ ft$) - (20$ k $\times 10$ ft$) = 325$ ft-k

$M_D = (75$ k $\times 10$ ft$) - 375$ ft-k $= 375$ ft-k

We see then that the location selection is correct, for no other moment is larger than M_p.

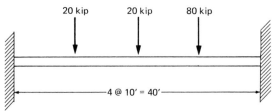

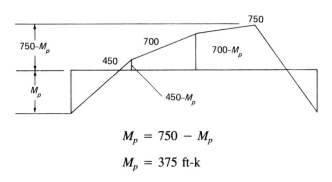

$M_p = 750 - M_p$

$M_p = 375$ ft-k

Example 10.10. Find the plastic moment for the given situation using a) virtual work and b) superposition.

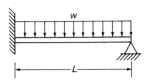

Solution. a) Since the hinge location is not known, the location must be found by the method of sections and assuming the location of zero shear and maximum moment equal to M_p.

PLASTIC DESIGN OF STEEL BEAMS

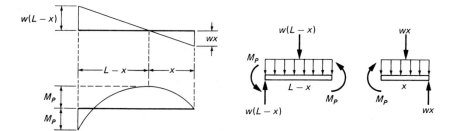

From the right section,

$$M_p + (wx)\left(\frac{x}{2}\right) - (wx)(x) = 0$$

$$M_p = \frac{wx^2}{2}$$

From the left section,

$$-2M_p + w(L-x)(L-x) - w(L-x)\left(\frac{L-x}{2}\right) = 0$$

$$M_p = \frac{w}{4}(L^2 - (2Lx) + x^2)$$

Equating the two equations,

$$\frac{wx^2}{2} = \frac{wL^2}{4} - \frac{wLx}{2} + \frac{wx^2}{4}$$

$$wx^2 + (2wLx) - wL^2 = 0$$

$$x = 0.414 L$$

$$L - x = 0.586 L$$

$$M_p = \frac{wx^2}{2} = \frac{w(0.414 L)^2}{2} = \frac{wL^2}{11.66}$$

$$M_p = \frac{wL^2}{11.66}$$

438 STEEL DESIGN FOR ENGINEERS AND ARCHITECTS

b)

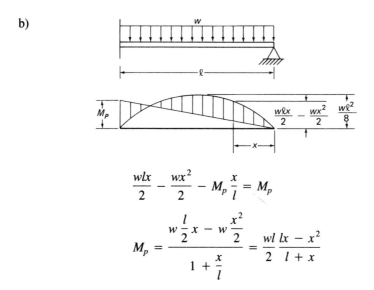

$$\frac{wlx}{2} - \frac{wx^2}{2} - M_p \frac{x}{l} = M_p$$

$$M_p = \frac{w\frac{l}{2}x - w\frac{x^2}{2}}{1 + \frac{x}{l}} = \frac{wl}{2} \frac{lx - x^2}{l + x}$$

x must be such that M_p is maximum. To maximize M_p, dM_p/dx must be equal to zero

$$(l - 2x)(l + x) - (lx - x^2) = 0$$
$$l^2 - 2lx + lx - 2x^2 - lx + x^2 = 0$$
$$x^2 + 2lx - l^2 = 0, \quad x = -l \pm \sqrt{l^2 + l^2}$$
$$x = 0.414\, l$$
$$M_p = \frac{wl^2}{2} \frac{0.414 - 0.171}{1.414} = 0.0858\, wl^2 = \frac{wl^2}{11.67}$$

10.4 PLASTIC DESIGN OF BEAMS

Plastic analysis enables the designer to calculate the maximum load that the structure can support just before it collapses; thus, the working loads must be multiplied by load factors to compensate for uncertainties in the loading and to provide the structure safety against collapse. The AISC calls for a load factor of 1.70, for dead and live loads only, or a load factor of 1.30, for dead, live, and wind or earthquake loads combined (AISCS N1). It may be noted that the load factor 1.70 is approximately equal to the average shape factor 1.12 (for sections having a compression and tension flange) times the safety factor of

PLASTIC DESIGN OF STEEL BEAMS 439

1.50 for a simply supported beam, and the load factor 1.30 is approximately equal to the load factor 1.70 reduced by 25%, which is equivalent to the 25% reduction in moment allowed in elastic design when wind or earthquake forces are taken into account.

To determine the required plastic moment capacity of any span of a structure by the method of virtual work:

1. Multiply loads by the appropriate load factors.
2. Assume the locations of the plastic hinges.
3. Introducing a virtual displacement in the structure, determine the rotations of the member at all plastic hinge locations and the displacements under all applied concentrated loads or at the interior hinge location for distributed loads.
4. Equating the internal virtual work to the external, calculate the required plastic moment.
5. Determine proper locations of hinges, and determine M_p.
6. Once the largest required plastic moment is obtained, determine Z required and select a satisfactory member from the AISCM (p. 2-15).

Note that steps 3 and 4 can be replaced by the method of superposition.

Up to now, only single-span beams were designed. The failure of a continuous beam is essentially the failure of one of its spans. Hence, in the case of a continuous beam, each of the spans is analyzed individually, and the largest moment M_p obtained becomes the design moment of the entire beam.

Example 10.11. Select the lightest beams according to elastic and plastic theories. Assume full lateral support and that the loads include the weight of the beam. Consider also all appropriate safety and load factors.

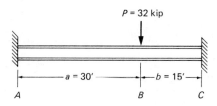

Solution. Elastic Approach

$$M_A = \frac{Pab^2}{L^2} = -106.7 \text{ ft-k}$$

440 STEEL DESIGN FOR ENGINEERS AND ARCHITECTS

$$M_B = \frac{2\,Pa^2b^2}{L^3} = 142.2 \text{ ft-k}$$

$$M_C = \frac{Pa^2b}{L^2} = -213.3 \text{ ft-k}$$

The moments can be redistributed in accordance with AISCS F1.1:

$$M_A = 0.9 \times -106.7 \text{ ft-k} = -96.0 \text{ ft-k}$$

$$M_C = 0.9 \times -213.3 \text{ ft-k} = -192.0 \text{ ft-k}$$

$$M_B = 142.2 \text{ ft-k} + \left(\frac{10.7 + 21.3}{2}\right) \text{ ft-k} = 158.2 \text{ ft-k}$$

$$S_{req} = \frac{M_{max}}{F_b} = \frac{192.0 \text{ ft-k} \times 12 \text{ in.}/\text{ft}}{0.66 \times 36.0 \text{ ksi}} = 97.0 \text{ in.}^3$$

Use W 18 × 55 $S_x = 98.3$ in.^{3}

Plastic Approach

$$\text{Factored load} = 1.7\,P = 1.7 \times 32.0 \text{ k} = 54.4 \text{ k}$$

Assume plastic hinges at A, B, and C

$$6\,\phi\,M_p = 54.4 \text{ k} \times 30\,\phi$$

$$M_p = \frac{1632\,\phi}{6\,\phi} = 272 \text{ ft-k}$$

$$Z = M_p/F_y = \frac{272 \text{ ft-k} \times 12 \text{ in.}/\text{ft}}{36.0 \text{ ksi}} = 90.7 \text{ in.}^3$$

The section must be chosen from the plastic design selection table (p. 2-22).

The procedure used to determine the most economical beam section is similar to the one used for the elastic analysis, as described in Chapter 2.

Use W 21 × 44 $Z_x = 95.4$ in.^{3}

Example 10.12. Select the lightest beams according to elastic and plastic theories. Assume full lateral support and that the loads include the weight of the beam.

PLASTIC DESIGN OF STEEL BEAMS 441

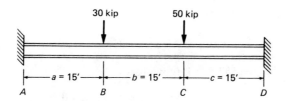

Solution. Elastic Approach

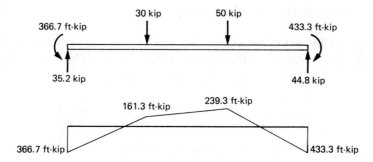

Redistributing moments in accordance with AISCS F1.1:

$$M_A = 0.9 \times -366.7 \text{ ft-k} = 330.0 \text{ ft-k}$$

$$M_D = 0.9 \times -433.3 \text{ ft-k} = 390.0 \text{ ft-k}$$

$$M_C = 239.3 \text{ ft-k} + \left(\frac{36.7 + 43.3}{2}\right) = 279.3 \text{ ft-k}$$

$$S_{req} = \frac{M_{max}}{F_b} = \frac{390.0 \text{ ft-k} \times 12 \text{ in./ft}}{0.66 \times 36.0 \text{ ksi}} = 195.0 \text{ in.}^3$$

Use W 24 × 84 $S_x = 196.0$ in.^{3}

Plastic Approach

Factored loads:

$$1.7 \times 30.0 \text{ k} = 51.0 \text{ k}$$

$$1.7 \times 50.0 \text{ k} = 85.0 \text{ k}$$

Assume hinges at A, B, and D

$$6 \phi M_p = (51.0 \text{ k} \times 30 \phi) + (85.0 \text{ k} \times 15 \phi)$$

$$M_p = 467.5 \text{ ft-k}$$

Assume hinges at A, C, and D

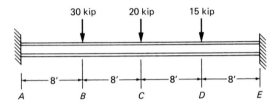

$6 \phi M_p = (51.0 \text{ k} \times 15 \phi) + (85.0 \text{ k} \times 30 \phi)$

$M_p = 552.5$ ft-k (governs)

$$Z = M_p/F_y = \frac{552.5 \text{ ft-k} \times 12 \text{ in./ft}}{36.0 \text{ ksi}} = 184.0 \text{ in.}^3$$

Use W 24 × 76 $Z = 200$ in.^{3}

Example 10.13. Select beams according to elastic and plastic theories. Assume full lateral support and that the loads include the weight of the beam.

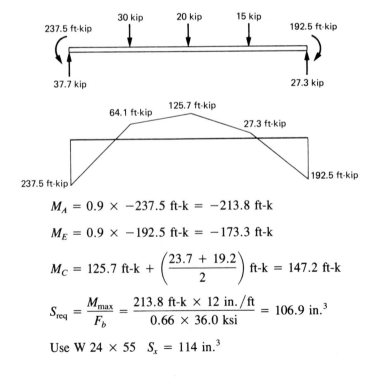

Solution. Elastic Approach

$M_A = 0.9 \times -237.5 \text{ ft-k} = -213.8 \text{ ft-k}$

$M_E = 0.9 \times -192.5 \text{ ft-k} = -173.3 \text{ ft-k}$

$M_C = 125.7 \text{ ft-k} + \left(\dfrac{23.7 + 19.2}{2}\right) \text{ft-k} = 147.2 \text{ ft-k}$

$$S_{\text{req}} = \frac{M_{\text{max}}}{F_b} = \frac{213.8 \text{ ft-k} \times 12 \text{ in./ft}}{0.66 \times 36.0 \text{ ksi}} = 106.9 \text{ in.}^3$$

Use W 24 × 55 $S_x = 114$ in.^{3}

PLASTIC DESIGN OF STEEL BEAMS

Plastic Approach

Factored loads:

$$1.7 P = 1.7 \times 30 \text{ k} = 51 \text{ k}$$
$$1.7 \times 20 \text{ k} = 34 \text{ k}$$
$$1.7 \times 15 \text{ k} = 25.5 \text{ k}$$

Assume hinges at A, B, and E:

$8 \phi M_p = (51 \text{ k} \times 24 \phi)$
$\qquad + (34 \text{ k} \times 16 \phi) + (25.5 \times 8 \phi)$

$M_p = 246.5$ ft-k

Assume hinges at A, C, and E:

$4 \phi M_p = (51 \text{ k} \times 8 \phi)$
$\qquad + (34 \text{ k} \times 16 \phi)$
$\qquad + (25.5 \times 8 \phi)$

$M_p = 289$ ft-k

Assume hinges at A, D, and E:

$8 \phi M_p = (51 \times 8 \phi) + (34 \times 16 \phi)$
$\qquad + (25.5 \times 24 \phi)$

$M_p = 195.5$ ft-k

Hinges at A, C, and E govern

$M_p = 289$ ft-k

$Z = M_p/F_y = 289 \text{ ft-k} \times 12 \text{ in.}/\text{ft } 36 \text{ ksi} = 96.3 \text{ in.}^3$

Use W 21 × 50 $Z = 110$ in.^{3}
or W 18 × 50 ($Z = 101.0$ in.^{3}) which has same weight but less depth.

Example 10.14. For the situation shown, design the lightest beam, using the plastic approach.

444 STEEL DESIGN FOR ENGINEERS AND ARCHITECTS

Solution.

Factored loads:

$1.7 P = 25 \times 1.7 = 42.5$ k

$1.7 \times 15 = 25.5$ k

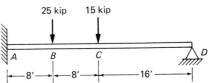

Assume plastic hinges at A and B, with an actual hinge at D.

$3 \phi M_p + 4 \phi M_p = (42.5 \times 24 \phi)$
$\qquad\qquad\qquad\quad + (25.5 \text{ k} \times 16 \phi)$

$M_p = 204$ ft-k

Assume plastic hinges at A and C, with an actual hinge at D.

$1 \phi M_p + 2 \phi M_p = (42.5 \text{ k} \times 8 \phi)$
$\qquad\qquad\qquad\quad + (25.5 \text{ k} \times 16 \phi)$

$M_p = 249$ ft-k (governs)

$Z = M_p/F_y = (249 \text{ ft-k} \times 12 \text{ in.}/\text{ft})/36 \text{ ksi} = 83 \text{ in.}^3$

Use W 21 × 44 $Z = 95.4$ in.^{3}

Example 10.15. Select, by plastic design, a continuous rolled section that is satisfactory for the entire structure.

Solution.

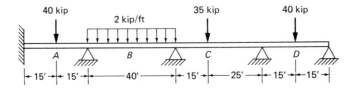

Since plastic hinges will form at the supports where negative moment is greatest, each span will be analyzed separately and the critical moment determined.

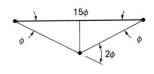

Span A $4 \phi M_p = 40 \text{ k} (1.7) 15 \phi$

$\qquad\qquad M_p = 255$ ft-k

Span B $M_p = w(L)^2/16 = \dfrac{2\,(1.7)\,(40)^2}{16}$

$M_p = 340$ ft-k (governs)

Span C $(80/15)\,\phi\,M_p = 35\,k\,(1.7)\,25\,\phi$

$M_p = 279$ ft-k

Span D $3\,\phi\,M_p = 40\,k\,(1.7)\,15\,\phi$

$M_p = 340$ ft-k (governs)

$Z = 340$ ft-k $\times$ 12 in./ft/36 ksi $= 113$ in.^{3}

Use W 24 $\times$ 55 $Z = 134$ in.^{3}

Example 10.16. Select, by plastic design, a continuous rolled section that is satisfactory for the entire structure.

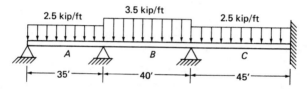

Solution.

Span A $M_p = w(L)^2/11.66 = \dfrac{2.5\,(1.7)\,(35)^2}{11.66}$

$M_p = 446.5$ ft-k

Span B $M_p = \dfrac{w(L)^2}{16} = \dfrac{3.5\,(1.7)\,(40)^2}{16}$

$M_p = 595.0$ ft-k (governs)

Span C $M_p = \dfrac{w(L)^2}{16} = \dfrac{2.5\,(1.7)\,(45)^2}{16}$

$M_p = 537.9$ ft-k

Maximum $M_p = 595$ ft-k

$Z = 595$ ft-k $\times$ 12 in./ft/36 ksi $= 198$ in.^{3}

Use W 24 $\times$ 76 $Z = 200$ in.^{3}

10.5 ADDITIONAL CONSIDERATIONS

Shear

The AISC Code requires, in Section N5, that the ultimate shear be governed by (N5-1):

$$V_u \leq 0.55 \, F_y t_w d \tag{10.6}$$

unless the web is reinforced. Nevertheless, the design of additional reinforcement for shear is quite involved, and it is suggested that the design of rolled sections meet the requirements of eqn. (10.6).

Minimum Thickness Ratio

Similar to the requirements for width-thickness ratios in AISCS B5, some rules have been established in Section N7 for beams. For most commonly used A36 steel, the following must be satisfied:

$$\frac{b_f}{2 \, t_f} \leq 8.5 \tag{10.7}$$

Also, the depth-thickness ratio of beam webs shall not exceed the value in (N7-1), shown below for the case when the factored axial load P is zero:

$$\frac{d}{t} = \frac{412}{\sqrt{F_y}} \tag{10.8}$$

Connections

In AISCM, Section N8, some rules for connections are established. The most important rule is that the connection elements shall resist the factored loads with a capacity equal to 1.7 times their respective stresses, given in AISCS Chapter J.

10.6 COVER PLATE DESIGN

It is generally uneconomical to use the same rolled section for all spans in a continuous multispan structure; the rolled section must be satisfactory for the span with the largest plastic moment, and, therefore, it will quite possibly be overdesigned for some if not all of the other spans. Usually in such cases, the rolled section is selected to satisfy the moment requirement of the span with the least moment, and cover plates are used at all other locations. Initially, the shear

PLASTIC DESIGN OF STEEL BEAMS 447

and moment diagrams of the structure are drawn, and the lengths and locations of the cover plates are determined for all portions of the beam exceeding in plastic moment requirement from that provided by the rolled section. Assuming a width b for the cover plate, the thickness t is then calculated.

A cost analysis has to be performed to determine if a single rolled section covering all spans is cheaper than the lighter section with cover plates (including the cost of welds).

Example 10.17. Using the plastic design method,

a) Design a single rolled section satisfactory for the entire structure.
b) Select a section for the span with the least moment, and design cover plates as necessary for the other spans.

Assume full lateral bracing.

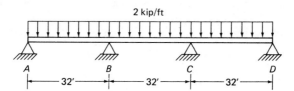

Solution. The plastic moments of span AB and CD are equal.

$$(M_p)_{AB} = (M_p)_{CD} = \frac{wL^2}{11.66} = \frac{(2 \times 1.7)(32)^2}{11.66}$$

$$(M_p)_{AB} = (M_p)_{CD} = 298.6 \text{ ft-k}$$

$$(M_p)_{BC} = \frac{wL^2}{6} = \frac{(2 \times 1.7)(32)^2}{16}$$

$$(M_p)_{BC} = 217.6 \text{ ft-k}$$

a) Governing $(M_p)_{AB\&CD} = 298.6$ ft-k

$$Z = M_p/F_y = \frac{298.6 \text{ ft-k} \times 12 \text{ in./ft}}{36.0 \text{ ksi}} = 99.5 \text{ in.}^3$$

Use W 18 × 50 $Z_x = 101$ in.^{3}

b) Least $(M_p)_{BC} = 217.6$ ft-k

$$Z = M_p/F_y = \frac{217.6 \times 12 \text{ in./ft}}{36.0 \text{ ksi}} = 72.53 \text{ in.}^3$$

Use W 16 × 40 $Z_x = 72.9$ in.^{3}

Section will carry plastic moment of span BC only. Cover plates need to be designed for plastic moments in other spans.

The plastic moment capacity of a W 16 × 40 beam is

$$M_p = \frac{Z_x F_y}{12 \text{ in./ft}} = \frac{72.9 \text{ in.}^3 \times 36 \text{ ksi}}{12 \text{ in./ft}} = 218.7 \text{ ft-k}$$

If W 16 × 40 is used for the entire structure, a plastic hinge of 218.7 ft-k moment will form at B, C, and at the midspan BC.

Summing moments about point A, the free body diagram of span AB becomes

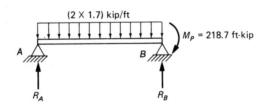

$$M = \left(32 \times 2 \times 1.7 \times \frac{32}{2}\right) + 219 - (R_B \times 32) = 0$$

$R_B = 61.24$ k

$R_A = 47.56$ k

$$M_{AB} = \frac{(47.56)^2}{2 \times 2 \times 1.7} = 332.6 \text{ ft-k}$$

$M_{CD} = M_{AB} = 332.6$ ft-k

PLASTIC DESIGN OF STEEL BEAMS

Moment and shear diagrams for the entire structure are

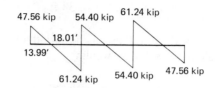

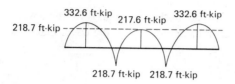

Cover plates must compensate for the deficient moment capacity of the W 16 × 40, as shown by the crosshatched area.

Since the change in moment equals the area under the shear diagram, R and S are found by letting $R = S$,

$$R \times \left(\frac{R}{13.99} \times 47.56\right) \times \frac{1}{2} = 113.9 \text{ ft-k}$$

$R = S = 8.19$ ft

$Q = 13.99 - 8.19 = 5.80$ ft

$T = 18.01 - 8.19 = 9.82$ ft

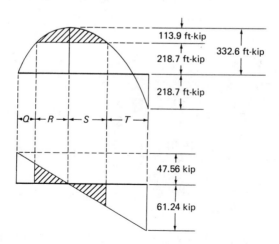

450 STEEL DESIGN FOR ENGINEERS AND ARCHITECTS

Theoretical length of cover plates

$$8.19 \text{ ft} \times 2 = 16.38 \text{ ft}$$

To design cover-plates for the beam, assume the width of the cover plate to be 4 in. to leave space along the sides for welding.

$$d = 16.01 \text{ in.}; \quad b_f = 6.995 \text{ in.}$$

Summing area moments about the neutral axis for the cover plates and equating them to the required moment capacity of the cover plates, we get

$$2 \times \left(\frac{16.01}{2} + \frac{t}{2} \right) \times 4 \times t = \frac{113.9 \times 12}{36}$$

$$t^2 + 16.01 t - 9.49 = 0$$

$t = 0.572$ in. Use $\frac{5}{8}$ in. $\times$ 4 in. cover plates

Connection of cover plates to flanges (AISCS N8)

Shear force at the theoretical end

$$8.19 \text{ ft} \times 2 \text{ k/ft} \times 1.7 = 27.85 \text{ k}$$

Shear flow

$$q = \frac{VQ}{I}$$

$$Q = \left(\frac{16.01}{2} + \frac{1}{2} \times \frac{5}{8} \right) \times 4 \times \frac{5}{8} = 20.79 \text{ in.}^3$$

$$I = 518 + 2 \times \left(\frac{16.01}{2} + \frac{1}{2} \times \frac{5}{8} \right)^2 \times 4 \times \frac{5}{8} = 863.9 \text{ in.}^4$$

$$q = \frac{27.85 \text{ k} \times 20.79 \text{ in.}^3}{863.9 \text{ in.}^4} = 0.67 \text{ k/in.}$$

Minimum size of E70 weld that can be used is $\frac{1}{4}$ in. (AISCS J2.2b)

Capacity of weld (AISC, Table J2.5)

$$0.30 \times \text{nominal tensile strength on throat} =$$

$$1.7 \times 0.30 \times 70 \times \frac{\sqrt{2}}{2} = 25.24 \text{ k/in.}^2$$

For $\frac{1}{4}$-in. weld, two-sided,

$$F_w = 2 \times 0.25 \times 25.24 = 12.62 \text{ k/in.}$$

If intermittent fillet weld is to be used,

$$l_{min} = 1.5 \text{ in. corresponds to } 1.5 \times 12.62 = 18.93 \text{ k/weld}$$

Spacing between welds

$$s = \frac{18.93 \text{ k/weld}}{0.67 \text{ k/in.}} = 28.3 \text{ in.}$$

Use at 12 in. on center.

Termination of welds (AISCS B10)

Force to be developed is MQ/I

$$\frac{218.7 \times 12 \times 20.79}{863.9} = 63.16 \text{ k}$$

Length of termination weld is 8 in.

$$F_w \times l = 12.62 \text{ k/in.} \times 8 \text{ in.} = 101 \text{ k} > 63.16 \text{ k} \quad \text{ok}$$

width/thickness ratio $= \dfrac{6.995}{2 \times 0.505} = 6.93 < 8.5 \quad \text{ok} \quad$ (AISCS N7)

depth/thickness ratio $= \dfrac{16.01}{0.305} = 52.49 < \dfrac{412}{\sqrt{F_y}} = 68.67 \quad \text{ok} \quad$ (AISCS N7)

Example 10.18. Using the plastic design method,

a) Design a single rolled section satisfactory for the entire structure.
b) Select a section for the span with the least moment, and design cover plates as necessary for the other spans.

452 STEEL DESIGN FOR ENGINEERS AND ARCHITECTS

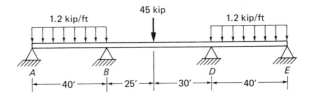

Solution. The plastic moments of span AB and DE are equal.

$$(M_p)_{AB} = (M_p)_{DE} = \frac{wL^2}{11.66} = \frac{(1.2 \times 1.7)(40)^2}{11.66} = 279.9 \text{ ft-k}$$

For M_p of span BD,

$$4.4 \phi M_p = 45 \text{ k} \times 1.7 \times 30 \phi$$

$$M_p = 521.6 \text{ ft-k for span } BD$$

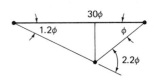

a) governing $(M_p)_{BD} = 521.6$ ft-k

$$Z = M_p/F_y = \frac{521.6 \text{ ft-k} \times 12 \text{ in./ft}}{36.0 \text{ ksi}} = 173.9 \text{ in.}^3$$

Use W 24 × 68 ($Z = 177$ in.^{3})

b) least $(M_p)_{AB\&DE} = 279.9$ ft-k

$$Z = M_p/F_y = \frac{279.9 \text{ ft-k} \times 12 \text{ in./ft}}{36.0 \text{ ksi}} = 93.3 \text{ in.}^3$$

Use W 21 × 44

M_p of W 21 × 44 beams is 286 ft-k ($Z = 95.4$ in.^{3})

The free body diagram of span BD gives

$$\Sigma M_D = -45 \times 1.7 \times 30 + 55 R_B = 0$$

$R_B = 41.7$ k

$R_D = 34.8$ k

PLASTIC DESIGN OF STEEL BEAMS

Moment and shear diagrams for the entire structure are

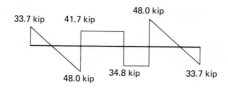

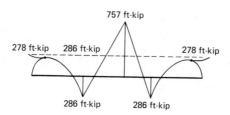

Cover plates must compensate for the deficient moment capacity of the W 21 × 44, as shown by the crosshatched area.

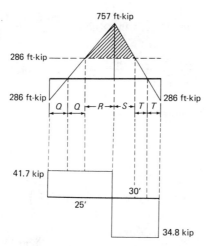

Since the change in moment equals the area under the shear diagram.

$$Q \times 41.7 = 286 \text{ ft-k}$$
$$Q = 6.85 \text{ ft}$$
$$R = 25 - 2(6.85 \text{ ft}) = 11.28 \text{ ft}$$
$$T = 286/34.8 = 8.22 \text{ ft}$$
$$S = 30 \text{ ft} - 2(8.22 \text{ ft}) = 13.56 \text{ ft}$$

454 STEEL DESIGN FOR ENGINEERS AND ARCHITECTS

Theoretical length of cover plate

$$R + S = 11.28 + 13.56 = 24.84 \text{ ft}$$

To design cover plates for the beam, assume the width of the cover plate to be 4 in. to leave space along the sides for welding.

$$d = 20.66 \text{ in.}, \quad b_f = 6.50 \text{ in.}$$

$$2 \times \left(\frac{20.66}{2} + \frac{t}{2}\right) \times (4\,t) \times F_y = M_{\text{req}} \times 12 \text{ in./ft}$$

$$(20.66 + t) \times (4\,t) \times 36 = (757 - 286) \times 12$$

$$t^2 + 20.66\,t - 39.25 = 0$$

$$t = 1.75 \text{ in.}$$

Checking the section,

$$Z = 95.4 \text{ in.}^3 + 4 \text{ in.} \times 1.75 \text{ in.} \times 2 \times \left(\frac{20.66}{2} + \frac{1.75}{2}\right) = 252.3 \text{ in.}^3$$

$$M_p = F_y \times Z = \frac{36 \text{ ksi} \times 252.3 \text{ in.}^3}{12 \text{ in./ft}} = 757 \text{ ft-k} \quad \text{ok}$$

Connection of plate to flange

Shear force

$$V_{\max} = 41.7 \text{ k}$$

$$I = 843 + 4 \times 1.75 \times \left(\frac{20.66}{2} + \frac{1.75}{2}\right)^2 \times 2 = 2601 \text{ in.}^4$$

$$Q = 4 \times 1.75 \times \left(\frac{20.66}{2} + \frac{1.75}{2}\right) = 78.44 \text{ in.}^3$$

$$q = \frac{VQ}{I} = \frac{41.7 \text{ k} \times 78.44 \text{ in.}^3}{2601 \text{ in.}^4} = 1.26 \text{ k/in.}$$

Design weld using $\tfrac{5}{16}$-in. E70 weld.
Capacity of weld

$$F_w = 1.7 \times 0.30 \times 70 \text{ ksi} \times \frac{\sqrt{2}}{2} \times \frac{5}{16} \text{ in.} = 7.89 \text{ k/in.}$$

Capacity of weld for two sides = 2 × 7.89 k/in. = 15.78 k/in.
Use intermittent fillet welds spaced at 12 in. on center.

$$q = 1.26 \text{ k/in.} \times 12 \text{ in./ft} = 15.12 \text{ k/ft}$$

$$l_{req} = \frac{15.12 \text{ k/ft}}{15.78 \text{ k/in.}} \cong 1 \text{ in. weld per foot of plate}$$

Use minimum weld length of $1\frac{1}{2}$ in.

Termination of welds

$$M = 286 \text{ ft-k}$$

$$H = \frac{MQ}{I} = \frac{286 \text{ ft-k} \times 12 \text{ in./ft} \times 78.44 \text{ in.}^3}{2601 \text{ in.}^4} = 103.5 \text{ k}$$

Try length of 8 in.

$$8 \text{ in.} \times 15.78 = 126.2 \text{ k} > 103.5 \text{ k} \quad \text{ok}$$

Check width/thickness and depth/thickness ratios (AISCS N7).

$$\frac{6.50 \text{ in.}}{2 \times 0.45 \text{ in.}} = 7.22 < 8.5 \quad \text{ok}$$

$$\frac{20.66 \text{ in.}}{0.35 \text{ in.}} = 59.0 < \frac{412}{\sqrt{F_y}} = 68.67 \quad \text{ok}$$

PROBLEMS TO BE SOLVED

10.1 through 10.5. Calculate the values S, Z, and shape factor for the following shapes with respect to a centroidal horizontal axis.

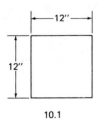

10.1

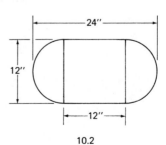

10.2

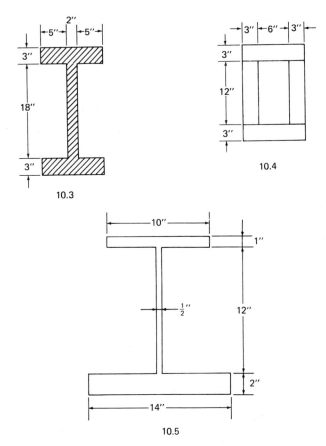

10.6 and 10.7. Draw the elastic and plastic moment diagrams for the following beams loaded as shown. Ignore load factors.

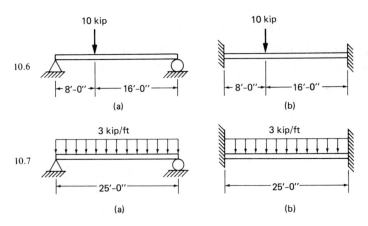

10.8. Calculate the plastic moment M_p for a beam loaded as shown, assuming (a) simple supports at both ends, (b) the left support hinged and the other fixed, and (c) both ends fixed.

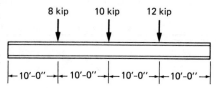

10.9. Calculate the plastic moment for the beam shown.

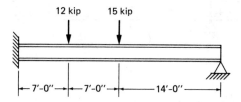

10.10. Calculate the plastic moment for the beam shown, using the principles of virtual work.

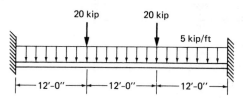

10.11 and 10.12. For the loadings shown, select the lightest beams according to elastic and plastic theories. Assume full lateral support, and neglect the weight of the beam.

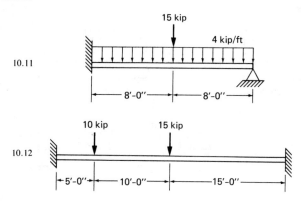

10.13 through 10.15. Select a continuous wide-flange section, using plastic design that is satisfactory for the entire structure.

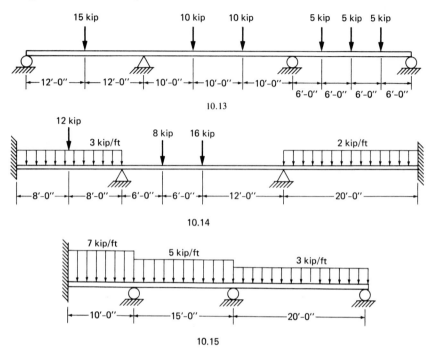

10.16 and 10.17. For the loadings shown determine:
a) A single rolled section satisfactory for this beam using elastic analysis and moment redistribution.
b) A single rolled section satisfactory for this beam using plastic analysis, and
c) A section for the span with the least moment, and design cover plates as necessary.

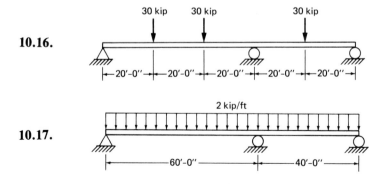

Appendix A
Repeated Loadings (Fatigue)

It has been shown experimentally that the failure stress of a steel member subjected to repeated loading and unloading depends on the number of load applications and the magnitude of the stress range (i.e., the algebraic difference between the maximum and minimum stresses). When the number of loading cycles anticipated over the lifetime of the member is below 20,000, the failure stress will have the same magnitude as the one obtained for the same member loaded statically. However, it has also been shown that when the number of cycles is over 20,000 and/or the difference between the maximum and minimum stresses is large, failure stress dramatically decreases. Furthermore, the configuration of the loaded member, as well as the type of connection (riveted, high-strength bolted, or welded), has a very strong influence on the magnitude of the failure load. When a member fails due to the application of repeated loading, this is commonly referred to as *fatigue failure*.

The provisions for fatigue design are given in AISCS Appendix K. The number of loading cycles expected to occur over the life of the member is divided into categories 1 through 4, as shown in Table A-K4.1. As noted above, fatigue need not be considered if the number of cycles is less than 20,000. In Table A-K4.2, various stress categories are established. Depending on the type of member, connections, and stress applications, a stress category ranging from A to F is assigned. Note that in all cases, the types of stress ranges included are tension, tension and compression, and shear. In no case is there a stress range completely in compression. Experiments have revealed that stress ranges that do not involve tensile or shear stresses cannot propagate cracks in the steel that would lead to eventual fracture (AISCS Commentary K4).[1] Figure A-K4.1 illustrates the different stress categories.

---

[1] When subjected to repeated cycles of tensile or shear stresses, fatigue cracks will eventually form in the microstructure of the steel. As the number of stress cycles increases, these cracks propagate and coalesce, which ultimately results in the overall fracture of the material.

460 STEEL DESIGN FOR ENGINEERS AND ARCHITECTS

Once a loading condition and a stress category are established, the allowable stress range (which includes a factor of safety) can be determined from Table A-K4.3. As can be seen from the table, the allowable stress range can be quite small. When designing the member, the maximum stress should not exceed the applicable basic allowable stress provided in AISCS Chapters D through K, and the maximum stress range should not exceed the one given in Table A-K4.3 (Appendix K4.2). The provisions for the tensile fatigue loading of A325 and A490 bolts are given in Appendix K4.3.

For structural members in conventional buildings, the number of load cycles (due to the change in the live load) is usually small, so that fatigue design is not required. However, if the member supports machinery, a crane, etc. (such as in an industrial building), fatigue may be a very important design consideration. Fatigue is also important in bridge design, where the number of load cycles may be very large. In general, when the ratio of dead load to live load is large, the effects of fatigue are not significant; however, if this ratio is small, the effects of fatigue may be large enough to govern the design. The following examples will illustrate some typical situations involving fatigue.

Example A.1. A 2 L 4 × 4 × $\frac{1}{2}$ is subjected to the loading shown. Determine the lengths l_1 and l_2 (neglect returns), and check the member for both static and fatigue loading if the number of cycles is 300,000 (i.e., 48 cycles/day, 250 working days per year, for a lifetime of 25 years).

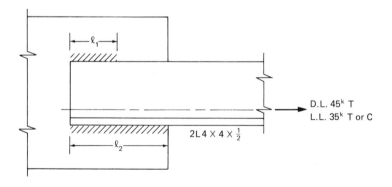

Solution.

- Static Loading

Maximum load = 45 k + 35 k = 80 k

For a ¼-in. weld, required total weld length $2l$ is

$$2l = \frac{80}{4 \times 0.928} = 21.55 \text{ in.}$$

$$l = 10.78 \text{ in. per angle}$$

Assuming $l_1 = l_2$, then $l_1 = l_2 = 10.78$ in./$2 = 5.39$ in.; say $5\frac{1}{2}$ in.

Note that the center of gravity of the welds does not coincide with the center of gravity of the member; for statically loaded members, the effects of this eccentricity may be neglected (AISCS J1.9).

Check the tensile capacity of the member:

$$A_g = 7.50 \text{ in.}^2; \quad A_e = 0.85 \times 7.5 = 6.38 \text{ in.}^2$$

$$P_{all} = \begin{cases} 0.6F_y A_g = 22 \text{ ksi} \times 7.50 \text{ in.}^2 = 165 \text{ k} \quad \text{(governs)} \\ 0.5F_u A_e = 29 \text{ ksi} \times 6.38 \text{ in.}^2 = 185 \text{ k} \end{cases}$$

$$P_{all} = 165 \text{ k} > 80 \text{ k} \quad \text{ok}$$

- Fatigue Loading

Balance the welds around the center of gravity of the angle for fatigue loading (AISCS J1.9).

Sum moments about the outstanding leg:

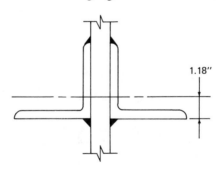

$$2 \times [1.18 \text{ in.} \times (l_1 + l_2)] = 2 \times [4 \text{ in.} \times l_1]$$

Loading condition 2 for 300,000 cycles (Table A-K4.1)

Stress category F for fillet welds (Table A-K4.2; see illustrative example no. 17)

462 STEEL DESIGN FOR ENGINEERS AND ARCHITECTS

Allowable stress range for weld = 12 ksi (Table A-K4.3)

$$F_w = 0.928 \times (12 \text{ ksi}/21 \text{ ksi}) = 0.530 \text{ k/in./sixteenth}$$

Design loads:

$$\text{Max. DL + LL} = 45 \text{ k} + 35 \text{ k} = 80 \text{ k} \quad T$$
$$\text{Min. DL + LL} = 45 \text{ k} - 35 \text{ k} = 10 \text{ k} \quad T$$
$$\text{Load range} = 80 \text{ k} - 10 \text{ k} = 70 \text{ k}$$

Use maximum weld size of $\frac{7}{16}$ in. to obtain minimum total weld length $2l$:

$$2l = \frac{70 \text{ k}}{0.530 \times 7} = 18.87 \text{ in.}$$

$$l = l_1 + l_2 = 9.44 \text{ in. per angle}$$

$$1.18 \text{ in.} \times 9.44 \text{ in.} = 4 \text{ in.} \times l_1 \rightarrow l_1 = 2.79 \text{ in.} \quad \text{say } 2\tfrac{7}{8} \text{ in.}$$

$$l_2 = 9.44 \text{ in.} - 2.79 \text{ in.} = 6.65 \text{ in.} \quad \text{say } 6\tfrac{3}{4} \text{ in.}$$

Check weld length for statically loaded angles:

$$2 \times (2.875 \text{ in.} + 6.75 \text{ in.}) \times 0.928 \times 7 = 125 \text{ k} > 80 \text{ k} \quad \text{ok}$$

Check base metal:

Stress category E for base metal (see illustrative example no. 17)

Allowable stress range for base metal = 13 ksi

$$f_t = \frac{70 \text{ k}}{0.85 \times 7.5 \text{ in.}^2} = 10.98 \text{ ksi} < 13 \text{ ksi} \quad \text{ok}$$

Example A.2. A 1 in. × 3 in. plate (A588 Gr 50) is welded to a heavy member, as shown. Determine the maximum allowable P that can be applied to the $\frac{5}{16}$-in. welds (E60 electrodes) for a) $\theta = 0°$, b) $\theta = 90°$, and c) $\theta = 30°$. In all cases, consider only fatigue loading. The number of load cycles is 300,000, and the load can range from $+P$ to $-P$.

REPEATED LOADINGS (FATIGUE)   463

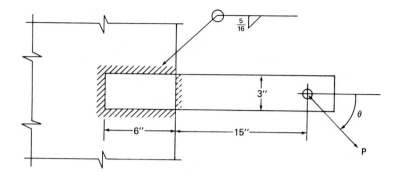

Solution.

a) $\theta = 0°$
   For the fillet weld:
   Loading Condition 2
   Stress Category F
   Allowable stress range = 12 ksi

$$F_w = \frac{12 \text{ ksi}}{0.3 \times 60 \text{ ksi}} \times 0.795 \text{ k/in./sixteenth} \times 5 = 2.65 \text{ k/in.}$$

$$P_{\text{range}} = P - (-P) = 2P$$

$$2P = 2.65 \text{ k/in.} \times 18 \text{ in.} = 47.7 \text{ k}$$

$$P = 23.85 \text{ k}$$

b) $\theta = 90°$
   Use elastic method (Alternate Method 1) and compare with Ultimate Strength Method (p. 4-57 AISCM).

Elastic method:
For a 1-in. weld,

$$I_x = 2\left[\frac{1}{12} \times (3 \text{ in.})^3 + 6 \text{ in.} \times (1.5 \text{ in.})^2\right] = 31.5 \text{ in.}^4$$

$$I_y = 2\left[\frac{1}{12} \times (6 \text{ in.})^3 + 3 \text{ in.} \times (3.0 \text{ in.})^2\right] = 90.0 \text{ in.}^4$$

$$J = I_x + I_y = 121.5 \text{ in.}^4$$

464 STEEL DESIGN FOR ENGINEERS AND ARCHITECTS

$$M = P_{\text{range}} \times 18 \text{ in.} = 2 \times P \times 18 \text{ in.} = 36 \times P$$

$$f_h = \frac{36P \times 1.5}{121.5} = 0.444P$$

$$f_v = \frac{2P}{18} + \frac{36P \times 3}{121.5} = 1.0P$$

$$f_r = \sqrt{(0.444P)^2 + (1.0P)^2} = 1.094P$$

$$2.65 \text{ k/in.} = 1.094P \rightarrow P = 2.42 \text{ k}$$

Ultimate Strength Method:
Use Table XXII, p. 4-78 AISCM with a = 3 and k = 0.5

$$C = \frac{12 \text{ ksi}}{18 \text{ ksi}} \times 0.330 = 0.22$$

$$2P = CC_1Dl = 0.22 \times 0.857 \times 5 \times 6 = 5.66 \text{ k}$$

$$P = 2.83 \text{ k}$$

c) $\theta = 30°$
Use elastic method of analysis (Alternate Method 1):

$$P_h = 0.866 \times P_{\text{range}} = 1.732P$$

$$P_v = 0.5 \times P_{\text{range}} = 1.0P; \quad M = 1.0P \times 18 = 18P$$

$$f_h = \frac{1.732P}{18} + \frac{18P \times 1.5}{121.5} = 0.3184P$$

$$f_v = \frac{1.0P}{18} + \frac{18P \times 3}{121.5} = 0.50P$$

$$f_r = \sqrt{(0.3184P)^2 + (0.50P)^2} = 0.5928P$$

$$2.65 \text{ k/in.} = 0.5928P \rightarrow P = 4.47 \text{ k}$$

Example A.3. A water tank is supported by two W 16 × 57 beams (A572 Gr 50), as shown. The beams were originally designed for a total weight of 68 kips (10 k for the weight of the empty tank and 58 k for the weight of the water). Due to factory upgrades, the existing tank is to be replaced with a new tank weighing 10 kips which has a capacity to carry water weighing 130 kips. The tank is to be filled and emptied 25 times per day. Determine the required reinforcement for the existing W 16 × 57 beams, assuming that the top flanges are adequately braced the full length of the beams. Assume 250 work days per year and a 25-year life for the structure.

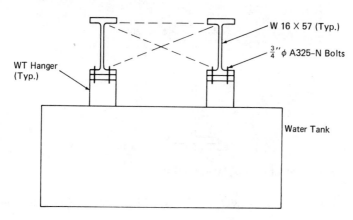

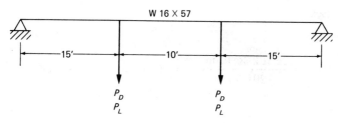

Solution.

- Static Loading:
Original loading:

$$P_D = 10 \text{ k}/4 = 2.5 \text{ k}$$

$$P_L = 58 \text{ k}/4 = 14.5 \text{ k}$$

$$M_{max} = (2.5 \text{ k} + 14.5 \text{ k}) \times 15 \text{ ft} = 255 \text{ ft-k}, \quad S = 92.2 \text{ in.}^3$$

$$f_b = \frac{255 \text{ ft-k} \times 12 \text{ in./ft}}{92.2 \text{ in.}^3} = 33.19 \text{ ksi} \cong 0.66 \times 50 \text{ ksi} = 33 \text{ ksi} \quad \text{ok}$$

New loading:

$$P_D = 10 \text{ k}/4 = 2.5 \text{ k}$$

$$P_L = 130 \text{ k}/4 = 32.5 \text{ k}$$

$$M_{max} = (2.5 \text{ k} + 32.5 \text{ k}) \times 15 \text{ ft} = 525 \text{ ft-k}$$

The existing beams must be reinforced with cover plates to carry the new loads. After some trials, assume the following cover plates:[2]

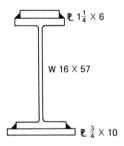

Section properties of cover-plated section:

$\bar{y} = 9.02$ in., $I = 1897.4$ in.^{4}, $S_t = 201.6$ in.^{3}, $S_b = 210.4$ in.^{3}

Check stresses:
Top:

$$f_b = \frac{525 \text{ ft-k} \times 12 \text{ in./ft}}{201.6 \text{ in.}^3} = 31.25 \text{ ksi} < 33 \text{ ksi} \quad \text{ok}$$

Bottom:

$$f_b = \frac{525 \text{ ft-k} \times 12 \text{ in./ft}}{210.4 \text{ in.}^3} = 29.94 \text{ ksi} < 33 \text{ ksi} \quad \text{ok}$$

Determine the theoretical cutoff points for the plates:

$$M_{\text{cutoff}} = 92.2 \text{ in.}^3 \times \frac{33 \text{ ksi}}{12 \text{ in./ft}} = 253.6 \text{ ft-k}$$

$$\frac{x}{253.6} = \frac{15}{525} \rightarrow x = 7.25 \text{ ft i.e., 7 ft, 3 in. from the support}$$

Determine required fillet welds:

$$q = \frac{VQ}{I}$$

---

[2]The widths of the new cover plates were selected as shown so that overhead welding to the existing beams would not be required.

$Q = (0.75 \times 10) \times (9.02 - 0.375) = 64.84$ in.^{3} (tension cover plate)

$$q = \frac{35 \text{ k} \times 64.84 \text{ in.}^3}{1897.4 \text{ in.}^4} = 1.2 \text{ k/in.}$$

For continuous $\frac{5}{16}$-in. fillet welds:

$$F_w = 2 \times 5 \times 0.928 = 9.28 \text{ k/in.} > 1.2 \text{ k/in.} \quad \text{ok}$$

(Note: An intermittent fillet weld is not used due to fatigue, as will be discussed below.)

Check the force in the termination region:

$$H = \frac{MQ}{I} = \frac{253.6 \text{ ft-k} \times 12 \text{ in./ft} \times 64.84 \text{ in.}^3}{1897.4 \text{ in.}^4} = 104 \text{ k}$$

Assume $a' = 2.0 \times 7.12$ in. $= 14.24$ in.   say 14.5 in.

$H_{\text{weld}} = (2 \times 14.5 \text{ in.}) \times 0.928 \times 5 = 134.6$ k

$H_{\text{weld}} > H$ ok

Total length of cover plate $= 40 - (2 \times 7.25) + \left[2 \times \left(\frac{14.5}{12} \right) \right]$

$= 27.92$ ft   say 28 ft centered on the span

Stress at end of cover plate:

$$M = (7.25 - 1.21) \times \frac{525}{15} = 211.4 \text{ ft-k}$$

$$f_b = \frac{211.4 \text{ ft-k} \times 12 \text{ in./ft}}{92.2 \text{ in.}^3} = 27.51 \text{ ksi}$$

- Design for Fatigue
  No. cycles = 25 cycles/day × 250 days/yr × 25 yr. life-time = 156,250
  From Table A-K4.1, this number of cycles corresponds to Loading Condition 2
  From Table A-K4.2:

For built-up member, base metal: stress category B (illust. ex. 5)

468 STEEL DESIGN FOR ENGINEERS AND ARCHITECTS

For built-up member, base metal at ends of partial length welded cover plates: stress category E′
For fillet weld metal: stress category F

Allowable stress ranges (Table A-K4.3):
  Category B: 29 ksi[3]
  E′: 9 ksi
  F: 12 ksi

Stress at midspan for full DL + LL: f_b = 29.94 ksi (see above)
Stress at midspan for DL only:

$$f_b = \frac{2.5 \text{ k} \times 15 \text{ ft} \times 12 \text{ in./ft}}{210.4 \text{ in.}^3} = 2.14 \text{ ksi}$$

Stress range at midspan = 29.94 ksi − 2.14 ksi = 27.80 ksi

27.80 ksi < Allowable stress range Category B = 29.0 ksi ok

$$\text{Stress range at end of plate} = 27.51 \text{ ksi} \times \frac{32.5 \text{ k}}{35.0 \text{ k}} = 25.55 \text{ ksi}$$

25.55 ksi > Allowable stress range Category E′ = 9.0 ksi N.G.

It is clear that the cover plate cannot be terminated at this point; thus, either the cover plate must be made much heavier or the cutoff point must be moved much closer to the support. Determine the cutoff point of the plate given an allowable stress range of 9 ksi:

$$S \text{ (W 16} \times \text{57)} = 92.2 \text{ in.}^3$$

$$M_{\text{range}} = \frac{92.2 \text{ in.}^3 \times 9 \text{ ksi}}{12 \text{ in./ft}} = 69.2 \text{ ft-k}$$

Let X be the distance from the support to the cutoff point.

$$32.5 \text{ k} \times X = M_{\text{range}} = 69.2 \text{ ft-k}$$

$$X = 2.13 \text{ ft from the support.}$$

Hence, extend the bottom cover plate the full length of the beam.

---

[3]This allowable stress range is for the case when continuous fillet welds are used to connect the plate to the beam. Intermittent fillet welds are not acceptable, since at the point where the weld is interrupted, the stress range goes to E or E′, which results in a significant decrease in the allowable value.

Check the continuous fillet weld:

$$F_w = 0.928 \times \frac{12 \text{ ksi}}{21 \text{ ksi}} = 0.53 \text{ k/in./sixteenth} \times 5$$

$$= 2.65 \text{ k/in.}$$

$$q_{LL} = \frac{V_{LL}Q}{I} = \frac{32.5 \text{ k} \times 64.84 \text{ in.}^3}{1897.4 \text{ in.}^4} = 1.11 \text{ k/in.}$$

2×2.65 k/in. $= 5.3$ k/in. > 1.11 k/in.   ok

Hanger design (see Chap. 7):

Design the hangers for the full static load and then check the provisions in AISCS Appendix K4.3. Assume A36 steel for the hangers.

Use $\frac{3}{4}$-in. A325-N bolts (minimum practical size for such elements).

Select a preliminary size of tee flange by using the table provided on p. 4-89 of the AISCM:

Try $b = 2$ in. and assume 7-in. fitting length.

$B = 19.4$ k (Table I-A, p. 4-3 AISCM)

$$\text{No. of bolts required} = \frac{35 \text{ k}}{19.4 \text{ k}}$$

$$= 1.8 \quad \text{use four bolts (minimum practical number)}$$

$$T = \frac{35 \text{ k}}{4} = 8.75 \text{ k} < 19.4 \text{ k} \quad \text{ok}$$

$$\text{Load per linear in.} = \frac{2 \times 8.75 \text{ k}}{(7 \text{ in.}/2)} = 5.0 \text{ k/in.}$$

From the preliminary selection table with $b = 2$ in. and load $= 5.0$ k/in., the thickness of the flange of the tee should be approximately 0.75 in. Tentatively select WT 9 $\times$ 32.5, $t_f = 0.75$ in., $t_w = 0.45$ in., $b_f = 7.59$ in. Assume 4-in. beam gage.

$$b = \frac{4 \text{ in.} - 0.45 \text{ in.}}{2} = 1.775 \text{ in.} > 1\frac{1}{4} \text{ in. tightening clearance}$$
$$\text{(see p. 4-137)}$$

$$a = \frac{7.59 \text{ in.} - 4 \text{ in.}}{2} = 1.795 \text{ in.}$$

$1.25b = 2.219$ in. > 1.795 in.   use $a = 1.795$ in.

$b' = 1.775$ in. $- (0.75$ in. $/2) = 1.400$ in.

$a' = 1.795$ in. $+ (0.75$ in. $/2) = 2.170$ in.

$p = 7.0$ in. $/2 = 3.5$ in.

$d' = 0.75$ in. $+ 0.0625$ in. $= 0.8125$ in.

$\delta = 1 - (0.8125$ in. $/3.5$ in.$) = 0.768$

$\rho = 1.40$ in. $/2.170$ in. $= 0.645$

$$\beta = \frac{1}{0.645}\left(\frac{19.4 \text{ k}}{8.75 \text{ k}} - 1\right) = 1.89 > 1 \rightarrow \alpha' = 1$$

$$t_{req} = \sqrt{\frac{8 \times 8.75 \text{ k} \times 1.40 \text{ in.}}{3.5 \text{ in.} \times 36 \text{ ksi}(1 + 0.768)}} = 0.66 \text{ in.} < 0.75 \text{ in.} \text{ ok}$$

Note that a WT 9×30 would probably be satisfactory in this case as well; however, the savings in weight for such small hangers would be negligible.

Use WT 9×32.5 hangers.

Determine the prying force Q:

$$t_c = \sqrt{\frac{8 \times 19.4 \times 1.40 \text{ in.}}{3.5 \text{ in.} \times 36 \text{ ksi}}} = 1.313 \text{ in.}$$

$$\alpha = \frac{1}{0.768}\left[\frac{8.75 \text{ k}/19.4 \text{ k}}{(0.75 \text{ in.}/1.313 \text{ in.})^2} - 1\right] = 0.498$$

$$Q = 19.4 \text{ k} \times 0.768 \times 0.498 \times 0.645 \times \left(\frac{0.75 \text{ in.}}{1.313 \text{ in.}}\right)^2 = 1.56 \text{ k}$$

Check the requirements of AISCS Appendix K4.3, p. 5-107:

Total load acting on one bolt $= 8.75$ k $+ 1.56$ k $= 10.31$ k

$$f_{bolt} = \frac{10.31 \text{ k}}{0.4418 \text{ in.}^2} = 23.34 \text{ ksi} < 40 \text{ ksi} \text{ ok}$$

Also, 1.56 k $< 0.60 \times 8.75$ k $= 5.25$ k ok

Note that in lieu of the above calculations, the prying effect can be eliminated by putting thick enough washers between the surfaces in contact. The four $\frac{3}{4}$-in.-diameter A325-N bolts determined above would work in this case as well.

Appendix B
Highway Steel Bridge Design

B.1. OVERVIEW OF HIGHWAY STEEL BRIDGE DESIGN AND THE AASHTO CODE

The main intention of this appendix is not to give a comprehensive coverage of highway steel bridge design but to familiarize the reader with the basic differences between building and bridge design. The code which will be considered here is the "Standard Specifications for Highway Bridges" of the American Association of State Highway and Transportation Officials (AASHTO). In general, the design provisions of the AISC and AASHTO are similar. However, some basic differences in allowable stresses, limiting thicknesses and sizes, etc. are evident. This section will present a general overview of these differences, and the order of presentation will follow Sect. 10 of the AASHTO code. The loads and their distribution (Sect. 3 of AASHTO) will be presented in Sect. B.1.5. Throughout this appendix, reference is made to the applicable section numbers and equations from the AASHTO code.[1]

B.1.1. General Requirements and Materials (AASHTO Sect. 10, Part A). The designations of the materials follows the AASHTO M listing given in Table 10.2A. The equivalent ASTM (AISC) designations for these materials are also given.

B.1.2. Design Details (AASHTO Sect. 10, Part B).

B.1.2.1. Repetitive Loading and Toughness Considerations (AASHTO 10.3). In AASHTO Table 10.3.2A, the number of stress cycles to be considered in the design are given, depending on the type of loading and the type of

---

[1]"Standard Specifications for Highway Bridges," Copyright 1989. The American Association of State Highway and Transportation Officials, Washington, D.C. Used by permission.

roadway. Table A-K4.3 of the AISC is almost identical to the allowable stress range Table 10.3.1A in AASHTO for redundant load path structures. Also given in Table 10.3.1A are the allowable stress ranges for nonredundant load path structures which are about 20% smaller than the previous ones. By definition, a nonredundant load path structure is one in which failure of a single element could cause collapse. Also, slight differences occur in the stress category classifications given in AISC Table A-K4.2 and AASHTO Table 10.3.1B.

Members subjected to tensile stress capacity require special Charpy V-Notch impact properties (AASHTO 10.3.3). No such requirement is given in the AISC.

B.1.2.2. Depth Ratios and Deflections (AASHTO 10.5, 10.6). In AISCS L3, suggested maximum live load deflections are provided to prevent cracking of plaster ceilings. Typically, the maximum live load deflection should not exceed $\frac{1}{360}$ of the span. According to AASHTO, the ratio of the depth of the beam to the length of the span should not be less than $\frac{1}{25}$ for beams or girders; for composite girders, the ratio of the overall depth (slab plus steel) to the length should preferably not be less than $\frac{1}{25}$, and the ratio of the steel girder alone to the length of the span should preferably not be less than $\frac{1}{30}$. In all cases, the deflection due to service live loads plus impact should not exceed $\frac{1}{800}$ of the span.

B.1.2.3. Limiting Lengths of Members (AASHTO 10.7). According to AASHTO, the limiting value of the slenderness ratio KL/r for compression members is 120 for main members and 140 for secondary members. In AISCS B7, this ratio is suggested to be limited to 200 for all compression members. As far as tension members are concerned, the ratio L/r is required to be less than 200 for main members in AASHTO 10.7.5, while AISCS B7 suggests a limit of 300.

B.1.2.4. Minimum Thickness of Metal (AASHTO 10.8). In general, the minimum thickness of metal is $\frac{5}{16}$ in. unless other considerations (e.g., minimum thickness ratio) control.

B.1.2.5. Effective Area of Angles and Tee Sections in Tension (AASHTO 10.9). For a tee section, a single angle, or a double angle section with outstanding legs connected together, the effective area is the net area of the connected leg or flange plus one-half of the area of the outstanding leg (AASHTO 10.9.1). This provision is quite different from that of the AISCS (AISCS B3) but may still give very similar results.

B.1.2.6. Outstanding Legs of Angles (AASTHO 10.10). The widths of outstanding legs of angles in compression should not exceed 12 times the thickness of the angle in main members nor 16 times the thickness in secondary members. The corresponding limitations in the AISC are given in Table B5.1.

B.1.2.7. Cover Plates (AASHTO 10.13). The maximum thickness of a cover plate shall be less than two times the thickness of the flange to which it is attached. AISCS B10 requires that for plates which are bolted or riveted to the beam flange, the total cross-sectional area of the cover plates should not exceed 70% of the total flange area.

B.1.2.8. Splices (AASHTO 10.18). According to the service load design method, splices should be designed for the average of the calculated design stress at the point of the splice and the allowable stress of the member at the same point, or 75% of the allowable stress in the member, whichever is larger. This AASHTO provision is quite different from the one given in AISCS J7, which requires that the splice develop the strength required by the stresses at the point of the splice. In the case of splices created by welding of material of different widths or thicknesses, transition details are given in AASHTO Fig. 10.18.5A.

B.1.2.9. Diaphragms and Cross Frames (AASHTO 10.20). Special provisions are made for the location and depth of diaphragms. Empirical formulas are given for the additional stresses induced in the main member's flanges due to the application of wind through the diaphragm. See Sect. B.2 for more details.

B.1.2.10. Fasteners (Rivets and Bolts) (AASHTO 10.24). Bearing-type connections using high-strength bolts are limited to members subjected to compression and to secondary members. All other high-strength connections shall be SC connections. This is very different from the provision of the AISCS, which allows bearing-type connections to be used in all situations except in the termination zone of partial length cover plates (AISCS B10). The AISCS permits the use of A307 bolts in limited cases. The spacing and edge distance requirements given in AASHTO 10.24.5 and 10.24.7 are somewhat larger than those given in AISCS J3.

B.1.3. Service Load Design Method (AASHTO Sect. 10, Part C).

B.1.3.1. Allowable Stresses in Steel (AASHTO 10.32.1). The allowable stresses are summarized as follows (see Table 10.32.1A):

- Tension: $0.55 F_y$ on gross area
  $0.46 F_u$ on net area
  These allowable values are about 10% less than those given in AISCS D1.

- Compression in a concentrically loaded column:

$$\text{when } \frac{KL}{r} \leq C_c: \quad F_a = \frac{F_y}{2.12}\left[1 - \frac{(KL/r)^2}{2C_c^2}\right]$$

$$\text{when } \frac{KL}{r} > C_c: \quad F_a = \frac{\pi^2 E}{2.12(KL/r)^2} \quad \text{where } C_c = \sqrt{\frac{2\pi^2 E}{F_y}}$$

These equations have a uniform factor of safety of 2.12, which is different from that of the AISCS, where the factor of safety varies from $\frac{5}{3}$ to $\frac{23}{12}$ (AISCS E2). Essentially, $(\frac{23}{12}) \times 1.10 = 2.11$, which indicates that although a uniform factor of safety has been given in AASHTO, it has also been increased by about 10%. Note that in situations where A36 steel is used, the slender column case cannot be reached, since KL/r is limited to 120, which is less than $C_c = 126.1$. Values for F_a are shown graphically in Appendix C of AASHTO.

- Compression in extreme fibers of flexural members:

full lateral support: $0.55\ F_y$ (far below $0.66\ F_y$ given in AISCS F1.1)
partial support or unsupported: a special formula taking into consideration lateral-torsional buckling is provided.

- Shear in girder webs on the gross section: $0.33\ F_y$ (compared to $0.40\ F_y$ given in AISCS F4)
- Bearing on milled stiffeners: $0.80\ F_y$ (compared to $0.90\ F_y$ given in AISCS J8)

B.1.3.2. Weld Metal (AASHTO 10.32.2). The provisions can be summarized as follows:

$$F_y(\text{weld metal}) \geq F_y(\text{base metal})$$

$$F_u(\text{weld metal}) \geq F_u(\text{base metal})$$

For fillet welds, F_v = allowable stress = $0.27\ F_u$, which is 10% less than the allowable stress given in AISCS Table J2.5.

B.1.3.3. Allowable Stress on Bolts (AASHTO 10.32.3). Bearing-type connectors have 10% less allowable capacity in shear and tension (Table 10.32.3B) than those given in the AISCS (Table J3.2). For SC connectors, the same 10% is true for shear.

B.1.3.4. Applied Tension, Combined Tension and Shear (AASHTO 10.32.3.3). For the combination of tension and shear in SC and bearing-type connectors, completely different expressions from those given in AISCS Table J3.3 are used. The use of the AASHTO equation for prying forces on a bolt is much simpler than the procedure given in the AISCM (see p. 4-89).

B.1.3.5. Fatigue (AASHTO 10.32.3.4). The allowable design stresses for tensile fatigue loading of high-strength bolts given in this section should be greater than the tensile stresses due to the application of the service load plus the prying force resulting from the application of the service load. Additionally, it is required that the prying force shall not exceed 80% of the externally applied load. Similar provisions are given in AISCS Appendix K4.3. The allowable stresses given are about 12% larger than the AASHTO ones, and the prying force should not exceed 60% of the externally applied load.

B.1.3.6. Bearing on Masonry (AASHTO 10.32.6). The allowable bearing pressure on the concrete is less than the one given in AISCS J8. Also, the same edge distances are recommended for the case of bearing on less than the full area of the concrete support.

B.1.3.7. Plate Girders (AASHTO 10.34). The design and sizing of plate girders follow requirements similar to the ones given in AISCS. In particular, limiting width-thickness ratios, minimum-web thickness, etc. are more conservative in AASHTO. Allowable stresses, as noted earlier, are more conservative as well.

B.1.3.8. Stiffeners (AASHTO 10.34.4, 10.34.6). The expressions for spacing and sizing both bearing and transverse stiffeners are similar to the ones given in AISCS K1; however, lower allowable stresses are required in AASHTO. Note that longitudinal stiffeners are treated in AASHTO 10.34.3.2 and are not used in the AISCS.

B.1.3.9. Combined Stresses (AASHTO 10.36). Equations (10-41) and (10-42) are identical to eqns. (H1-1) and (H1-2) in the AISCS, with one exception: the denominator of the first term in (10-42) is $0.472 \, F_y$ compared to $0.60 \, F_y$ in (H1-2). Note that eqn. (H1-3) has been omitted in AASHTO. The term F'_e in eqn. (10-43) is the same as the one given in AISCS, except that the factor of safety is 2.12 instead of $\frac{23}{12}$. The C_m coefficients are tabulated in Table 10.36A.

B.1.3.10. Composite Girders (AASHTO 10.38). The values of the modular ratio n are specified in integers for various ranges of the concrete compressive strength (AASHTO 10.38.1.3). The effect of creep is specifically considered in

Sect. 10.38.1.4, where it is stated that the modular ratio n must be multiplied by a factor of 3. Elaboration on this increase in n will be given in Sect. B.2.

The effective flange width of the concrete may be governed by the thickness of the slab (AASHTO 10.38.3); in the AISCS, the effective width does not depend on the slab thickness (AISCS I1).

Because fatigue loading is always present, the shear connectors must be verified for both fatigue and ultimate strength (AASHTO 10.38.5), while in the AISCS it is required that only the ultimate capacity be checked (AISCS I4).

Stresses and deflections must be checked for the following cases: 1) dead load (the steel alone resisting its own weight plus the weight of the wet concrete), 2) superimposed dead load (the composite section, based on $3n$, resisting the superimposed loads), and 3) live load (the composite section, based on n, resisting the live loads, which include the impact factor, the distribution factor, and the multiple loaded lanes reduction factor). See Sect. B.2 for more detailed information.

B.1.4. Strength Design Method (AASHTO Sect. 10, Part D). This method can be compared to the Load and Resistance Factor Design (LRFD) method; for this reason, it will not be discussed here.

B.1.5. Loads (AASHTO Sect. 3). In general, a bridge should be designed for dead loads, live loads (including impact), wind loads, and other applicable forces, as discussed in AASHTO 3.2. Only the dead and live loads given in AASHTO 3.3 and 3.4, respectively, will be discussed here, since the effects of the other loads are beyond the scope of this text. The dead load consists of the weight of the entire structure, which should include the roadway (deck), future resurfacing, sidewalks, supporting members, and any other permanent attachments to the bridge. The live loads, which are very different from those encountered in building design, consist of truck loading and lane loading (AASHTO 3.7).

Four standard classes of highway live loadings are given in AASHTO 3.7.2: H20-44, H15-44, HS20-44, and HS15-44. The H loading consists of a truck with two axles or a corresponding lane load. The number following the H is the gross weight in tons of the standard truck, and the second number refers to the year 1944, when these trucks were standardized. The truck loadings for the H type trucks are shown in Fig. B.1a. The HS loading is a tractor truck with a semitrailer having a total of three axles or a corresponding lane loading. For the HS truck loading, the number which follows the HS indicates the gross weight in tons of the tractor truck (i.e., the first two axles). The third axle is assumed to carry a load equal to 80% of the weight of the tractor truck, which is the weight of the second axle (see Fig. B.1b).

The lane loadings associated with the four standard classes are shown in Fig. B.2. A lane loading, consisting of a uniform load per linear foot of lane and a single concentrated load for a simple span,[2] simulates a truck train and approximately represents the fully loaded bridge. As shown in the figure, the magnitude of the concentrated load depends on whether shears or moments are computed; it is to be placed so as to produce maximum stresses.

According to AASHTO 3.6, standard truck or lane loadings are assumed to occupy a width of 10 ft. These loads are to be placed in 12-ft-wide design traffic lanes, which are to be spaced across the entire bridge width (measured between curbs) in number and position so as to produce the maximum stress in the member under consideration.

The type of live loading (truck or lane) which should be used is the one which produces the maximum stress on the member. Considering that all of the lanes will not be fully loaded at the same time (statistically speaking), the governing live load can be reduced by 10% for a bridge with three lanes and by 25% for one with four or more lanes loaded simultaneously (AASHTO 3.12). In Appendix A of the AASHTO code, tables are provided which give the unreduced maximum reactions and maximum moments corresponding to the governing truck or lane loading for simple spans (one lane).

To account for the increase in stresses (i.e., impact effect) due to the sudden application of the moving loads, the applicable live load must be increased by the factor $(I + 1)$, with I given in AASHTO eqn. (3-1):

$$I = \frac{50}{L + 125} \leq 0.30$$

where I = impact factor and L = length in feet of the portion of the span that is loaded to produce the maximum stress in the member.

When a concentrated load from a truck wheel acts on a bridge deck, this load is distributed over an area larger than the contact area between the wheel and the deck. The distribution of the load described above is given in AASHTO 3.23. For shear, the lateral distribution of the wheel load shall be obtained by assuming that the flooring acts as a simple span between stringers. For the bending moments of an interior stringer, the fraction of a wheel load (both front and rear) shall be applied to the stringer. It shall be $S/7.0$ (where S is the stringer spacing in feet) for one traffic lane or $S/5.5$ for two or more traffic

---

[2]For a continuous beam, the lane loading should consist of the uniform distributed load and the two concentrated loads (one per span), all positioned so as to produce the maximum negative moment at an interior support. To obtain the maximum positive moment, the uniform load and one concentrated load should be positioned accordingly.

478 STEEL DESIGN FOR ENGINEERS AND ARCHITECTS

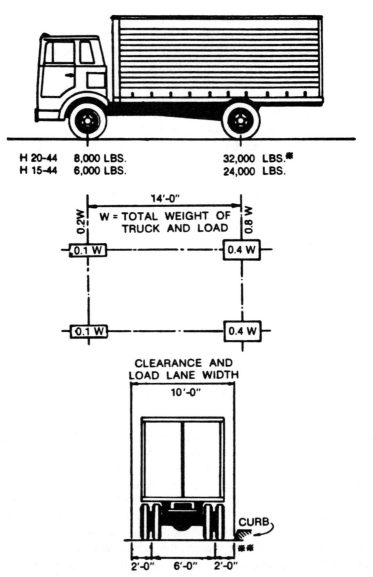

*In the design of timber floors and orthotropic steel decks (excluding transverse beams) for HS 20 loading, one axle load of 24,000 pounds or two axle loads of 16,000 pounds each, spaced 4 feet apart may be used, whichever produces the greater stress, instead of the 32,000-pound axle shown.

**For slab design, the center line of wheels shall be assumed to be 1 foot from face of curb. (See Article 3.24.2.)

a.

Fig. B.1. Standard truck loading: (a) H type; (b) HS type (AASHTO Figs. 3.7.6A and 3.7.7A; "Standard Specifications for Highway Bridges," Copyright 1989. The American Association of State Highway and Transportation Officials, Washington, D.C. Used by permission).

HIGHWAY STEEL BRIDGE DESIGN 479

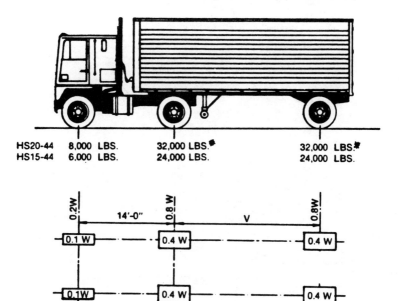

W = COMBINED WEIGHT ON THE FIRST TWO AXLES WHICH IS THE SAME AS FOR THE CORRESPONDING H TRUCK.
V = VARIABLE SPACING — 14 FEET TO 30 FEET INCLUSIVE. SPACING TO BE USED IS THAT WHICH PRODUCES MAXIMUM STRESSES.

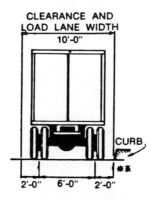

*In the design of timber floors and orthotropic steel decks (excluding transverse beams) for H 20 loading, one axle load of 24,000 pounds or two axle loads of 16,000 pounds each spaced 4 feet apart may be used, whichever produces the greater stress, instead of the 32,000-pound axle shown.

**For slab design, the center line of wheels shall be assumed to be 1 foot from face of curb. (See Article 3.24.2.)

b.

Fig. B.1. (*Continued*)

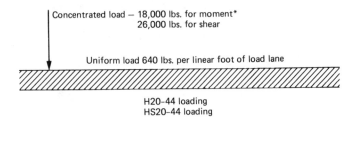

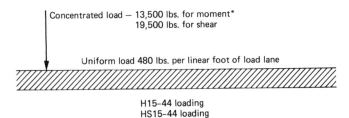

*For the loading of continuous spans involving lane loading refer to Article 3.11.3 which provides for an additional concentrated load.

Fig. B.2. Standard H and HS lane loading (AASHTO Fig. 3.7.6B; "Standard Specifications for Highway Bridges," Copyright 1989. The American Association of State Highway and Transportation Officials, Washington, D.C. Used by permission).

lanes. For an exterior stringer, the distribution factor is $S/5.5$ when $S \leq 6$ ft or $S/(4.0 + 0.25S)$ when 6 ft $< S <$ 14 ft. In no case shall an exterior stringer have less capacity than an interior one.

B.2 DESIGN OF A SIMPLY SUPPORTED COMPOSITE HIGHWAY BRIDGE STRINGER

In this section, only the essential aspects of the design of simply supported composite stringers will be discussed. As with composite beams in buildings, it is also generally required that the stringer and the concrete deck work together so that a stiffer and shallower structure can be obtained. All of the appropriate section and equation numbers referenced in this section will be from the AASHTO code.

The Service Load Design Method, Part C of AASHTO Section 10, will be used here to proportion the composite member. The allowable stresses given in this method are at least 10% less than those given in the AISCS. Considering the type of loading which acts on a bridge, fatigue must also be considered. The details for the design of a typical composite bridge beam is given in the following example, with reference to the AASHTO code wherever applicable.

HIGHWAY STEEL BRIDGE DESIGN

Example B.1. Determine an interior composite stringer for the 88-ft-long, simply supported highway bridge shown below. The deck is 37 ft, 3 in. wide curb-to-curb and is divided into three lanes, each 12 ft, 5 in. wide. The highway live loading is HS20-44, with a traffic frequency greater than 2,000,000 cycles/year. Use M270 Gr 36 steel (equivalent to A36 steel) and a normal-weight concrete deck $7\frac{1}{2}$ in. thick with $f'_c = 3500$ psi. Unshored construction.

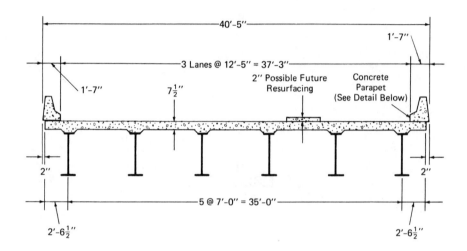

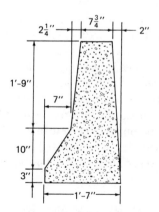

Typical parapet (Jersey type) as per Illinois Department of Transportation (IDOT)

Solution. After some trials, assume the following section:

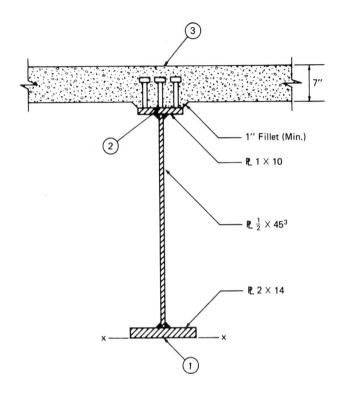

A more economical composite member is obtained by using a welded steel section with the bottom flange larger than the top flange, as shown. Also note that when computing section properties, the concrete of the 1-in. fillet is assumed to be negligible, and the $7\frac{1}{2}$-in. concrete deck is reduced to a thickness of 7 in. to account for any loss in wearing surface caused by the traffic.[4]

The stresses in the steel at points 1 and 2 are due to the combination of the dead load, the superimposed dead load, and the highway live load. Since the beam is unshored, the steel alone will carry the dead load, while the composite section will carry the superimposed dead load and live load. The stress in the concrete at point 3 is due to the superimposed dead load and the live load; this stress is resisted by the composite section. Thus, section properties are required

---

[3]Some savings in poundage could be obtained by using a $\frac{3}{8}$-in.-thick web; however, the labor that would be required for cutting and welding the extra stiffeners would sharply increase the overall cost.

[4]This $\frac{1}{2}$-in. reduction in the concrete deck, although not required by AASHTO, is often used by many designers to provide additional safety.

for the steel alone for the dead load, the composite section for the superimposed dead load, and the composite section for the live load.

In general, when determining section properties, it is often convenient to take the reference axis at the bottom of the section, and to tabulate all of the data for each of the components in the section with respect to the reference axis, as shown below. The notation used in the tables is as follows:

A_i = area of the component under consideration, in.^{2}

y_i = distance from the reference axis to the centroid of the component under consideration, in.

$(I_o)_i$ = moment of inertia of the component under consideration about its own centroid, in.^{4}

$(I_x)_i$ = moment of inertia of the component under consideration about the reference axis, in.^{4}

The location of the neutral axis for the entire section and the corresponding moment of inertia can then be easily determined from statics.

- Section properties for the steel alone (reference axis taken at the bottom of the section):

Component	A_i (in.^{2})	y_i (in.)	$A_i y_i$ (in.^{3})	$A_i y_i^2$ (in.^{4})	$(I_o)_i$ (in.^{4})	$(I_x)_i$ (in.^{4})
Pl 2 × 14	28.0	1.0	28.0	28.0	—^{a}	28.0
Pl ½ × 45	22.5	24.5	551.3	13,505.6	3797.0	17,302.5
Pl 1 × 10	10.0	47.5	475.0	22,562.5	—^{a}	22,562.5
Σ	60.5		1054.3			39,893.0

^{a}Negligible.

$$y_s = \frac{1054.3 \text{ in.}^3}{60.5 \text{ in.}^2} = 17.43 \text{ in. measured from the bottom of the section to the neutral axis}$$

$$I_s = 39{,}893.0 \text{ in.}^4 - 60.5 \text{ in.}^2 \times (17.43 \text{ in.})^2 = 21{,}520 \text{ in.}^4$$

$$S_1 = \frac{21{,}520 \text{ in.}^4}{17.43 \text{ in.}} = 1234.7 \text{ in.}^3$$

$$S_2 = \frac{21{,}520 \text{ in.}^4}{48 \text{ in.} - 17.43 \text{ in.}} = 704.0 \text{ in.}^3$$

- Section properties of the composite section for the superimposed dead loads:

484 STEEL DESIGN FOR ENGINEERS AND ARCHITECTS

When dead load acts on the composite section, the effects of creep must be considered (AASHTO 10.38.1.4). To account for creep, the ratio of the modulus of elasticity of the steel to that of the concrete n must be multiplied by 3.[5]

$$\text{For } f'_c = 3500 \text{ psi}, \quad n = 9 \quad \text{(AASHTO 10.38.1.3)}$$

Therefore, for creep, base transformed section properties of the concrete on $n = 3 \times 9 = 27$.

Effective flange width of the concrete (AASHTO 10.38.3):

$$b = \text{lesser of} \begin{cases} \text{Span}/4 = 88 \text{ ft} \times 12 \text{ in.}/\text{ft}/4 = 264 \text{ in.} \\ \text{Center-to-center spacing of girders} \\ \quad = 7 \text{ ft} \times 12 \text{ in.}/\text{ft} = 84 \text{ in.} \\ 12 \times \text{slab thickness} = 12 \times 7 \text{ in.} = 84 \text{ in.} \end{cases}$$

Thus, $b = 84$ in.

$$b_{tr} = \frac{84 \text{ in.}}{27} = 3.11 \text{ in.}$$

$$A_{ctr} = 3.11 \text{ in.} \times 7 \text{ in.} = 21.8 \text{ in.}^2$$

$$I_{ctr} = \frac{1}{12} \times 3.11 \text{ in.} \times (7 \text{ in.})^3 = 89 \text{ in.}^4$$

Component	A_i (in.^{2})	y_i (in.)	$A_i y_i$ (in.^{3})	$A_i y_i^2$ (in.^{4})	$(I_o)_i$ (in.^{4})	$(I_x)_i$ (in.^{4})
Steel stringer	60.5	17.43	1054.5	18,380	21,520	39,900
Transformed concrete	21.8	52.50	1144.5	60,086	89	60,175
Σ	82.3		2199.0			100,075

$$y_{tr} = \frac{2199.0 \text{ in.}^3}{82.3 \text{ in.}^2} = 26.7 \text{ in.}$$

$$I_{tr} = 100{,}075 \text{ in.}^4 - 82.3 \text{ in.}^2 \times (26.7 \text{ in.})^2 = 41{,}319 \text{ in.}^4$$

$$S_1 = \frac{41{,}319 \text{ in.}^4}{26.7 \text{ in.}} = 1547.5 \text{ in.}^3$$

---

[5]Since the superimposed dead load is always present on the composite section, the concrete will undergo some creep. This results in a change in the modular ratio from n to $3n$.

$$S_2 = \frac{41{,}319 \text{ in.}^4}{48 \text{ in.} - 26.7 \text{ in.}} = 1940.0 \text{ in.}^3$$

$$S_3 = \frac{41{,}319 \text{ in.}^4}{56 \text{ in.} - 26.7 \text{ in.}} = 1410.0 \text{ in.}^3$$

- Section properties of the composite section for the live load: Transformed section properties of the concrete are determined for $n = 9$.

$$b_{tr} = \frac{84 \text{ in.}}{9} = 9.33 \text{ in.}$$

$$A_{ctr} = 9.33 \text{ in.} \times 7 \text{ in.} = 65.3 \text{ in.}^2$$

$$I_{ctr} = \frac{1}{12} \times 9.33 \text{ in.} \times (7 \text{ in.})^3 = 267 \text{ in.}^4$$

Component	A_i (in.^{2})	y_i (in.)	$A_i y_i$ (in.^{3})	$A_i y_i^2$ (in.^{4})	$(I_o)_i$ (in.^{4})	$(I_x)_i$ (in.^{4})
Steel stringer	60.5	17.43	1054.5	18,380	21,520	39,900
Transformed concrete	65.3	52.50	3428.3	179,983	267	180,250
Σ	125.8		4482.8			220,150

$$y_{tr} = \frac{4482.8 \text{ in.}^3}{125.8 \text{ in.}^2} = 35.6 \text{ in.}$$

$$I_{tr} = 220{,}150 \text{ in.}^4 - 125.8 \text{ in.}^2 \times (35.6 \text{ in.})^2 = 60{,}408 \text{ in.}^4$$

$$S_1 = \frac{60{,}408 \text{ in.}^4}{35.6 \text{ in.}} = 1697.0 \text{ in.}^3$$

$$S_2 = \frac{60{,}408 \text{ in.}^4}{48 \text{ in.} - 35.6 \text{ in.}} = 4872.0 \text{ in.}^3$$

$$S_3 = \frac{60{,}408 \text{ in.}^4}{56 \text{ in.} - 35.6 \text{ in.}} = 2961.0 \text{ in.}^3$$

Loading:

Dead load:

$$\text{Beam weight} = 60.5 \text{ in.}^2 \times \frac{3.4}{1000} = 0.206 \text{ klf}$$

Details, diaphragms, etc. (15% of beam weight) = 0.031 klf

$$\text{Slab} = \frac{7.5 \text{ in.}}{12 \text{ in./ft}} \times 0.145 \text{ k/ft}^3 \times 7 \text{ ft} = 0.634 \text{ klf}$$

$$\text{Fillet} = \frac{1 \text{ in.}}{12 \text{ in./ft}} \times 0.145 \text{ k/ft}^3 \times \frac{10 \text{ in.}}{12 \text{ in./ft}} = 0.010 \text{ klf}$$

$$w_D = 0.206 + 0.031 + 0.634 + 0.010 = 0.881 \text{ klf}, \quad \text{say } 0.90 \text{ klf}$$

Superimposed dead load:

Concrete parapets:

$$2 \times 0.145 \text{ k/ft}^3 \times 2.83 \text{ ft}^2 = 0.82 \text{ klf}$$

$$\text{Resurfacing} = \frac{2 \text{ in.}}{12 \text{ in./ft}} = 0.145 \text{ k/ft}^3 \times 37.25 \text{ ft} = 0.90 \text{ klf}$$

The stiff reinforced concrete deck and the existence of the diaphragms forces all of the stringers to deflect the same amount; thus, they will all carry the same moment (AASHTO 10.6.4). Consequently, the superimposed dead load is to be divided by the number of stringers when computing the moment in a stringer caused by this loading.

$$w_{SD} = (0.82 + 0.90) = 1.72 \text{ klf}/6 \text{ beams} = 0.287 \text{ klf}$$

Maximum moments:

$$M_D = \frac{0.90 \text{ klf} \times (88 \text{ ft})^2}{8} = 871.2 \text{ ft-k}$$

$$M_{SD} = \frac{0.287 \text{ klf} \times (88 \text{ ft})^2}{8} = 277.8 \text{ ft-k}$$

From Appendix A of the AASHTO code, the live load moment is 1308.5 ft-k (from interpolation). This moment is to be modified by the impact factor, the distribution factor, and the multiple loaded lanes reduction factor.

$$I = \frac{50}{88 + 125} = 0.235 < 0.30$$

$$\text{Distribution factor} = \frac{7.0 \text{ ft}}{5.5} = 1273 \times \tfrac{1}{2} \text{ (per wheel line)}$$

$$= 0.636$$

$$\text{Reduction factor} = 1 - 0.1 = 0.9$$

Therefore,

$$M_{L+I} = 1308.5 \text{ ft-k} \times (1 + 0.235) \times 0.636 \times 0.9 = 925.0 \text{ ft-k}$$

Maximum stresses (AASHTO 10.38.4):

$$f_1 = \frac{871.2 \times 12}{1234.7} + \frac{277.8 \times 12}{1547.5} + \frac{925.0 \times 12}{1697.0}$$

$$= 8.47 + 2.15 + 6.54$$

$$= 17.16 \text{ ksi} < 20 \text{ ksi (AASHTO Table 10.32.1A)} \quad \text{ok}$$

$$f_2 = \frac{871.2 \times 12}{704.0} + \frac{277.8 \times 12}{1940.0} + \frac{925.0 \times 12}{4872.0}$$

$$= 14.85 + 1.72 + 2.28 = 18.85 \text{ ksi} < 20 \text{ ksi} \quad \text{ok}$$

$$f_3 = \frac{277.8 \times 12}{1410.0 \times 27} + \frac{925.0 \times 12}{2961.0 \times 9} = 0.09 + 0.42$$

$$= 0.51 \text{ ksi} < 0.4 \times 3.5 = 1.4 \text{ ksi} \quad \text{ok}$$

We can see that the bending stresses due to fatigue (i.e., $LL + I$) are quite small and under control.

For this composite girder, the ratio of the top compression flange width to thickness $b/t = 10/1 = 10$. The limiting value of b/t is given in AASHTO eqn. (10-20) as follows:

$$\frac{b}{t} = \frac{3860}{\sqrt{f_{2,DL}}}$$

where $f_{2,DL}$ = compression stress in the top flange due to the noncomposite dead load; in this case, $f_{2,DL}$ is 14,850 psi (see maximum stress calculations above). Thus, maximum $b/t = 3860/\sqrt{14,850} = 31.7 > 10$, which is ok.

Check shear stresses (AASHTO 10.38.5.2):

$$V_D = \tfrac{1}{2} \times 0.9 \text{ klf} \times 88 \text{ ft} = 39.6 \text{ k}$$

$$V_{SD} = \tfrac{1}{2} \times 0.287 \text{ klf} \times 88 \text{ ft} = 12.6 \text{ k}$$

$$V_{L+I} = 64.4 \text{ k (from Appendix A of the AASHTO code)} \times (1 + 0.235)$$

$$\times 0.636 \times 0.9 = 45.5 \text{ k}$$

488 STEEL DESIGN FOR ENGINEERS AND ARCHITECTS

$V_{\text{total}} = 39.6 + 12.6 + 45.5 = 97.7 \text{ k}$

$$f_v = \frac{97.7 \text{ k}}{45 \text{ in.} \times 0.5 \text{ in.}} = 4.34 \text{ ksi} < \frac{F_y}{3} = \frac{36 \text{ ksi}}{3} = 12 \text{ ksi} \quad \text{ok}$$

Check for intermediate stiffeners (AASHTO 10.34.4):

Transverse intermediate stiffeners are not required if the web thickness is not less than $D/150$ and f_v is less than the value given in AASHTO eqn. (10-25):

$$F_v = \frac{7.33 \times 10^7}{(D/t_w)^2} \leq \frac{F_y}{3}$$

where D = superimposed depth of the web plate between flanges, in.
t_w = web thickness, in.
F_v = allowable shear stress, psi

In this case, $D/150 = 45/150 = 0.3$ in. < 0.5 in., which is provided. Also,

$$F_v = \frac{7.33 \times 10^7}{(45/0.5)^2} = 9049 \text{ psi} < 12{,}000 \text{ psi}$$

$f_v = 4340 \text{ psi} < 9049 \text{ psi}$

Since both conditions are satisfied, no intermediate stiffeners are required.

Shear connectors (AASHTO 10.38.2):

To resist the horizontal shear at the interface between the slab and the steel girder, mechanical shear connectors must be provided (AASHTO 10.38.5.1). The shear connectors must be designed for fatigue and then checked for ultimate strength.

Fatigue (AASHTO 10.38.5.1.1):

The range of horizontal shear shall be computed by AASHTO eqn. (10-57):

$$S_r = \frac{V_r Q}{I}$$

where S_r = range of horizontal shear at the slab-girder interface at the point in the span under consideration, k/in.
V_r = range of shear due to live loads and impact; at any section, the

range of shear is the difference in the minimum and maximum shear envelopes (excluding the dead load), kips.[6]

Q = statical moment about the neutral axis of the composite section of the transformed compressive concrete area, in.^{3}

I = moment of inertia of the transformed composite girder, in.^{4}

The allowable range of horizontal shear on an individual shear connector is given in AASHTO eqn. (10-59) for welded studs:

$$Z_r = \alpha d^2$$

where Z_r = allowable range of horizontal shear, lb
α = constant that depends on the number of cycles
d = diameter of the stud, in.

Once S_r and Z_r are obtained, the required pitch of the shear connectors is determined by dividing the allowable range of horizontal shear of all connectors at one transverse girder cross-section by the quantity S_r.

To begin the analysis here, calculate the maximum and minimum values of the live load shear (including impact) at the following sections along the beam: 1) at the support, 2) at 0.25 the span length, and 3) at 0.50 the span length. It can be shown that the truck live loading controls over the lane live loading in this case. Note that the truck loads shown in Fig. B.1 must be multiplied by the impact, distribution, and reduction factors, i.e.,

$$R = \overline{R} \times (1 + 0.235) \times 0.636 \times 0.9$$

where R = axle load multiplied by all of the appropriate factors and $\overline{R}$ = 8 kips or 32 kips for the HS20-44 truck loading, which gives R = 5.7 k and R = 22.6 k, respectively.

1) At the support, the maximum shear V_{max} is obtained by arranging the loads on the span as follows:

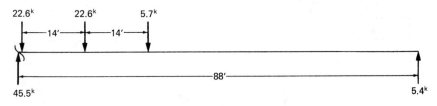

---

[6]For information on fatigue, see Appendix A.

Summing moments results in the following:

$$V_{max} = \frac{22.6 \text{ k} \times 88 \text{ ft} + 22.6 \text{ k} \times 74 \text{ ft} + 5.7 \text{ k} \times 60 \text{ ft}}{88 \text{ ft}} = 45.5 \text{ k}$$

The minimum shear V_{min} is obtained when the truck has passed over the bridge with only its last wheel at the support. Therefore,

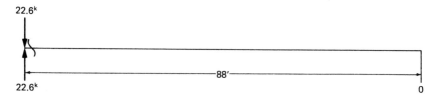

$$V_{min} = 22.6 \text{ k} \times 88 \text{ ft} - 22.6 \text{ k} \times 88 \text{ ft}$$
$$= 0 \therefore V_r = 45.5 - 0 = 45.5 \text{ k}$$

2) At $(0.25 \times 88 \text{ ft}) = 22$ ft from the support, V_{max} is obtained when the truck loads are arranged as shown below:

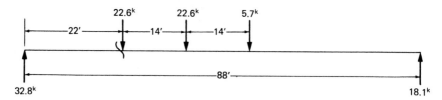

$$V_{max} = \frac{22.6 \text{ k} \times 66 \text{ ft} + 22.6 \text{ k} \times 52 \text{ ft} + 5.7 \text{ k} \times 38 \text{ ft}}{88 \text{ ft}} = 32.8 \text{ k}$$

For V_{min}, the loads should be arranged as follows:

$$V_{min} = -\left(\frac{22.6 \text{ k} \times 8 \text{ ft} + 22.6 \text{ k} \times 22 \text{ ft}}{88 \text{ ft}}\right) = -7.7 \text{ k}$$

$$V_r = V_{max} - V_{min} = 32.8 - (-7.7) = 40.5 \text{ k}$$

3) At $(0.5 \times 88 \text{ ft}) = 44 \text{ ft}$,

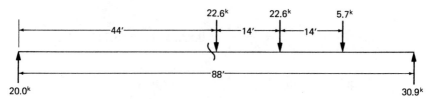

$$V_{max} = -(V_{min}) = \frac{22.6 \text{ k} \times 44 \text{ ft} + 22.6 \text{ k} \times 30 \text{ ft} + 5.7 \text{ k} \times 16 \text{ ft}}{88 \text{ ft}}$$

$$= 20.0 \text{ k}$$

$$V_r = 20.0 - (-20.0) = 40.0 \text{ k}$$

The values of V_r along the span length are plotted below.

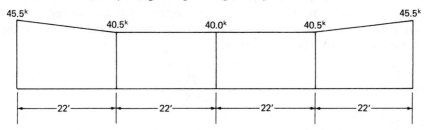

Using $\frac{7}{8}$-in.-diameter shear studs, the required connector spacing will be determined for both the first and second quarters of the beam span.

$$Q = 7 \text{ in.} \times \frac{84 \text{ in.}}{9} \times (49 \text{ in.} + 3.5 \text{ in.} - 35.6 \text{ in.}) = 1104 \text{ in.}^3$$

$$I = 60{,}408 \text{ in.}^4$$

For a stud subjected to over 2,000,000 cycles, $\alpha = 5500$.

For the first quarter of the beam span $(0 - 22 \text{ ft})$:

$$Z_r = 5500 \times (0.875)^2 = 4211 \text{ lb}$$

If three studs are used per line, $Z_r = 3 \times 4211 = 12{,}633 \text{ lb}$

$$S_r = \frac{45.5 \text{ k} \times 1104 \text{ in.}^3}{60{,}408 \text{ in.}^4} = 0.832 \text{ k/in.}$$

$$\text{Required spacing} = \frac{12.63 \text{ k}}{0.832 \text{ k/in.}} = 15.2 \text{ in.} < 24 \text{ in. max.} \quad \text{ok}$$

492 STEEL DESIGN FOR ENGINEERS AND ARCHITECTS

For the second quarter of the beam span (22 − 44 ft):

$$S_r = \frac{40.5 \text{ k}}{45.5 \text{ k}} \times 0.832 \text{ k/in.} = 0.741 \text{ k/in.}$$

$$\text{Required spacing} = \frac{12.63 \text{ k}}{0.741 \text{ k/in.}} = 17.0 \text{ in.}$$

If we use a spacing of 15 in. for the first quarter and a spacing of 17 in. for the second quarter, the total number of studs required for half of the beam length is 53 + 47 = 100.

Ultimate strength (AASHTO 10.38.5.1.2):

The number of shear connectors that were provided for fatigue must now be checked to ensure that the ultimate strength requirements are satisfied. The required number of shear connectors for ultimate strength shall be at least the value given in AASHTO eqn. (10-60):

$$N_1 = \frac{P}{\phi S_u}$$

where N_1 = number of connectors required between points of maximum positive moment and adjacent end supports

S_u = ultimate strength of the shear connector, lb
    = $0.4 d^2 \sqrt{f'_c E_c}$ for welded studs (AASHTO eqn. (10-66))
d = diameter of stud, in.
f'_c = compressive strength of the concrete at 28 days, psi
E_c = modulus of elasticity of the concrete, psi
    = $w^{3/2} 33 \sqrt{f'_c}$ (AASHTO eqn. (10-67))
w = unit weight of the concrete, lb/ft^3
ϕ = reduction factor = 0.85
P = force in the slab = P_1 or P_2 (whichever is smaller), lb
P_1 = $A_s F_y$ (AASHTO eqn. (10-61))
P_2 = $0.85 f'_c b t_s$ (AASHTO eqn. (10-62))
A_s = total area of the steel including cover plates, in.^{2}
F_y = specified minimum yield stress of the steel, psi
b = effective flange width (AASHTO 10.38.3), in.
t_s = thickness of the concrete slab, in.

In this case,

$$E_c = (145 \text{ pcf})^{3/2} \times 33 \times \sqrt{3500 \text{ psi}} = 3{,}409{,}000 \text{ psi}$$

$$S_u = 0.4 \times (0.875 \text{ in.})^2 \times \sqrt{3500 \text{ psi} \times 3{,}409{,}000 \text{ psi}} = 33{,}450 \text{ lb}$$

$$P_1 = 60.5 \text{ in.}^2 \times 36{,}000 \text{ psi} = 2{,}178{,}000 \text{ lb}$$

$$P_2 = 0.85 \times 3500 \text{ psi} \times 84 \text{ in.} \times 7 \text{ in.} = 1{,}749{,}300 \text{ lb (governs)}$$

$$N_1 = \frac{1{,}749{,}300 \text{ lb}}{0.85 \times 33{,}450 \text{ lb}} = 61.5 \text{ studs on half of the beam}$$

Since this number is smaller than the one obtained for fatigue, the number of studs required for fatigue governs.

Bearing stiffeners (AASHTO 10.34.6):

At the ends of the composite beam where it bears on supports, bearing stiffeners are required. Typically, the stiffeners are provided on both sides of the web plate and extend as close to the outer edges of the flange plate as possible. The bearing stiffeners are designed as columns; the connection of the stiffener to the web should be designed to transmit the entire reaction to the bearings. Both the allowable compressive stress and the bearing pressure should not exceed the values given in AASHTO 10.32. In general, the thickness of the bearing stiffener plate should not be less than the one given in AASHTO eqn. (10-34):

$$t = \frac{b'}{12} \sqrt{\frac{F_y}{33{,}000}}$$

where b' = width of the stiffener, in.
F_y = specified minimum yield stress of the stiffener, psi

In this case, if we use $b' = 6$ in., then the minimum thickness is

$$t = \frac{6 \text{ in.}}{12 \text{ in.}} \sqrt{\frac{36{,}000 \text{ psi}}{33{,}000}} = 0.52 \text{ in.} \quad \text{Try } \frac{9}{16} \text{ in.}$$

For stiffeners consisting of the two plates, the column section is taken as the two stiffener plates plus a centrally located strip of the web plate whose width is equal to not more than 18 times its thickness. The stiffener location and the column section are shown below.

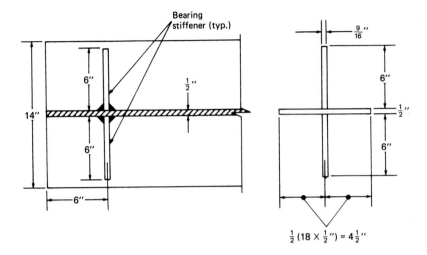

Section properties for the column section are as follows:

$$A = 9 \text{ in.} \times 0.5 \text{ in.} + 2 \times 6 \text{ in.} \times 0.56 \text{ in.} = 11.25 \text{ in.}^2$$

$$I = \frac{1}{12} \times (0.56 \text{ in.}) \times (12.5 \text{ in.})^3 = 91.6 \text{ in.}^4$$

$$r = \sqrt{\frac{91.6 \text{ in.}^4}{11.25 \text{ in.}^2}} = 2.85 \text{ in.}$$

$$\frac{KL}{r} = \frac{1.0 \times 45 \text{ in.}}{2.85 \text{ in.}} = 15.79$$

From AASHTO Table 10.32.1A, the allowable compressive stress F_a for M270 Gr 36 steel is ($KL/r = 15.79 < C_c = 126.1$)

$$F_a = 16{,}980 - 0.53(15.79)^2 = 16{,}848 \text{ psi}$$

Maximum end reaction is 97.7 kips. Therefore, maximum axial stress is

$$f_a = \frac{97.7 \text{ k}}{11.25 \text{ in.}^2} = 8.68 \text{ ksi} < 16.85 \text{ ksi} \quad \text{ok}$$

Check bearing of stiffener on bottom flange:

Assume that the stiffeners are ground to fit with the corner clipped as shown.

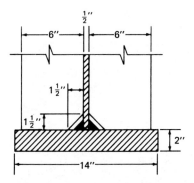

The load per stiffener is 97.7 k/2 = 48.9 k. The bearing stress is then

$$f_p = \frac{48.9 \text{ k}}{(6 \text{ in.} - 1.5 \text{ in.}) \times (0.56 \text{ in.})} = 19.4 \text{ ksi}$$

From AASHTO Table 10.32.1A, $F_p = 0.8 F_y = 29$ ksi > 19.4 ksi ok

Connection of bearing stiffener to web:

Assume $\frac{5}{16}$-in. E70 weld.

$$F_v = 0.27 F_u = 0.27 \times 70 \text{ ksi} = 18.9 \text{ ksi} \quad (\text{AASHTO eqn. (10-12)})$$

For welds on both sides of the stiffener, the required weld length is

$$l = \frac{48.9 \text{ k}}{2 \times 0.707 \times 0.3125 \text{ in.} \times 18.9 \text{ ksi}} = 5.86 \text{ in.}$$

Therefore, use two bearing stiffeners each end, $\frac{9}{16} \times 6 \times 2$ ft, 11 in.

Connection of flange plates to web (assume that the entire transformed section carries all the shear):

Top flange:

$$Q = (10 \text{ in.} \times 1 \text{ in.}) \times (48 \text{ in.} - 0.5 \text{ in.} - 35.6 \text{ in.})$$
$$+ (7 \text{ in.} \times 9.33 \text{ in.}) \times (56 \text{ in.} - 3.5 \text{ in.} - 35.6 \text{ in.})$$
$$= 119 + 1104 = 1223 \text{ in.}^3$$

$$I = 60{,}408 \text{ in.}^4$$

$$q = \frac{97.7 \text{ k} \times 1223 \text{ in.}^3}{60{,}408 \text{ in.}^4} = 1.98 \text{ k/in.}$$

For a $\frac{3}{8}$-in. E70 weld on each side,

$$F_w = 2 \times 0.27 \times 70 \text{ ksi} \times 0.707 \times 0.375 \text{ in.}$$
$$= 10 \text{ k/in.} > 1.98 \text{ k/in. ok}$$

Bottom flange:

$$Q = (14 \text{ in.} \times 2 \text{ in.}) \times (35.6 \text{ in.} - 1 \text{ in.}) = 969 \text{ in.}^3$$

$$q = \frac{97.7 \text{ k} \times 969 \text{ in.}^3}{60,408 \text{ in.}^4} = 1.57 \text{ k/in.} < 10 \text{ k/in.}$$

Therefore, use $\frac{3}{8}$-in. continuous fillet weld on both the top and bottom flange.

Diaphragms (AASHTO 10.20):

When a girder which has its top flange continuously supported is subjected to wind loading, an additional stress is induced in its bottom flange. Diaphragms are to be provided along the length of the girder, and are to be spaced at intervals such that the additional stress in the bottom flange due to the wind is kept to a minimum. The maximum induced stress F can be computed from AASHTO eqn. (10-5):

$$F = RF_{cb}$$

where $R = (0.2272L - 11)S_d^{-2/3}$ when no lateral bracing is provided
$R = (0.059L - 0.64)S_d^{-1/2}$ when bottom lateral bracing is provided

$F_{cb} = \dfrac{72M_{cb}}{t_f b_f^2}$, psi

$M_{cb} = 0.08 \ WS_d^2$, ft-lb
W = wind loading along the exterior flange, lb/ft
S_d = diaphragm spacing, ft
L = span length, ft
t_f = thickness of bottom flange, in.
b_f = width of bottom flange, in.

In our case, assume a wind load of 50 psf and diaphragms spaced 22 ft, 0 in. o.c. Also, assume that no lateral bracing on the bottom is provided.

$$M_{cb} = 0.08 \times \left(50 \text{ psf} \times \frac{48 \text{ in.}}{12 \text{ in./ft}} \times \frac{1}{2}\right) \times (22 \text{ ft})^2 = 3872 \text{ ft-lb}$$

$$F_{cb} = \frac{72 \times 3872 \text{ ft-lb}}{2 \text{ in.} \times (14 \text{ in.})^2} = 711 \text{ psi}$$

$$R = (0.2272 \times 88 \text{ ft} - 11)(22 \text{ ft})^{-2/3} = 1.15$$

$$F = 1.15 \times 711 \text{ psi} = 818 \text{ psi}$$

The total stress on the bottom flange is then 17.16 ksi + 0.818 ksi

$$= 18.0 \text{ ksi} < 20 \text{ ksi} \quad \text{ok}$$

It is important to note that the diaphragm depth should be at least one-half and preferably three-quarters of the girder depth. A W 24 × 55 diaphragm, spaced at 22 ft, 0 in. on center and positioned near the bottom flange of the girder, would be sufficient in this case. The diaphragms at the ends of the girder should be framed near the top flange, and should be sized to carry a live load moment (including impact) caused by the load of one wheel acting at its midspan. The factored load in this case would be 1.3 × 16 k = 20.8 k.

$$\text{Max. } M = \frac{20.8 \text{ k} \times 7 \text{ ft}}{4} = 36.4 \text{ ft-k}$$

Use W 12 × 35.

A minimum 3-in. fillet should be provided for these end members.

Deflection (AASHTO 10.6):

For the dead load:

$$w_D = 0.90 \text{ klf}/12 \text{ in.}/\text{ft} = 0.075 \text{ k/in.}$$

$$L = 88 \text{ ft} \times 12 \text{ in.}/\text{ft} = 1056 \text{ in.}$$

$$I = 21{,}520 \text{ in.}^4$$

$$\Delta_D = \frac{5w_D L^4}{384EI} = \frac{5 \times 0.075 \times (1056)^4}{384 \times 21{,}520 \times 29{,}000} = 1.95 \text{ in.}$$

In order to follow the profile of the roadway, a large camber between the crown and the two supports may be needed. This distance and the dead load deflection will make up the total camber that must be provided for the beam.

For the superimposed dead load:

$$w_{SD} = 0.287 \text{ klf}/12 \text{ in.}/\text{ft} = 0.0239 \text{ k/in.}$$

498 STEEL DESIGN FOR ENGINEERS AND ARCHITECTS

$$\Delta_{SD} = \frac{5 \times 0.0239 \times (1056)^4}{384 \times 41{,}319 \times 29{,}000} = 0.32 \text{ in.}$$

For the live load:

Conservatively assume that the load $P = 6 + 32 + 32 = 72$ kips acts at the midspan of the bridge. As noted previously, the stiff concrete deck and the diaphragms force the composite girders to deflect the same amount when subjected to the superimposed dead and live loads. This was taken into consideration in the above computations for the superimposed dead load deflection by using $w_{SD} = 1.72$ klf/6 girders $= 0.287$ klf. For the live load, the load per girder can be obtained as follows:

$$\frac{3 \text{ lanes} \times 72 \text{ k} \times (1 - 0.235) \times 0.90}{6 \text{ girders}} = 40.0 \text{ k}$$

$$\Delta_{LL} = \frac{PL^3}{48EI} = \frac{40.0 \times (1056)^3}{48 \times 29{,}000 \times 60{,}408} = 0.56 \text{ in.}$$

$$\text{Max. allowable } \Delta_{LL} = \frac{L}{800} = \frac{88 \times 12}{800} = 1.32 \text{ in.} > 0.56 \text{ in.} \quad \text{ok}$$

Note that for composite girders, the ratio of the overall depth of the girder (concrete slab + steel girder) to the length of the span preferably should not be less than $\frac{1}{25}$ (AASHTO 10.5). In our case,

$$\frac{(7.5 + 1 + 48)}{88 \times 12} = \frac{1.34}{25} > \frac{1.0}{25} \quad \text{ok}$$

Also, the ratio of the depth of the steel girder alone to the length of the span preferably should not be less than $\frac{1}{30}$:

$$\frac{48}{88 \times 12} = \frac{1.36}{30} > \frac{1.0}{30} \quad \text{OK}$$

Note that a steel girder depth of $88 \times 12/30 \cong 35.5$ in. would have resulted in a less stiff bridge.

Index

Allowable stress design, xxxi
Allowable stress range, 460, 472
American Association of State Highway and Transportation Officials (AASHTO), xxxi, 471
American Institute of Steel Construction (AISC), xxx
American Railway Engineering Association (AREA), xxxi
American Society for Testing and Materials (ASTM), xxxii
American Society of Civil Engineers (ASCE), xxviii
American Welding Society (AWS), 231
Amplification factor, 163
Anchor bolts, 303
Angle of rotation. See Angle of twist
Angle of twist, 324, 331, 350
Arc welding, 231
Axial compression, allowable stress, 147–149, 153, 163, 474. See also Columns

Base plates, 148, 178
  subject to moment, 302–305
Beam-columns, 163
  C_m factor, 165
  interaction formulas, 164
Beams, 39
  allowable stress, 39
    bending, 41, 45, 56, 58, 108, 123, 474
    shear, 69, 108, 124, 474
  bearing plates, 80
  biaxial bending, 64, 164
  built-up, 107, 114, 123, 215
  compact section, 41, 56, 58
  composite. See Composite beams
  continuous, 60
    plastic design, 439, 446
  deflection, 84
    composite beams, 369, 396
  holes in, 72, 99
  lateral instability, 43
  lateral support, 42, 114
  plastic design method, 420, 423, 436, 439
  plastic hinge, 422, 427, 431, 439
  plastic modulus, 423
  plastic moment, 422, 427, 431, 439, 446, 447
  plastic neutral axis, 423
  plate girder. See Plate girder
  principal axes, 64
  section modulus, 40
  shape factor, 423, 438
  shear, 68
  shear center, 64, 324, 345
  shear stress, 68, 108, 487
  splices, 311, 473
  torsion. See Torsion
  typical types, 39
  web crippling, 75, 81
  web tearout. See Block shear
  web yielding, 75, 81
Bearing length, 81
Bearing plates, 80
Bearing stiffeners, 77, 112, 124, 475, 493
Bearing stresses, 80
  allowable, 81, 179, 221, 474
Bearing-type connections, 187, 189, 473

499

INDEX

Biaxial bending, 64, 164
Block shear, 219–222
Bolted connections, 187
  concentrically loaded, 189, 474
  eccentrically loaded, 196
    subject to shear and tension, 208, 475
    subject to shear and torsion, 196
    ultimate strength methods, 198, 199, 266
  failure modes, 189, 190
  framed-beam, 222, 223, 283
  types, 187
Bolts, 29, 97, 473
  allowable shear stress, 187, 190, 474
  allowable tensile stress, 187, 188
  bearing-type connections, 187
  combined shear and tension, 208, 475
  combined shear and torsion, 196
  failure modes, 189, 190
  high-strength, 187, 473
  in combination with welds, 187, 238
  pretension force, 209
  slip-critical type, 187, 473
  unfinished, 187, 473
Braced frames, 148, 165
Bracing, 42, 160
Bredt's formula, 350
Buckling, 41, 136, 139
Built-up beams, 107, 114, 123, 215
Built-up members, 29, 106, 161, 215

Chain of holes, 4
Column base plates, see Base plates
Columns, 136, 153
  allowable stress, 147–149, 153, 163, 474
  effective length, 143, 153
  effective length factor, 144–148, 153, 157, 165
  Euler's column, 143, 145
  Euler curve, 140, 149
  Euler load, 138
  Euler stress, 138, 163, 165
  slenderness ratio, 139, 147, 472
  SSRC curve, 140, 149
  stiffeners, 282, 284–286
Combined bending and axial load, see Beam-columns
Combined stress, 162, 176, 302, 475
  bolts, 196, 208, 475
  welds, 257, 263
Compact section, 41, 56, 58

Composite beams, 363, 475
  advantages, 363
  bridge stringer, 475, 483–485
  cover plates, 363, 392
  deflections, 369, 396
  design tables, 399
  effective moment of inertia, 384
  effective section modulus, 384
  effective width of concrete, 365, 387, 476, 484
  encased beam, 363, 377
  service load stresses, 369, 370
  shear connectors, 363, 365, 377–379, 386, 476, 488–493
  shored construction, 369, 396
  transformed section properties, 369, 384
  unshored construction, 369, 396, 482
Connections, 186, 222, 274, 446, 459
  angles, 223
  beam splices, 311, 473
  beam-column, 274
    end plate, 283
    moment resisting, 282
    rigid, 274, 282
    semi-rigid, 274, 282
    single shear plate, 277
    single tee, 279
  bearing type, 187, 189, 473
  bolted. *See* Bolted connections
  column bases. *See* Base plates
  construction types
    rigid frame (Type 1), 274
    semi-rigid frame (Type 3), 274
    simple frame (Type 2), 274
  framed-beam, 222, 283
  hanger type, 317
  moment resisting, 282
  N or X type, 187, 189, 473
  slip-critical type, 187, 189, 473
  special, 274
  stress reversals, 187, 475
  subject to combined stress. *See* Combined stress
  ultimate strength methods, 199, 266
  welded, 97, 231, 474
Connectors, shear. *See* Composite beams
Continuous beams. *See* Beams, continuous
Cover plates, 93, 97, 99, 446, 473
  connectors for, 97–99
  cut-off point, 99, 392

INDEX 501

Creep, 475, 484
Critical load, 136, 143

Dead load, xxviii, xxix, xxxii, 369
Deflections, 64, 84
   bridge stringers, 472, 497, 498
   composite beams, 369, 396
   floor beams and girders, 84
   roof purlins, 84
Diaphragms, 473, 486, 496
Distribution factor, 476, 486
Ductility, xxvii

Earthquake loads, xxviii, xxxii
Eccentric loads
   on bolted connections, 196, 208, 475
   on welded connections, 257, 263
Effective area of concrete, 368
Effective length, xxxii, 143, 153
Effective length factor, 144–148, 153, 157, 165
Effective net area, 1, 8, 13, 19, 472
Effective moment of inertia, 384
Effective section modulus, 384
Effective width, concrete slab, 365, 387, 476, 484
Encased beams, 363, 377
Euler theory. *See* Columns
Eyebars, 25

Factor of safety, xxx, 420, 438
Fasteners, 1, 20, 97, 223, 473. *See also* Connections
   for eccentric loads, 196, 223, 257
   for horizontal shear, 215
Fatigue, 34, 99, 459, 460
   failure, 459
   load, 99, 476
Fillet welds. *See* Welds, types
Fireproofing, xxviii
Flange area method, 107
Formed steel deck, 385–388
Fracture, 13, 459
Framing angles, 223

Gage, 4, 19
Girder, 39. *See also* Plate girder
Gross area, 1, 13, 23

Highway bridge design, 471, 481
Highway live loads, 476–482

High-strength bolts. *See* Bolts
Holes, 4
   chain, 4
   diameter, 19
   effect on beams, 72, 99
Hybrid beams, 60, 112, 123

Impact, 476, 486
Inspection of welds, 233
Interaction formulas. *See* Beam-columns
Intermediate connectors, 160
Intermittent welds, 29, 236, 246

Joists, 39, 124

Lacing members. *See* Tie plates
Lap joint, 236
Lateral bracing. *See* Bracing *and also* Lateral support
Lateral instability, 43
Lateral support, 42, 114
Length, effective. *See* Effective length
Length, unsupported. *See* Unsupported length
Lintel, 39
Live loads, xxix, 369
Load cycles, 459
Load factors, 420, 438
Loads, xxviii
   concentrated, 75
   critical. *See* Critical load
   dead, xxviii, xxix, xxxii. *See also* Dead load
   earthquake (seismic), xxviii, xxxii
   highway, 476–482
   impact, 476, 486
   live, xxviii, xxix, xxxii. *See also* Live load
   snow, xxviii
   transverse, 39, 42
   wind, xxviii, xxxii

Manual of Steel Construction, Allowable Stress Design, xxxi
Membrane analogy, 358
Modular ratio, 368
Modulus
   elasticity, 23, 367
      concrete, 367, 368
      shear, 325
   plastic, 423
   section, 40

INDEX

Moment of inertia, 40, 84, 93
  effective. *See* Composite beams
  polar, 325
  transformed. *See* Composite beams
Moment of inertia method, 168
Moment, redistribution of, 60, 275

Net area, 2, 6, 19, 25, 220, 472
  effective. *See* Effective net area
  stagger on, 4
Net width, 2
Neutral axis, plastic, 423
Noncompact members, 45, 57, 59

Open web steel joist, 134

Parapet, bridge, 481
Pin-connected members, 13, 25
Pitch, 4, 19
Plastic design, 420, 423, 438
Plastic hinge, 422, 427, 431, 439
Plastic moment, 422, 427, 431, 439, 446
Plastic neutral axis, 423
Plastic section modulus, 423
Plate girder, 107, 123, 475
  allowable bending stress, 123
  allowable shear stress, 124
  stiffeners, 123
  tension field action, 124
  web slenderness ratio, 123
Plug welds, 236
Poissons ratio, 325
Polar moment of inertia, 325
Product of inertia, 64
Prying force, 317
Purlin, 107

Radius of gyration, 23, 138
Redistribution of moments, 428
Reduction factor (multiple loaded lanes), 476, 486
Repeated loads. *See* Fatigue
Repetitive loading, 471
Residual stress, 139
Rivets, 29, 97, 186, 473
Safety factor. *See* Factor of safety
Section modulus, 40
  composite beam, 369
  effective, 384

  plastic, 423
  transformed, 369, 384
Shape factor, 423, 438
Shear, 67
  beams, 68
  bolted joints, 187
  center, 64, 324, 345
  connectors. *See* Composite beams
  flow, 97, 215, 335
  force, 68
  horizontal, 97
  modulus, 325
  stress, 68, 108, 487
  with torsion, 324, 331, 358
Shear lag principle, 1
Shored construction, 369, 396
Shoring, 364, 369
Single shear plate connection, 277
Single tee connection, 279, 280
Slenderness ratio
  compression, 139, 147, 153, 472
  effective, 149
  tension, 23, 30, 472
  web, 29, 107, 123
Slot weld, 236
Splices, 311, 473
Stagger of holes, 4
Stainless steel, 25
Stiffeners
  bearing, 77, 112, 124, 475, 493
  column, 282, 284–286
  intermediate, 108–112, 123, 475, 488
Strain, 23
Stress categories, 459, 472
Stress cycles, 471
Stress range, 459, 472
Stress reversal, 34, 187
Stress-strain relationship, xxvii, 420
Structural Stability Research Council (SSRC), 140, 149
Structural steel
  advantages, xxviii
  grades, xxxii
St. Venant torsion, 324

Tension field action, 108, 124
Tension members, 1
  allowable stress, 12, 14, 25, 177, 220, 473
  effective net area, 1, 8, 13, 19, 472
  net area, 2, 6, 19, 25, 220, 472

reduction coefficient for (U), 8, 19
single angle, 10, 72
slenderness ratio, 23
Termination zone, 99
Theoretical cut-off point, 99, 392
Threaded rods, 14
Throat, weld, 232, 234
Tie plates, 30
Torque, 324
Torsion, 324
angle of twist, 324, 331, 350
of axisymmetric members, 324
of closed sections, 350
of open sections, 335
of solid rectangular sections, 330
St. Venant, 324
Toughness considerations, 471
Transformed section properties. *See*
  Composite beams

U (reduction coefficient), 8, 19
Ultimate stress, xxv, xxxi, 12
Unbraced frames, 148, 165
Unbraced length. *See* Unsupported length
Unshored construction, 369, 396, 482
Unsupported length, 43, 108, 153

Virtual displacement, 431, 439
Virtual work, 431, 439

Warping, 324, 330
Web crippling, 75, 81

Web tearout. *See* Block shear
Web yielding, 75, 81
Welds, 97
allowable stress, 234, 474
capacity, 238
combined shear and bending, 257, 263
combined shear and torsion, 257
effective and required dimensions, 234
in combination with bolts, 187, 238
inspection, 233
intermittent, 29, 236, 246
length, 238
longitudinal, 8, 9
reinforcement, 232
returns, 236, 238
size, 231, 236
strength, 99, 238
subject to repeated stresses, 255
symbols, 231
throat, 232, 234
transverse, 8, 10
types, 231
ultimate strength method, 266
Width, net, 6
Width-thickness ratio
beams, 57, 108
intermediate stiffeners, 112
plastic design, 446
Wind load, xxviii, xxxii

Yield stress, xxv, xxxi, 12, 420